INTRINSIC TIME GEOMETRODYNAMICS
At One With The Universe

內蘊時間幾何動力學：
與宇宙同行

INTRINSIC TIME GEOMETRODYNAMICS

At One With The Universe

內蘊時間幾何動力學:
與宇宙同行

Chopin Soo

National Cheng Kung University, Taiwan

Hoi Lai Yu

Academia Sinica, Taiwan

NEW JERSEY · LONDON · SINGAPORE · BEIJING · SHANGHAI · HONG KONG · TAIPEI · CHENNAI · TOKYO

Published by

World Scientific Publishing Co. Pte. Ltd.

5 Toh Tuck Link, Singapore 596224

USA office: 27 Warren Street, Suite 401-402, Hackensack, NJ 07601

UK office: 57 Shelton Street, Covent Garden, London WC2H 9HE

Library of Congress Control Number: 2022945081

British Library Cataloguing-in-Publication Data
A catalogue record for this book is available from the British Library.

INTRINSIC TIME GEOMETRODYNAMICS
At One with the Universe

ISBN 978-981-126-359-0 (hardcover)
ISBN 978-981-126-360-6 (ebook for institutions)
ISBN 978-981-126-361-3 (ebook for individuals)

For any available supplementary material, please visit
https://www.worldscientific.com/worldscibooks/10.1142/13062#t=suppl

Desk Editor: Nur Syarfeena Binte Mohd Fauzi

In memory of our mothers
who taught us to tell right from wrong

Preface

We were first approached by Dr. Lim Swee Cheng (SC Lim) of World Scientific Publishing Company in 2017 to undertake a book project on "Intrinsic Time Gravity". It is the title of one of our articles, in a series of related works on geometrodynamics *sans* the paradigm of four-covariance; the particular paper was in collaboration with Niall Ó Murchadha. Having not written any book before, we demurred at that time on such an ambitious project — a discourse on Time, Gravitation, and the Universe. SC contacted us again with the same proposal two years later; in finally agreeing to this undertaking we definitely have his kind invitation, persistence, and patience to thank for. By then we had come to realize a book would indeed be the apposite format to allow us to present the framework, main theses and complementary arguments more cogently, without losing the subtleties and supplementary details that are necessary in making the case. It also allows us the luxury and the benefit of hindsight to revisit, refine, and enhance previous arguments and derivations; and to include new insights and results. We are also grateful to Ms. Nur Syarfeena Binte Mohd Fauzi, our desk-editor at World Scientific, for her professionalism in taking care of the finer details of publishing a book, and especially for her patience in our missed deadlines.

Little did we know at the time we took on the task that the last few years would be so challenging and so trying, not just for us personally, but for everyone else who would be affected by the global pandemic in one way or another. Besides the loss of his beloved mother to cancer during this period, CS is saddened by the passing, from COVID-related complications, of Lay Nam Chang, a lifelong friend and mentor. He will recall with special fondness the happy memories shared with Prof. Chang and treasure the nuggets of wisdom on life, physics, and the universe that are learned.

Niall passed away last year. It was from him that ITG, "Intrinsic Time Gravity" (or "Intrinsic Time Geometrodynamics") received its label and brand name (more of that in Chapter 4. We would like to dedicate the Chapter to the memory of our friend and collaborator). Eyo Eyo Ita of the US Naval Academy joined us after the first few articles on ITG have been completed. He has been an invaluable collaborator since, and a tireless worker in our joint publications; we appreciate his participation in the ITG program, and acknowledge without reservations his input and contributions. CS would like to record his sincere appreciation and his deep sense of privilege in mentoring and working with several outstanding graduate students at the National Cheng Kung University. Huei-Chen Lin (林暉程), a Ph.D. student; and three M.Sc. students, Wei-Chen Chen (陳威辰), Yung-Hui Chang (張永輝) and Dian-Heng Li (李典恆) deserve special mention. Their contributions and related results, some of which appear in one form or another in this book, are gratefully acknowledged; indeed some expositions in the book are in response to the questions they raised and to their personal struggles in understanding certain technical details and conceptual issues in General Relativity — fresh graduate students have yet to sign on to conventional wisdoms and established paradigms!

The final title of the book, *Intrinsic Time Geometrodynamics: At One With The Universe*, emphasizes several things: Gravitation is a theory of the whole universe in which the spatial metric (or dreibein when chiral fermions are also to be included) is the fundamental dynamical variable. This bequeaths the theory with several advantages, including a well-defined concept of a spacelike hypersurface — a notion, in Wheeler's words, of "an instance of simultaneity" — that is needed in imposing the microcausality of spacelike local commutativity or anticommutativity of quantum fields. Time is inherent in the framework, rather than external; it comes from a degree of freedom of the spatial metric or the intrinsic geometry, as opposed to one arising from extrinsic curvature or extrinsic geometry. The book aims to explicitly demonstrate, affirm and substantiate, the fundamental reality of time in Einstein's theory and its extensions; to emphasize it is the universe itself that plays the role of the all-encompassing and infinitely robust clock that renders "time" in gravitation concrete, universal, and comprehensible, whilst its own dynamics, origin and destiny are dictated by the laws of gravitation and quantum mechanics. Einstein's geometrodynamics is the dynamics of three-geometry, not four-geometry; quantum gravity spells the demise of the classical ideal of a four-dimensional space-time. Instead of a theory obfuscated by four-covariant paradigm, Schrödinger-Heisenberg quantum evolution and causal temporal ordering with respect to the cosmic clock of our expanding closed universe is well-defined in a theory of gravitation with just spatial diffeomorphism invariance. The paradigm shift prompts modifications of Einstein's theory to include a Cotton-York term for power-counting ultraviolet renormalizability, in addition to infrared convergence from spatial compactness. The physical contents of Einstein's General Relativity are captured at low frequencies and low curvatures; except in the very early Cotton-York dominated era, the time-dependent physical Hamiltonian ensures primacy of Einstein's theory and its observational consequences afterwards and at late times. This very time dependence permits the universe to begin very differently from how it will end. An exact Chern-Simons Hartle-Hawking state is advocated as the initial state of the universe. At the level of expectation values, it is in accord with Penrose's Weyl Curvature Hypothesis, of a smooth Robertson-Walker, but also hot beginning due to Euclidean instanton tunneling; the gravitational arrow of time of increasing intrinsic volume concurs with the thermodynamic Second Law arrow of increasing entropy. The state also exhibits quantum gravity metric fluctuations, with, at the lowest approximation, scale-invariant two-point correlations as one of its defining characteristics.

This book presents gravitation almost exclusively, but unapologetically, from the canonical point of view. This is for consistency and out of necessity, given the nature and theses of the discourse, and the paradigm shift that is involved. We appreciate that quantum physics and gravitation have been enriched by different perspectives and formulations, and are cognizant of complementary methodologies. Be that as it may, demonstration of the fundamental reality of cosmic time and causation in Quantum Geometrodynamics is best fulfilled with Hamiltonian mechanics, Dirac's analyses of constrained systems, and Schrödinger-Heisenberg pictures.

Chopin Soo (許祖斌) and *Hoi Lai Yu* (余海禮)

June 2022

"The difficulty, as in all this work, is to find a notation which is both concise and intelligible to at least two people of whom one may be the author."
—in P. T. Matthews and Abdus Salam, *The Renormalization of Meson Theories*, Rev. Mod. Phys. **23** (1951) 311.

A few words on notation and conventions

Ideally notation should be intuitive, rather than pedantic.

Spatial components are denoted by lower case Latin letters from i onwards. $SO(3,C)$ Lorentz indices are labeled by lower case Latin letters from a to h. Unless indicated otherwise in the text, Greek letters denote spacetime indices.

$\tilde{\pi}^{ij}$, of density weight 1, is the canonical conjugate momentum to the spatial metric, which is denoted by q_{ij}. Its trace is written as $\tilde{\pi} := q_{ij}\tilde{\pi}^{ij}$. Spatial dreibein e_{ai} carry one Lorentz and one spatial index. Its inverse, the "triad", is denoted by E^{ia}. Unimodular metric and dreibein entities are denoted as barred objects: $\bar{e}_{ai} := e^{-\frac{1}{3}}e_{ai}$, with e denoting the determinant of e_{ai}; and $\bar{q}_{ij} := q^{-\frac{1}{3}}q_{ij}$ is the unimodular spatial metric, with $q = e^2$ being the determinant of the spatial metric. The symbols for their respective inverses are $\bar{q}^{ij}$ and $\bar{E}^{ia}$. These carry weight $q^{\frac{1}{3}}$ and $e^{\frac{1}{3}} = q^{\frac{1}{6}}$ respectively.

It is not the convention in this book to denote tensor densities by the number of tildes or bars above or below the objects in accordance of their weights, nor is it practical to do so given the fractional weights that are encountered. One may quibble about the inverse of $\bar{q}_{ij}$ being written as $\bar{q}^{ij}$, instead of, say, $\underline{q}^{ij}$; we opt for the simpler notation $\bar{q}^{ij}$, contravariant placement of the indices suffices to distinguish it as the inverse to the covariant $\bar{q}_{ij}$. Traceless momentum entities associated with $\bar{q}_{ij}$ and $\bar{e}_{ai}$ are denoted respectively by $\bar{\pi}^{ij}$ and $\bar{\pi}^{ai}$. They transform, respectively, as weight $q^{\frac{1}{2}+\frac{1}{3}}$ and $q^{\frac{1}{2}+\frac{1}{6}}$ tensor densities, so $\bar{q}_{ij}\bar{\pi}^{kl}$ and $\bar{e}_{ai}\bar{\pi}^{ib}$ remain of weight 1 i.e. of the same weight as $\tilde{\pi}^{ij}$. Classically $\bar{q}_{ij}\bar{\pi}^{kl}$ is equal to the momentric variable which is denoted as $\bar{\pi}^i_j$.

Yang-Mills indices are written in sans-serif fonts e.g. a to differentiate from $SO(3,C)$ Lorentz index a; so $A_{ia}dx^i$ is a Yang-Mills connection 1-form on a spatial hypersurface. A 2-component Weyl fermion is denoted simply as ψ which is 2-component column matrix, and π_ψ stands for its conjugate momentum. On using the same symbol for different entities: for instance, ϕ denotes a scalar field and also the conformal factor of the spatial metric; but the context the symbol is invoked should quell any potential confusion.

Contents

List of Figures

List of Tables

Chapter 1

Gravitation and quantum mechanics: A match made in heaven

*"Truth is ever to be found in simplicity,
and not in the multiplicity and confusion of things."*
—Isaac Newton

*"A man may imagine things that are false,
but he can only understand things that are true,
for if the things be false,
the apprehension of them is not understanding."*
—Isaac Newton

1.1 Introduction

1.1.1 *Newton, the apple, and the moon*

No other theory in physics can boast of such a memorable fable of its origin. It is most certainly a myth that an apple hitting Newton's head was the Eureka moment of the birth of the theory of gravity. But there is basis to the fable that the falling apple was among Newton's inspirations. Voltaire, wrote in his *Essay on the Civil Wars* (1727) of "Sir Isaac Newton walking in his garden had the first thought of his system of gravitation upon seeing an apple falling down from the tree"; he might possibly have heard of the apple story from Newton's niece. William Stukeley, Newton's friend and one of his first biographers; and John Conduitt, Newton's assistant and the husband of his niece, confirmed the "falling apple" inspiration, even if it did not actually hit Newton on the head. Conduitt also wrote, "whilst he was musing in a garden it came into his thought that the same power of gravity (which made an apple fall from the tree to the ground) was not limited to a certain distance from the Earth but must extend much farther than was usually thought — why not as high as the Moon said he to himself and if so that must influence her motion and perhaps retain her in her orbit, ... He began his calculations anew and found it perfectly agreeable to his theory." From the point of view of science and the progress of human thought, the more significant (and romantic) conception involved the moon. That the calculations agree "pretty nearly" (as Newton said) for the motion of the moon and objects near the Earth's surface, carries the profound implication the same laws (of mechanics and of gravitation) that govern the apple also govern the moon. Newton's law of gravity is thus called the Universal Law of Gravitation. For the first time, precise quantitative laws, discovered by man, which dictate the fate of material bodies on the earth extend to the heavens! The workings of the cosmos can be comprehensible to mere mortals.

The completion of the Law of Gravitation in the set of ten-component Einstein's field equations with energy-momentum tensor as source of the metric tensor was to take some two and a half centuries after Newton made that momentous leap for mankind. The Equivalence Principle implies gravitation is unlike any other force in that its effects at a particular point can be compensated, or conversely, mimicked, by a choice of coordinates. Einstein struggled with a question: What formulation of the Law of Gravitation could by choice of coordinates at a particular point render the equations of physics to take their form as in his Special Theory of Relativity? He had the profound insight to come to the conclusion that gravitation must be the consequence of spacetime curvature. In a curved spacetime, the tangent plane at any particular point is isomorphic to flat Minkowski spacetime; thus a choice of local coordinates will make the Laws of Physics there agree with Special Relativity (SR) (to take a simple example, at any point on the curved surface of a football the tangent plane is a flat 2-dimensional surface; but it changes from point to point and there is no single global flat surface which is the tangent plane of all the points. The obstruction to achieving this is the curvature of the ball's surface). The problem and mathematical description of curvature in arbitrary number of dimensions had already been completely solved by the German mathematician Georg Friedrich Bernhard Riemann. "From him I learned for the first time about Ricci and later Riemann. So I asked him whether my problem could be solved by Riemann's theory ..." said Einstein. The "him" was Einstein's dear friend, classmate, collaborator, and mathematician Marcel Grossmann. Named in gratitude by the gravitation community, the Marcel Grossmann Meetings (since 1975) is the world's biggest conference devoted to the topic of gravitation.

1.1.2   *Einstein's Nobel prize*

Einstein was not present at the 1922 Nobel award ceremony. His achievements were presented by the famous Swedish physical chemist Svante Arrhenius. The Nobel committee grudgingly, and cryptically, acknowledged Einstein's profound achievements in relativity in the first half of his award citation: "for his services to Theoretical Physics, and especially for his discovery of the law of the photoelectric effect." Ironically, it too was a "consolation prize" in that the committee had first decided none of 1921's nominations met the standards. According to the rules, the prize can in such a case be reserved until the following year, and this statute was applied to Einstein who therefore received his Nobel one year later. That such a statute was applied to the greatest physicist of the last century was one of the injustices of the Nobel awards. The Nobel Committee rewarded him for what Einstein must have judged to be the lesser of his achievements. "Being too remote from Sweden, Professor Einstein could not attend the ceremony" (on December 10, 1922) was the officially stated reason. Einstein was indeed not yet back from a long cruise to the Far East with his wife Elsa, with Japan as the final destination.

The official Nobel prize website still lists his Gothenburg Lecture of 1923 as the "Nobel lecture". But Einstein was, on that occasion, in no mood to speak on the photoelectric effect. "Fundamental ideas and problems of the theory of relativity" was his chosen title; and, perhaps to underscore his irritation and the irony of his award citation, the title was tagged with Einstein's defiant footnote: "The lecture was not delivered on the occasion of the Nobel prize award, and did not, therefore, concern the discovery of the photoelectric effect". The lecture's final sentence read "Should the form of the general equations someday, by the solution of the quantum problem, undergo a change however profound, even if there is a complete change in the parameters by means of which we represent the elementary process, the relativity principle will not be relinquished and the laws previously derived therefrom will at least retain their significance as limiting laws." Despite hackneyed claims of his staunch opposition to certain aspects of quantum mechanics, Einstein was clearly willing to accept that, "solution of the quantum problem" could alter the final form of his General Relativity (GR).

Current understandings of quantum field theories and GR seem to be odds; there are deep technical and conceptual issues to resolve when trying to formulate a theory which combines the two. One of the great dividing issues concerns time and how it is treated in quantum mechanics and GR. Can we really understand the cosmos right down to the microscopic level, explain the universe and trace its very origin all the way to the domain of quantum mechanics when deficiencies of even Einstein's GR become manifest? Commenting on the laws of the electromagnetic field, Einstein in the final paragraph of his Gothenburg Lecture of 1923 said "it should not be forgotten that a theory relating to the elementary electrical structures is inseparable from the quantum theory problems. So far also relativity theory has proved ineffectual in relation to this most profound physical problem of the present time." It was to take more than two decades before full agreement between quantum mechanics and SR was finally achieved by Dyson, Feynman, Schwinger and Tomonaga in the completion of quantum electrodynamics as a relativistic quantum field theory.

1.1.3   *Why quantize gravity?*

The completion of a successful quantum theory of gravitation is considered by many as the preeminent challenge of the physics of our time. But is it really necessary to quantize gravity? Hackneyed responses run from "consistency of physics requires quantum matter fields and gravity to be treated in the same framework" to "why not?" Another more specific response is Einstein's field equations equates geometry to energy-momentum tensor, with matter and Yang-Mills fields on the R.H.S. subject to the laws of quantum physics; ergo the need to quantize the L.H.S. as well, the detraction that Einstein's equations are classical equations notwithstanding.

It is more illuminating to start by asking "why quantum mechanics?" Besides its demonstrated validity in countless experiments and its technological upshots, what is it about quantum mechanics that makes it universal? What are its defining qualities? It is, first of all, the supreme unification! More so than Grand Unified Theories (GUTs), Supersymmetry, and anything else we know so far. A postulate of quantum mechanics says all that can be known about a closed system is contained in the quantum state. One quantum state describes all; regardless of whether there is or is not a final unification of forces, or for that matter, a unification of forces and matter. One wave function to rule them all!

Instead of an equation for each particle (of which there are more than 10^{80}), or one for each type of field (of which there are several dozens in the Standard Model of particle physics), one and only equation is needed for the quantum wave function of the entire universe. How dynamics is prescribed in quantum mechanics is breathtakingly simple. Three centuries of mechanics, from Newton, to Euler-Lagrange, to Hamilton, to Schrödinger-Heisenberg finally settled on that simplest of descriptions (see Sec. 1.8.2 for more discussions). The rate of change of the state is given by $\frac{d\Psi}{dT} = \frac{1}{i\hbar}\hat{H}_{phys}\Psi$; linearity of the equation implies superposition of solutions are solutions, self-adjointness of the physical Hamiltonian guarantees unitarity or conservation of total probability.

Not only is all of dynamics contained in a single equation, the equation always has a formal solution! Given an initial $\Psi(T_0)$, the state at any later time T is $\Psi(T) = U(T,T_0)\Psi(T_0)$, in which the time-development operator is $U(T,T_0) = \mathcal{T}[\exp(-\frac{i}{\hbar}\int_{T_0}^{T}\hat{H}_{phys}\,dT)]$. In this we come to the third property of quantum physics: non-commutativity. Unlike classical physics which deals with c-numbers, two quantum operators do not in general commute with each other. So in the solution of the dynamical equation, an ordering has emerged! And it is an ordering in time T (the $\mathcal{T}$ in the solution stands for time-ordering). Its precise operational formula, and further discussions, can be found in later sections and chapters. To date, all our successful quantum descriptions, whether in the Schrödinger or Heisenberg picture (the former evolves states; the latter, operators), is founded upon this concept of time-ordered evolution. However, time in the above scheme is a c-number (see the discussion on Pauli's theorem [1] in Sec. 1.3). There is no quantum fuzziness in time. Quantum mechanics assumes and needs a global c-number time parameter, its values arranged on a line, for the entire framework to function. From Schrödinger-Heisenberg quantum mechanics, to semiclassical Hamilton-Jacobi theory, to the derived classical Hamilton's equations, it is the same T that appears.

What then is time? In his *Confessions*, one of the wisest and deepest of men, Saint Augustine of Hippo, whose writings influenced Christianity and western philosophy, admitted "What then is time? If no one asks me, I know what it is. If I wish to explain it to him who asks, I do not know." Is not time well understood in physics? A layperson might suspect a physics student who is adept at solving problems on trajectories of particles would know a thing or two about time. Alas, it was none other than Newton himself who proclaimed, "I do not define time, space, place, and motion, as they are well known to all. Absolute space by its own nature, without reference to anything external, always remains similar and unmovable." When an entry makes its appearance in first-year undergraduate physics textbooks it is considered well-understood and noncontroversial. Almost invariably in the very first chapter, we find the entry *one second is the time taken by*

9 192 631 770 oscillations of the light of a specified wavelength emitted by cesium-133 atom. This very standard of time reveals how much, or how little, physicists really know about time. Time is defined by a number of cycles; so how do we know 100 billion oscillations "have passed"? How is, say, the experience and enjoyment of music — an art in time — possible from a mere cyclical understanding of time? The oft-quoted response is there is an arrow of time, one that is encapsulated in the second law of thermodynamics, that entropy of a closed system always increases, which differentiates the future from the past. However, this statistical law is not inscribed in the fundamental dynamics of microscopic systems. This does not imply it is not valid for macroscopic systems, but how is it related to the time variable that is in the aforementioned Schrödinger-Heisenberg mechanics? Is there a fundamental arrow of time in the equation? And does it agree with the thermodynamic arrow of time?

What about time in gravitation? Einstein's theories of relativity, both the Special and the General Theory, had dismantled the Newtonian view of absolute time and space. The mathematician and logician Kurt Gödel was troubled by the implications of SR, by the relativity of simultaneity. However, in SR with Minkowski spacetime, the time sequence of *timelike* separated events is absolute and Lorentz-invariant. So long as there is no faster-than-light motion, the ordering of the events is absolute even if the time durations between events are not. GR poses a much greater challenge, and threat, to our intuitive understanding of time and to temporal ordering. Einstein's field equations are known to have solutions which exhibit non-standard, and even bizarre, causal structures. On Einstein's 70th birthday, Gödel, Einstein's friend and colleague at the Institute of Advanced Study in Princeton, contributed an essay in which he presented an exact solution [2,3] of Einstein's equations, a solution that would later come to be known as the Gödel universe. It was the first exact solution to allow *closed timelike curves*; it revealed a universe in which a traveller can, without superluminal motion, choose to end up anywhere at any time, including in the past (see Fig. 1.1). For Gödel, the possibility and proof of traveling backward in time implies that the past is always there, and the flow and passage of time is an illusion (see Chapter 8 for Gödel's exact words).

In his response to Gödel, Einstein acknowledged "...the problem here involved disturbed me at the time of the building up of the general theory of relativity without my having succeeded in clarifying it...". Later on, upon learning of the death of his close friend Michele Besso, Einstein wrote "Now he has departed from this strange world a little ahead of me. That

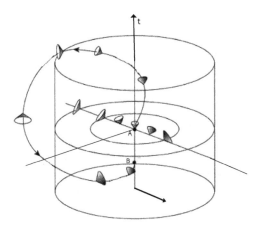

Fig. 1.1   Timelike path from point A to point B (which is in the past of A).

means nothing. People like us, who believe in physics, know that the distinction between past, present, and future is only a stubbornly persistent illusion." Perhaps those were consolatory remarks, not necessarily scientific pronouncements out of conviction. The first sentence was sadly prescient; Einstein was to pass away a month later.

Consider then the forbidding task and the practicality of consistently accounting for the causal relationships of bodies in motion in all possible solutions of GR. Attempts have been made to protect GR with conjectures [4]; but there is more to Einstein's reply to Gödel's provocative solution. After conceding his failure, Einstein ended with "It will be interesting to weigh whether these are not to be excluded on physical grounds." To the great logician's challenge, the great physicist had appealed to nature, as the final arbiter, for the resolution. Einstein's faith was not misplaced. The universe has to have a say on what time is; and part of the answer lies within Einstein's own GR, but buried under misunderstanding and prejudice. One can argue that of all the possible solutions, there is only one classical spacetime that matters and only one that exists, and that is our all-encompassing universe. A central thesis of our work is that it is not incidental all observational data points to an ever-expanding [5–7], and even spatially closed [8], universe; and unless and until the theory of gravitation is synthesized with this precise input it cannot be logically and physically sound.

Why quantize gravity then? The short answer is gravitation needs quantum mechanics to make sense of time and causality! This may seem paradoxical; would not the quantization of gravity lead to "quantum fuzziness" of spacetime, and hence of time as well? This book affirms and substantiates that it will not (see Sec. 1.3 and later sections of this chapter). Quantum gravity has no dominion over the cosmic clock. In Schrödinger-Heisenberg quantum mechanics is to be found the temporal ordering, the reality of the relations "earlier than" and "later than" that is essential to the notion and reality of time in the description of the physical world. In an ever-expanding spatially closed universe is to be found the arrow of time and the cosmic clock of Schrödinger-Heisenberg quantum mechanics (the ticks of the cosmic clock is the fractional change in its spatial volume); and in the Hamiltonian H_{phys} is to be found that pervasive time asymmetry and the clues to the origin of the universe.

1.1.4 *Why geometrodynamics?*

Geometrodynamics has the spatial metric and its conjugate momentum as fundamental variables. The fact that quantum mechanics owes its well-being to gravity is not often pointed out, because quantum mechanics is often formulated on a fixed background spacetime. Quantum field theories are founded on the bedrock of microcausality. However, microcausality, the spacelike local commutativity or anti-commutativity of field operators in quantum field theories, is often postulated in the form of "equal-time" relations, even though simultaneity is already relative in SR. What is really needed is a spacelike hypersurface, on which these fundamental quantum relations are imposed; commutation relations for bosonic fields and anti-commutation relations for fermionic ones. These relations are preserved under Heisenberg evolution of field operators (induced by similarity transformations with $U(T, T_0)$); and the commutativity(anti-commutativity) rules are precisely those that ensure the viability and well-being of the time-ordering that is needed for the framework. But a notion of distance is required in the concept of a "spacelike hypersurface"; it requires a metric, and one which is positive-definite everywhere on the hypersurface. Geometrodynamics bequeathed with spatial metric as its fundamental variable supplies and guarantees this spacelike separation. A spacelike hypersurface allows, to quote Wheeler's words [9], "a notion of 'simultaneity' and a common moment of a rudimentary 'time'". From this spatial surface must come a value of time T for quantum mechanics. What can it be other than being related to the spatial volume of the universe? That this emerges from the unification of GR with quantum mechanics will be substantiated and elucidated. What happens when gravity is quantized and four-dimensional spacetime loses its validity? The quantum wave function is the amplitude of the probability of different three-geometries; but any weighted distribution of spacelike surfaces will still be spacelike. So there is no fundamental contradiction in imposing the microcausality relations in the quantization of all fields, including the gravitational field.

$$|\Psi(T)\rangle = \mathcal{T}[\exp(-\frac{i}{\hbar}\int_{T_0}^{T} H_{Phys}(T')dT')]|\Psi(T_0)\rangle$$

Fig. 1.2 The elephant that is Quantum Gravity.

The road to quantum gravity is obstructed by formidable conceptual and technical challenges. The beauty of the elephant that is quantum gravity is often in the eye of the beholder. The standard Arnowitt-Deser-Misner (ADM) [10] canonical formulation of Einstein's theory reveals a total Hamiltonian that is, apart from a boundary term (we shall ignore the boundary term (it does not arise in a closed universe)), made up solely of constraints. There is no ready generator of "time" translation; and no immediate variable to be identified as "time". GR is a theory about spacetime; yet, a straightforward notion "time" does not come readily; it is even plagued by the "problem of time" [11]. "The label x^0 itself is irrelevant, and 'time' must be determined intrinsically" was DeWitt's conclusion, and suggestion [12]. A degree of freedom (d.o.f.) of the theory should be identified as "time" to deparameterize the dynamics. DeWitt's suggestion has an additional technical meaning. In his seminal work [12], the d.o.f. chosen as the time is $q^{\frac{1}{4}}$ where q is the determinant of the spatial metric. (3+1)-decomposition of spacetime separates four-geometry into intrinsic entities pertaining to a spatial hypersurface, and extrinsic quantities which describe how a spatial hypersurface is curved in spacetime; so De-Witt's choice is "intrinsic". In ITG, "intrinsic time" is advocated in the same context of arising from the spatial metric. Canonical Dirac quantization of the Hamiltonian constraint of GR leads to the Wheeler-DeWitt equation [9, 12]. But this does not bring gravity into the fold of quantum mechanics (see Sec. 1.8.1). The intrinsic time variable here is also spatially dependent, thus many-fingered [13]. Instead of well-defined one-parameter time-ordering in Schrödinger-Heisenberg quantum mechanics, mutual consistency of "time-orderings" and consistent interpretation of "causation" cannot be guaranteed due to the freedom in choosing any of the infinitude of ordered paths in a multidimensional time parameter space (see, for instance, Fig. 1.3). The ordering of paths in multidimensional parameter space is not time-ordering. As illustrated in Fig. 1.3, between the end points, path p_1 is path-ordered to be always increasing in t_1; likewise for path p_2 with respect to t_2. However, even if the integral along two different paths were to give the same numerical value, there is a section of p_2 in

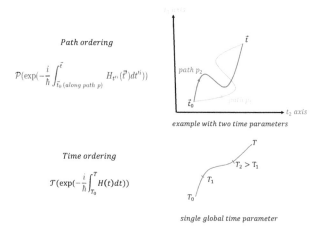

Fig. 1.3   Existence of more than one time variable. There is no mutual consistency of "time ordering" and "causation".

which t_1 time decreases; likewise for p_1 with respect to t_2 time. There is no mutual consistency of "time-ordering", of "earlier than" and "later than". This does not happen with a single global time parameter.

To make gravity compatible with quantum mechanics as we know it, first has to come a renouncement: four-dimensional spacetime is a classical ideal of "limited applicability" in quantum gravity. There is a photo of John Wheeler standing in front of the chalkboard with two figures [14]; one of a sharp trajectory joining two points A and B, the other of propagating wave fronts. "CLASSICAL" is written with bold letters on top of the first, and "QUANTUM" on top of the second (see Fig. 1.4). "There is no such thing as 4-geometry in quantum geometrodynamics, and for a simple reason. No probability amplitude function $\psi(^3\mathcal{G})$ can propagate through superspace as an infinitely sharp wave packet." [9] The technical statement is the spatial metric and its conjugate momentum do not commute; hence they are subject to quantum Heisenberg uncertainty relations. This demise of spacetime is not predicated upon quantum geometrodynamics with the metric as fundamental variable; any other set of fundamental variables used to describe quantum gravity will be subject to quantum non-commutativity and uncertainty.

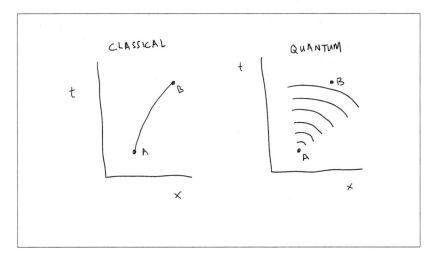

Fig. 1.4   Classical spacetime vs quantum propagation; x represents the dynamical variable in GR which is 3-dimensional space.

1.1.5   *Dirac's clarion call to abandon four-covariance*

An unequivocal call to abandon four-covariance was sounded by Dirac, one of the founding fathers of quantum mechanics, in Ref. [15]. Einstein's theory describes the dynamics or the change in time of three-geometry, not four-geometry. Simplicity of the Hamiltonian analysis and the fact that only the spatial metric, rather than the full spacetime metric, is dynamical led Dirac to conclude and declare "four-dimensional symmetry is not a fundamental symmetry of the physical world." An exemplar of succinctness in his writings and a private person of few words, Dirac nevertheless made the call in no less than five instances in the paper: once in the abstract, and four times in the conclusion. Two of these instances were even highlighted in italics!

In the Wheeler-DeWitt equation, the Hamiltonian constraint is turned into a dynamical quantum equation for three-geometry (equivalence class of spatial metric modulo spatial diffeomorphisms); the absence of four-covariance in this formulation of quantum geometrodynamics is fairly clear. The lapse function also makes no appearance at all in quantum gravity. Dirac however was also declaring, with reasons and apparently with conviction, the absence of four-covariance even in classical GR. Here the matter is more subtle; the lapse function is the Lagrange multiplier for the Hamiltonian constraint; thus it is *a priori* arbitrary in value, and it appears in the classical equations of motion (EOM) of the dynamical variables. As the total Hamiltonian is made up solely of first class constraints, the question is whether, with the arbitrariness of the multipliers, true physical evolutions rather than gauge histories are generated. The latter would be in line with the claim, and Gödel's position, that the passage of time is an illusion. As will be elucidated, the resolution lies in the distinction between the *a posteriori* physical nature of the lapse function and its *a priori* arbitrariness [16]; and true physical evolutions (modulo only spatial diffeomorphisms) occur in GR. Other clues that indicate four-covariance is neither present nor needed in GR can be found in the initial data formulation of York [17]. Therein, the Hamiltonian constraint is treated as an equation to uniquely fix the conformal factor when the trace of the momentum is made spatially constant and adopted as the time variable (this is the formalism of extrinsic time. See Chapters 4 and 8 for the comparison with ITG). Thus, instead of a generator of "gauge symmetry" the Hamiltonian constraint is used to eliminate a d.o.f. from the theory. ITG parallels this, but with a time variable that is intrinsic, spatial diffeomorphism invariance leads to an integrated physical Hamiltonian generating global cosmic time translation. The comparison of the local gauge symmetries in GR with local Yang-Mills [18] gauge invariance is discussed at length in Sec. 1.7.

1.1.6　*Quantum gravity and the cosmic clock*

What is perfectly sensible, at the very least in the quantum context, is the renouncement of classical spacetimes must come with the abandonment of four-covariance. Instead of puzzling over the merits and demerits of equivalent, or inequivalent, infinitude ways of slicing up all possible four-dimensional spacetimes, quantum gravity sets the tasks of figuring out the initial state of the universe and following its dynamical evolution with Schrödinger-Heisenberg mechanics. It was Poincaré who said, albeit before the advent of quantum mechanics, "Time should be so defined that the equations of mechanics may be as simple as possible." The paradigm shift makes the whole framework of Schrödinger-Heisenberg mechanics compatible with the true local gauge invariance of GR; whereas four-dimensional invariance excludes the possibility of a meaningful clock.

ITG affirms and substantiates the fundamental reality of time in Einstein's theory and its extensions, and of temporal ordering. It affirms and actualizes "the singular universe and the reality of time" of Smolin and Unger [19]. Gravitation and quantum mechanics are united in time by the cosmic clock which makes time universal, concrete, and comprehensible. Not only does gravity no longer stand apart from the rest; all are at one with the universe. Quantum gravity spells the demise of spacetime, but not of the reality of time and temporal ordering.

To bring gravity into the fold of quantum field theories further demands must be met — the requirements of renormalizability, unitarity, and absence of gauge anomalies that have been learned from decades of study. But the paradigm shift from four-covariance to only spatial diffeomorphism symmetry opens up a hitherto unexplored vista to overcome these challenges. This is recounted in later sections of this chapter, as well as in other places in the book. The foundation of gravitation on fundamental metric variable remains untouched by the shift, but its workings can be reconciled with the dynamics and triumphs of modern quantum field theories. ITG solves the Hamiltonian constraint exactly to gain a physical Hamiltonian which is time-dependent. This time asymmetry permits the universe to begin very differently from how it will end; except in the very early Cotton-York [17,20] dominated era, the time-dependent Hamiltonian ensures primacy of Einstein's theory and its observational consequences at late times.

It is the quantization of gravity that leads to universal Schrödinger-Heisenberg mechanics and fundamental temporal ordering for all. The union of gravity and quantum mechanics is needed to address the physics of the early universe, and to elucidate the microscopic beginning of our universe in conditions of extreme temperatures and curvatures. In the Chern-Simons Hartle-Hawking state [21, 22] is to be found the quantum origin which yields the semiclassical conditions compatible with Penrose's Weyl Curvature Hypothesis [23–25] — the low entropy beginning which makes the gravitational arrow of time of increasing volume concur with the second law of thermodynamics when the entropy increases with the formation of structures, galaxies, stars, black holes and gigantic black holes as our universe expands and ages.

1.2   Problem(s) of time

The "problem of time" in classical and quantum gravity is a multifaceted, rather than a single conceptual bafflement to be resolved, or a single technical hurdle to overcome. There are at least three difficulties which must be tackled:

1) Problem of time in Einstein's GR in which, apart from a boundary term, the Hamiltonian consists solely of constraints, with the entailing guise of a canonical framework frozen in time. Chapter 2 addresses this issue and its resolution in detail; other chapters also provides complementary perspectives. The resolution in ITG involves a paradigm shift from four-covariance to just spatial diffeomorphism invariance.

2) Problem of many-fingered (or Schwinger-Tomonaga [13]) time, in which the chosen time variable is spatially or even spacetime dependent. This complication arises naturally when a d.o.f. in a *field theory* is selected as the time variable to deparametrize the theory. As pointed out earlier, having many-fingered time, or for that matter more than one time variable, inevitably gives rise to the problem of the mutual compatibility and consistency of "time evolution(s)" along different paths in a multidimensional (even infinite-dimensional in the field theory context) time parameter space. This is compounded by that fact that along different paths the values of the time parameters are not necessarily monotonic to one another; so the time-ordering(s) are path dependent and even mutually inconsistent. "What happens next is without meaning" [9] and denying the physicality of any fundamental time variable is a way out; but this is not what is advocated in ITG which contends that the universe itself plays the role of

the preferred clock. After all, the ultimate temporal ordering is that of the wave function of the universe.

3) Problem of time as an operator in quantum mechanics, in which non-commutativity of the time operator and the Hamiltonian results in undesirable consequences in quantum mechanics; as asserted in Pauli's theorem [1], this leads to the Hamiltonian with an unbounded and continuous spectrum.

In ITG, $\delta \ln q^{\frac{1}{3}}$ the variable for intrinsic time interval comes from a field d.o.f., q, which has nontrivial commutation with $\tilde{\pi}$. Thus all three problems listed above must be addressed. Resolution of the first problem is delegated to Chapter 2, with complementary and supplementary discussions in the rest of the book. Remarkably, the second problem owes its deliverance to spatial diffeomorphism invariance! The gauge invariant part of $\delta \ln q^{\frac{1}{3}}$, and the physical intrinsic time interval, is dT which is spatially independent [16, 26]. It serves as the global one-parameter time interval; furthermore $dT = \frac{2}{3} d \ln V$ with V being the finite spatial volume. It is invariant under spatial diffeomorphisms. A spatially closed universe [8] is part and parcel of the framework of ITG. In such an ever expanding universe, the cosmic clock of expanding spatial volume is the very physical entity that makes time, and time-ordering of quantum evolution well-defined, universal, and comprehensible. On the third potential problem, the Hamiltonian of ITG is *not* afflicted with the consequences of Pauli's theorem.

1.3 Pauli's theorem on time, and the cosmic clock of ITG

Wolfgang Pauli in his *Handbuch der Physik* article [1] discussed a theorem which essentially points out that time should not be an operator in quantum mechanics. More precisely, if a self-adjoint Hamiltonian $\hat{H}$ is semibounded, then there does not exists a self-adjoint time operator $\hat{T}$ such that

$$[\hat{T}, \hat{H}] = i\hbar I, \tag{1.1}$$

is satisfied. The relation Eq. (1.1) is sometimes naively thought of as a natural "covariant extension" of $[\hat{x}^i, \hat{p}_j] = i\hbar \delta^i_j$ to include the time variable x^0 and Hamiltonian p_0. Pauli's theorem points out the price and futility of this route. If this relation holds, it follows that [27]

$$\exp\left(\frac{i}{\hbar}\hat{T}\epsilon\right) \hat{H} \exp\left(-\frac{i}{\hbar}\hat{T}\epsilon\right) \Psi = (\hat{H} - \epsilon I)\Psi. \tag{1.2}$$

Thus $\hat{H}$ and $(\hat{H} - \epsilon I)$ are unitarily equivalent and have the same spectrum for all real ϵ, which means that the spectrum of $\hat{H}$ must range over the whole real line. Since ϵ is arbitrary, the resultant spectrum of the Hamiltonian operator must also be continuous, in contradiction with the observed discrete energy levels in many quantum mechanical physical systems. Pauli concluded that time as an operator in quantum mechanics must be abandoned; and time can only be regarded as an ordinary commuting number (or "c-number") and $[\hat{T}, \hat{H}] = 0$ instead.

In standard quantum mechanics time, unlike the position of the particle, has no quantum fuzziness or quantum fluctuations, because time is not an operator but an ordinary number. The direct result of this is the nonexistence of a fundamental uncertainty such as $\Delta E \Delta T \gtrsim \frac{\hbar}{2}$; unlike the uncertainty relation $\Delta x \Delta p \gtrsim \frac{\hbar}{2}$ which is a consequence of the canonical commutator of position and momentum. In quantum mechanics, the Schrödinger and Heisenberg pictures treat time as an ordinary number; and the resultant time-ordering of states and time development operator as well-defined. Were time to be an operator rather than a simple single parameter, temporal ordering of quantum evolution would not be viable, at the very least it would bewildering.

In ITG, q, has nontrivial commutation with $\tilde{\pi}$, but not with anything else. $\tilde{\pi}$ is proportional to the Hamiltonian density $\bar{H}$ of the theory. However, the Hamiltonian of ITG, $H_{phys} = \frac{1}{\beta} \int \bar{H} d^3x$, commutes with q, and hence with the spatial volume of the expanding universe. This fortuitous circumstance comes about because the Hamiltonian constraint in Einstein's theory and its extensions allows $\tilde{\pi}$ to be solved or eliminated in terms of other variables which all commute with q and hence with global intrinsic time variable T which is thus effectively a "c-number" despite its origin in a field d.o.f. $\bar{H}$ and the physical Hamiltonian are independent of $\tilde{\pi}$; this remains true even when other fields, especially Weyl fermions, are also incorporated into the theory. Chapter 9 gives a further account of this remarkable happy circumstance. In ITG the only variable, $\tilde{\pi}$, that has nontrivial commutation with the cosmic clock is eliminated, through the Hamiltonian constraint, in terms of other variables.[1] As a result the cosmic clock is quantum mechanically robust, it is not subject to any quantum fluctuations and uncertainty relations! To all this is owed the viability and actualization of temporal causation, quantum time-ordering and time-development of our universe with respect to the cosmic clock.

---

[1] The emergence of H_{phys} and the reduced phase space is covered in detail in Chapter 8, and it is also discussed in complementary ways in other chapters.

1.3.1 *Mandelstam-Tamm time-energy uncertainty relation*

It should be pointed out the absence of fundamental uncertainty in cosmic time and its trivial commutation with the Hamiltonian do not rule out quantum fuzziness in effective timescales of operators. For instance, the Mandelstam-Tamm time-energy uncertainty relation [28], $\Delta E \Delta t_O \gtrsim \frac{\hbar}{2}$, follows from considering an operator timescale defined by

$$\Delta t_O := \left| \frac{d\langle \hat{O} \rangle}{dT} \right|^{-1} \Delta O. \tag{1.3}$$

This is the time duration required for a significant change of the expectation value of an observable $\hat{O}$. The uncertainty relation that arises is a consequence of the EOM,

$$\frac{d\langle \hat{O} \rangle}{dT} = \frac{i}{\hbar} \langle [H, \hat{O}] \rangle, \tag{1.4}$$

which leads to $\Delta E \Delta O \gtrsim \frac{\hbar}{2} |\frac{d\langle \hat{O} \rangle}{dT}|$ and thus to Eq. (1.3). Be that as it may, the existence of this uncertainty for Δt_O (which has been observed and confirmed in experiments) does not in any way negate the fundamental nature of the cosmic clock of ITG and the consequences and boon of its trivial commutation with H_{phys}.

1.4 Dirac's derivation of the Schrödinger equation from an extended phase space

There is a close parallel between origin of the Schrödinger equation in ITG and Dirac's "derivation" of the equation [29]. Venturing beyond quantum mechanics as eigenvalue problem, Dirac toyed with a derivation of the *time dependent* Schrödinger equation which circumvented Pauli's no-go theorem. He introduced an extended phase space, with an additional conjugate pair $(T, -W)$ on top of the original canonical variables (q^r, p_r). In this extended phase space the Poisson bracket is

$$\{A, B\}_{P.B.} = \left(\frac{\partial A}{\partial q^r} \frac{\partial B}{\partial p_r} - \frac{\partial A}{\partial p_r} \frac{\partial B}{\partial q^r} \right) - \frac{\partial A}{\partial T} \frac{\partial B}{\partial W} + \frac{\partial A}{\partial W} \frac{\partial B}{\partial T}. \tag{1.5}$$

Dirac further imposes a constraint

$$H(q^r, p_r, T) - W = 0. \tag{1.6}$$

If Hamiltonian $\hat{H}$ is independent of W, then T commutes with it; the afflictions of Pauli's theorem are absent because T does not generate translations of $\hat{H}$ as in Eq. (1.2). Furthermore, the constraint translates in the quantum context to

$$\left(\hat{H} + \frac{\hbar}{i} \frac{\partial}{\partial T} \right) \Psi = 0, \tag{1.7}$$

which is the Schrödinger equation.

1.4.1   *Parallels with ITG*

What if we take this idea with utmost seriousness? That the fundamental constraint governing the universe is of the form of Eq. (1.6); with the same independence of the Hamiltonian of the conjugate to T. The result is the circumvention of the troubles in Pauli's theorem *despite* the origin of time as a d.o.f. in an extended phase space. But Einstein's theory is precisely dictated by such a Hamiltonian constraint! Even when other fields are incorporated, the Hamiltonian constraint with all those desired properties remains. Would not the accompanying Schrödinger equation govern all that matters in the universe?

Following and extending Dirac's paradigm, we extrapolate that

- Time in a quantum field theory of the universe can be traced back to an extra conjugate pair.
- This extra pair must be universal to all systems since we apply the fundamental Schrödinger equation with the same time variable T when we are studying different subsystems of interest.
- This extra pair should come naturally from field variables which determine the structure of space i.e. from gravitation; and it is universal to all theories since gravity interacts with all fields.
- There is indeed a marvellous exact parallel between the pair $(\tilde{\pi}, \ln q^{1/3})$ extracted from the initial data of gravitation and Dirac's extended variables $(t, -W)$; wherein q is the determinant of spatial metric fields and $\tilde{\pi}$ is the trace of its canonical momentum. This extra pair $(\tilde{\pi}, \ln q^{1/3})$ can be regarded as being part of the "extended phase space" $(q_{ij}, \tilde{\pi}^{ij})$ of Geometrodynamics. The analog of (q^r, p_r) is the unimodular spatial metric and the traceless part of the momentum, $(\bar{q}_{ij}, \bar{\pi}^{ij})$, albeit that these are not strictly canonical (see Chapter 2). The Hamiltonian constraint is expressible as in Eq. (1.6); and q does Poisson commute with the analog of the Hamiltonian above, and with all other variables. It will be shown also in the later chapters that not just the Poisson, but the Dirac

bracket [30], $\{\ln(\frac{q}{q_{ref}})^{\frac{1}{3}}, \bar{H}\}_{D.B.} = 0$ holds. This important result confirms there is no quantum fluctuation in the time variable in the framework of Dirac's second class constrained system in which it is the Dirac, rather than Poisson brackets than are promoted to quantum commutation relations.

It must also be pointed out that it is *same time variable T* that is in the equations when one proceeds from the Schrödinger to Hamilton-Jacobi to Hamilton's equations; quantum, semiclassical and classical "time" are the same.[2] Note also that while quantum unitary is guaranteed by self-adjointness of the Hamiltonian, there is no fundamental reason for the Hamiltonian to be time-independent. This is true of the physical Hamiltonian, H_{phys}, of ITG; and the time dependence permits different physics to dominate and come into play at different eras of the universe, without the need of postulating ever-changing fundamental laws.

While the correspondence is remarkable, there is the caveat that the "time" variable, $\ln q^{\frac{1}{3}}$, as a field d.o.f. is many-fingered; but this is where spatial diffeomorphism invariance comes in. Instead of a Schrödinger equation which dictates the dynamics of the universe w.r.t. to a gauge-dependent many-fingered Schwinger-Tomonaga "time", the physical question is how the quantum state of universe evolves with respect to gauge-invariant expanding intrinsic spatial volume. If full four-dimensional spacetime diffeomorphism symmetry holds, rather than spatial diffeomorphism invariance, an expanding spatial volume would fail to be gauge invariant. A paradigm shift in symmetry is part and parcel of the framework which affirms the fundamental nature of time.

1.5 From many-fingered to gauge invariant global cosmic time

The last sections have made it clear time must not only commute with the Hamiltonian, it must be a simple global (spatially independent) parameter rather than many-fingered.

$\tilde{\pi}$ and $\sqrt{q}$ are weight one scalar densities; a non-dynamical reference or fiducial q_0 can be introduced so that $\ln(\frac{q}{q_{ref}})^{\frac{1}{3}}$ transforms as a scalar under spatial coordinate transformations (the factor $\frac{1}{3}$ which is kept throughout this book for accounting purposes is not physically significant). They obey

---

[2]Section 2.9.1 contains further related comments.

$\{\tilde{\pi}(x), \frac{1}{3}\ln(\frac{q}{q_{ref}})(x')\}_{P.B.} = \delta^3(x, x')$; so strictly speaking it is this pair that is canonical. But a non-changing[3] $q_{ref}(x)$ has no effect on the intrinsic time interval which involves the change in the logarithmic function, since $\delta\ln(\frac{q}{q_{ref}})^{\frac{1}{3}} = \delta\ln q^{\frac{1}{3}}$.

Even in Galilean-Newtonian physics, it is time interval, and not absolute time, that is physically relevant. Nevertheless, $\delta\ln q^{\frac{1}{3}}(x)$ is spatially dependent; but there is a further ace in the logarithm function[4]: under spatial diffeomorphisms, the Lie derivative of $\ln q^{\frac{1}{3}}$ is $\mathcal{L}_{\vec{N}}\ln q^{\frac{1}{3}} = \frac{2}{3}\nabla_i N^i$; while a Hodge decomposition of a zero form (see also Chapters 2 and 3) yields

$$\delta\ln q^{\frac{1}{3}}(x) = \delta T + \nabla_i \delta Y^i(x); \tag{1.8}$$

wherein dT is spatially independent[5] while $\nabla_i \delta Y^i$ is precisely on the form of a Lie derivative of $\ln q^{\frac{1}{3}}$ which can be gauged away (by choosing the shift function $N^i = \frac{3}{2}\delta Y^i$; the notation δY^i merely reflects that it is infinitesimal quantity entity) completely with a spatial diffeomorphism. Thus the gauge invariant change is dT which is global (spatially independent on each Cauchy hypersurface); furthermore, this is related to the change in the spatial volume V of our universe by $dT = \frac{2}{3}d\ln V = \frac{2}{3}dV/V$. This dimensionless fractional change in the volume of the universe is the ticking of the cosmic clock. The dynamics and evolution of our expanding universe is with respect to its own all-encompassing intrinsic clock. Equally astonishing is an apparent *raison d'etre* for spatial diffeomorphism invariance: it collapses many-fingered time to global cosmic time.

That time is dimensionless should not be disconcerting, in Robertson-Walker cosmology with scale factor a, how long ago an astronomical event took place is often measured by the dimensionless redshift parameter z, with $(1 + z) = a_{now}/a$, and in this particular case $dT = \frac{2}{3}d\ln V = 2d\ln a$. Even without homogeneity the fractional change in the volume of the universe remains a well-defined and spatial diffeomorphism invariant concept. What is paramount in the temporal ordering of quantum evolution is not the physical dimension of the time variable, but the availability of an invariant

---

[3]In later chapters we shall advocate that a spatially closed simply-connected conformally flat beginning and the Poincaré conjecture suffice to pin down the initial $q_0(x)$ to be proportional to the determinant of the standard S^3 round metric; and this q_0 can be adopted as q_{ref}. The EOM for $\ln q$ with T as time variable yields the solution $q(x, T) = q_0(x)e^{3(T-T_0)}$.

[4]This advantage is another reason for choosing $\delta\ln q^{\frac{1}{3}}$ instead of DeWitt's $\delta q^{\frac{1}{4}}$ [12] as the intrinsic time interval.

[5]Since this variation is spatially independent, we use δT and dT interchangeably.

global parameter, infinitely robust and all-encompassing, to order the wave function of the universe. Temporal ordering of events has no fundamental and invariant meaning in quantum gravity; the physical state of the universe is the fundamental gauge invariant entity that is causally time-ordered. In it lies our shared past, present, and future.

1.6   Horava gravity theories: problems and promise

1.6.1   *Conflict between renormalizability and unitary in four-covariant theories of gravitation*

Petr Horava pointed out a key obstacle to the viability of Einstein's theory and its four-covariant extensions as renormalizable perturbative quantum field theories: it lies in the deep conflict between unitarity and spacetime general covariance [31]. Renormalizability of GR can be improved and achieved through the introduction of higher derivative terms, but space-time covariance also requires higher time derivatives, as well as spatial derivatives, of the same order; thus compromising the stability and unitarity of the theory. It is understood through explicit cases (see for instance Eq. (10.4)) that introduction of higher derivatives of the metric can tame the divergences in the theory and even render the theory perturbatively renormalizable; however, it is also known since the late nineteenth century that theories with higher than two time derivatives in the action are usually unstable because the corresponding Hamiltonian will be unbounded from below. In a quantum theory, unitary or conservation of probability is a reasonable demand, particularly when a single wave function is employed to describe all. On another hand, the paradigm of four-covariance is long-cherished in gravitation. Horava boldly proposes to keep unitarity at the expense of relinquishing full space-time covariance, to retain only spatial diffeomorphism symmetry at the fundamental level. The proposed modification to Einstein's theory is to introduce higher spatial, but not time, derivatives of the metric to improve ultraviolet convergence in the effective graviton propagator and achieve power-counting renormalizability. Without the complication of higher time derivatives, unitary can be protected, with General Relativity recovered at low curvatures and low frequencies.

In a theory with only spatial diffeomorphism invariance, the preeminent term to add [31] to the Hamiltonian involves the trace of the square of Cotton-York tensor [17, 20]. This tensor is transverse and traceless, and contains up to three spatial derivatives of the metric. It is special to three-dimensions, and it would never have been introduced had four-covariance been the overriding concern. ITG is a framework which recognizes the primacy of the Cotton-York tensor in overcoming the many technical challenges of quantum gravity; the ramifications it brings are studied and recounted in this book. In the history of science, tension between assumed basic principles often gave birth to breakthroughs and the overthrow of long-cherished paradigms. For instance, conflict between action-at-a-distance in Newtonian gravitation and the finite speed of light led to the birth and discovery of Einstein's theories of relativity.

1.6.2 *Problems with Horava gravity*

In the canonical ADM formulation [10] of GR, apart form a possible boundary term, the total Hamiltonian consists of a combination of two types of constraints: a vector constraint (usually called the momentum constraint) which implements spatial diffeomorphism invariance of the theory; and a constraint is a scalar (or scalar density) constraint (called the Hamiltonian constraint $H(x) = 0$) with the lapse function, N, as its Lagrange multiplier. The role of the Hamiltonian constraint, which is quadratic in the momentum conjugate to the metric, is contentious even in GR, although it has been advocated that together with the momentum constraints it enforces on-shell four-covariance. What is clear is all the constraints are needed to capture the physical content of Einstein's theory.

In Horava gravity theories the nature of the Hamiltonian constraint was the subject of many investigations; and Horava gravity theories can be divided into two broad categories: the "projectable" and "non-projectable" versions.

- The projectable version; in which the lapse function $N(t)$ is spatially independent and a function only of ADM coordinate time. This implies an integrated global Hamiltonian constraint, $\int H(x)\, d^3x = 0$, rather than a constraint at each spatial point. This resultant constraint algebra is self-consistent in a trivial way (this global constraint commute with itself, and with the momentum constraint if $H(x)$ is spatial tensor density of weight one). However, the theory does not reproduce GR, which needs all four constraints to be local to reduce to two field d.o.f. (the perturbative quantum excitation should be a purely transverse spin 2 graviton). In this type of Horava gravity theory, simple analysis, such as in Ref. [32], indicates the extraneous d.o.f. will not completely decouple from the theory.

There are too few constraints at each spatial point to generically eliminate any spurious pathological mode.

- The non-projective version; in which the spacetime dependent lapse function leads to a Hamiltonian constraint $H(x) = 0$ at each point. There is no lack of constraints; but as these are modifications of Einstein's theory, the resultant constraint algebra cannot be the same as the Dirac algebra of GR. Hojman, Kuchar and Teitelboim [33] had proven an essentially uniqueness theorem in their *Geometrodynamics Regained* program: Einsteinian geometrodynamics is the only (time-reversible) canonical representation of the Dirac algebra if the spatial metric and its conjugate momentum are the sole canonical variables. In a generic Horava gravity theory, neither the potential nor the DeWitt supermetric take the form as in Einstein's theory, so it is not surprising the constraint algebra ran into apparent inconsistencies [34,35].

1.6.3 *ITG formulation of Horava gravity: new vista to address the challenges of classical and quantum gravity*

The intriguing question is what replaces the Dirac algebra if one tries to modify Einstein's theory; in particular, what would be a consistent constraint algebra for Horava's program. More precisely, how would one formulate a consistent modification of Einstein's theory *within the confines of a first class constraint system*. By casting the Hamiltonian constraint as a master constraint [36–38], a mathematically consistent first class algebra (see Chapter 2, Sec. 2.4) emerges, *regardless of the actual form of the Hamiltonian constraint*. Through Eq. (2.22) one sees that the Hamiltonian constraint generates *no* extra symmetry. It is totally dynamical in content, provided a d.o.f. in the theory is adopted as the inherent time variable and a corresponding Hamiltonian is identified. This implies even Einstein's theory can be so casted without losing any of its physical contents as all the constraints are still present. But it would also carry the implications the Hamiltonian constraint generates no extra symmetry and there is only spatial diffeomorphism invariance; and that four-covariance in GR is a red herring. This is supported by the revelation that while the lapse function (which is the Lagrange multiplier for the Hamiltonian constraint of Einstein's theory) is *a priori* arbitrary, its *a posteriori* value is in fact determined by the EOM and when the Hamiltonian constraint is solved.[6] This substantiates true physical evolution in classical GR despite the *a priori* arbitrariness of the lapse function. The distinction, between the *a priori* and *a posteriori* lapse function, that is pointed out in ITG removes a mental block on the true nature of the Hamiltonian constraint. It is clearly not a generator of any local gauge symmetry.

With the paradigm shift from four-covariance, a global time variable invariant under just spatial, but not four-dimensional diffeomorphisms, becomes well-defined; it is now consistent with the gauge symmetries of the theory. This time variable does not commute with the Hamiltonian constraint of GR; but the constraint is no longer regarded as a generator of symmetry (the *a posteriori* lapse function demonstrates that it generates true dynamical evolution, rather than gauge transformation). Instead, the Hamiltonian associated with many-fingered intrinsic time translation can be read off from the master constraint; and the gauge invariant physical Hamiltonian generating global cosmic time translation can be inferred. With respect to the cosmic clock of our expanding universe, temporal causation and time-ordering of quantum evolution are now all gauge invariant, and reassuringly consistent with the spatial diffeomorphism symmetry of the theory. *Were four-covariance to be insisted upon, time and temporal ordering would have no invariant meaning; and the divide between such a theory of quantum gravity and well-established understandings of time-ordered quantum theories in the rest of physics would be insurmountable.*

All this is recounted in Chapter 2; the analogy of with the simple relativistic point particle in Sec. 2.11 is a quick route to comprehend the main ideas of the path from GR as a first class system to the final physical Hamiltonian via the provisional master constraint setup (the *a posteriori* vs *a priori* value of the Lagrange multiplier therein is also noteworthy). Chapter 3 discusses how the Hamiltonian constraint can be dispensed with; but the same *a posteriori* value of the lapse function in Einstein's theory is reproduced by a true Hamiltonian which generates global intrinsic time evolution. The paradigm shift means that higher spatial derivative terms can be added to the formalism without disturbing the consistency and physical gauge symmetry of the theory. Horava's program can now be consistently realized in the canonical framework that is erected [36]; with the reassurance that it will yield agreement — this was also one of the challenges of Horava's original formulation — with Einstein's theory in the appropriate limits. As discussed in this book, this requires H_{phys} to assume a

---

[6]The situation is akin to Lagrange multipliers in simple mechanical systems taking on physical significance and fixed values when the EOM and constraints are solved exactly.

square root form.[7] An additional Cotton-York term in the Hamiltonian is a salient feature; its far-reaching consequences,[8] including its similarity to the magnetic field in renormalizable Yang-Mills theory, are discussed in other chapters.

If the requirement of formulating the theory as a first class system is dropped, Einstein's theory and its Horava extensions can be consistently framed as Dirac second class systems [30] by introducing appropriate supplementary conditions, one to each constraint (this doubling results in a new total set of constraints. In second class systems, these are strongly vanishing c.f. weakly vanishing in first class systems). This is addressed in detail in Chapter 8. In a second class system all variables have identically zero Dirac brackets with all the constraints and supplementary conditions. So all entities on the total constrained surface are bona fide observables; in first class systems, the task of identifying observables which are required to Poisson commute (weakly) with all the constraints is formidable.

Horava's bold proposal triggered substantial initial interest among particle physicists and GR researchers. For GR practitioners, consistency problems described earlier dampened the initial interest. When Petr visited Taiwan in early 2014, we shared our own enthusiasm and tried to convince him, without much success, that his breakthrough proposal is correct and ought to be pursued vigorously within the framework of a consistent canonical theory.

ITG actualizes, in a new vista to overcome the many conceptual and technical challenges of quantum gravity, Dirac's conviction that four-dimensional symmetry is not a fundamental symmetry of nature and Horava's proposal of modifying Einstein's theory. The transverse-traceless Cotton-York tensor contains two d.o.f.; and it can be shown (see Chapter 10) that its changes are one-to-one with physical transverse-traceless excitations of the spatial metric. It can be thought of as the non-perturbative manifestation of the d.o.f. in gravitation; indeed at the ultraviolet end its contribution dominates the graviton propagator, and the it takes over as the two dynamical d.o.f.; moreover, since there is no mismatch in the number of d.o.f, there should not be a phase transition. As is well-known (and as discussed in Chapter 10), the Cotton-York tensor density $\tilde{C}^i_j$ is conformal invariant; its dominance of the Hamiltonian in the ultraviolet regime is a indicative of the existence of an ultraviolet fixed point in the

---

[7]The kinetic operator of the Hamiltonian has a new interpretation in terms of Klauder's momentric variables [39]. This is addressed in Chapter 5.

[8]We shall discuss in Chapter 10 its role in very origin of our universe.

renormalization group flow. A theory with ultraviolet fixed point is well-defined at arbitrarily small scales, raising the possibility that it may an ultimate, rather than an effective, field theory! Infrared divergences are tamed by spatial compactness; a spatially closed universe is part and parcel of ITG.

On the quantization of the gravitational field, eminent quantum field theorist Steven Weinberg had questioned [41] "Why in the world should anyone take seriously Einstein's original theory,[9] with just the Einstein-Hilbert action in which only two derivatives act on metric fields? Surely that's just the lowest order term in an infinite series of terms with more and more derivatives. In such a theory, loops are made finite by counter terms provided by the higher order terms in the Lagrangian." The first part of the contention is reasonable; but a Lagrangian theory of gravitation predicated upon the four-covariant paradigm would neither consider nor incorporate the Cotton-York tensor[10] as a saving grace, nor would it regard a spatial diffeomorphism invariant Hamiltonian formulation which requires no four-dimensional constructs to be the final word.

1.7   Gravitation and local gauge invariance

The discovery of local gauge symmetries shaped the development of modern physics in the latter half of the twentieth century; and the Standard Model of particle physics which incorporates matter and forces electromagnetic, weak and strong within the framework of a renormalizable quantum field theory is its crowning achievement [42]. In Maxwell's theory it was already noted the gradient of a function could be added to the vector potential without affecting the magnetic field. "From the long view of this history of mankind — seen from, say, 10,000 years from now — there can be little doubt that the most significant event of the 19th century will be judged as Maxwell's discovery of the laws of electromagnetism". This was Feynman's assessment of Maxwell's achievement.

Invariance of the action under general coordinate transformations was a guiding principle in Hilbert's 1915 derivation of the Einstein field equations from an action. In 1918 Hermann Weyl advocated the idea [43]

---

[9]The Loop Quantum Gravity community by and large believes in and works on completing this non-perturbatively.

[10]In ITG, every R-term of Einstein's theory is tempered by $\tilde{C}^i_j \tilde{C}^j_i$ which contains six derivatives of the metric. The Hamiltonian is also of the square root form; technically, through power series expansion there are "more and more derivatives".

of gauging a scale locally at each point by extending the metricity condition to $\nabla_\alpha g_{\beta\gamma} = A_\alpha g_{\beta\gamma}$ (this can be thought of as extending the covariant derivative ∇_α to include the connection A_α). The condition is invariant under transformation to $g'_{\beta\gamma} = e^\eta g_{\beta\gamma}, A'_\alpha = A_\alpha + \partial_\alpha \eta$; as is the curvature $F_{\mu\nu} = \partial_\mu A_\nu - \partial_\nu A_\mu$. Weyl's unification of electromagnetism and gravitation with the simplest invariant action of the form $S = \int (R^2 + \mathbf{g} F_{\mu\nu} F^{\mu\nu}) \sqrt{-g}\, d^4x$ led to fatal flaws noted by Einstein: proper time in clocks would depend upon location and history, and the broadening of spectral lines would occur, contradicting the sharp spectral lines that are observed.

After the development of quantum mechanics, Weyl (as well as Vladimir Fock and Fritz London) modified local scaling to the phase of the complex wave function, $\psi(x) \rightarrow e^{i\eta(x)}\psi(x)$; this symmetry is associated with the conservation of electric charge. Today, quantum electrodynamics is referred to as $U(1)$ "gauge invariant", after Weyl's "*Eichinvarianz*". In the effort to describe weak forces in elementary particles, Chen Ning Yang and Robert Mills in 1954 generalised the principle of local gauge invariance to non-abelian[11] $SU(2)$ [18]. Non-abelian Yang-Mills gauge theories are part and parcel of the Standard Model and its success as a renormalizable quantum field theory. A principal fiber bundle is the mathematical structure associated with the connection, curvature and the gauging of Lie groups.

In gravitation similarities and subtle, but profound, differences with Yang-Mills theories are present. Is Einstein's theory just a Yang-Mills theory, with gauge invariance under general coordinate transformations? Is it just a principal frame bundle i.e. a Yang-Mills theory on (pseudo)Riemannian base manifolds with the additional structure of the freedom of choosing at each point an orthonormal basis for the tangent space? What is the "correct" gauge group[12] of gravitation? Is gravitation even a fiber bundle theory? Judging from the number of articles that has been produced, the last question may seem provocative; but rather than focusing on a particular Lie group or being subject to the mathematical straightjacket of a fiber bundle, it is essential to note that *local gauge invariance* means there are redundant or unphysical d.o.f. in the *field* variables that are used to describe the theory; so *local gauge invariance* is associated with *field theories*; and in the quantum context, with *quantum field theories*

---

[11]Ronald Shaw working under Abdus Salam independently made the same proposal in his doctoral thesis.

[12]A survey and collection of reprints of original articles, with elucidating comments, on various models of gravitation as a gauge theory can be found in Ref. [44].

and not quantum mechanics. In the context of a field theory the generators of the symmetry are not mere Lie algebra elements of a finite-dimensional group, but they are infinite-dimensional generators of field transformations at each point. For instance, local gauge invariance of the pure Yang-Mills field theory is encapsulated in the Gauss Law constraint $\mathcal{G}^a := D_i \tilde{\pi}^{ia} = 0$, with $\tilde{\pi}^{ia}$ being the conjugate momentum of the gauge connection; it is $\mathcal{G}^a(x) = D_i \tilde{\pi}^{ia}(x)$ which generate the transformation of the fields at each point, and not Lie algebra elements of the fiber. The latter trivially commutes with the field variables. Also, the local gauge transformations that are generated through commutation with $\mathcal{G}^a$ are *infinitesimal gauge transformations*.

Often there is confusion that the distinction between gravity as a gauge theory and ordinary Yang-Mills theory is that gauge transformations in Yang-Mills are "internal" transformations, such as a phase multiplication, at the same point; whereas in gravitation they are "external" transformations comparing the fields at different spacetime points associated with external spacetime transformations e.g.

$$g'^{\mu\nu}(x') = \frac{\partial x'^\mu}{\partial x^\alpha} \frac{\partial x'^\nu}{\partial x^\beta} g^{\alpha\beta}(x). \tag{1.9}$$

This distinction is an unfortunate fallacy, which will be revealed by the symmetry generators and constraints of the theories involved. To add to this confusion, it is often argued there is an "active versus passive" view: that for infinitesimal coordinate transformations

$$x^\mu \to x'^\mu = x^\mu - \xi^\mu(x), \tag{1.10}$$

$$g'^{\mu\nu}(x') = g^{\mu\nu}(x) - \xi^\nu_{,\alpha} g^{\mu\alpha}(x) - \xi^\mu_{,\alpha} g^{\alpha\nu}(x) + \mathcal{O}(\xi^2);$$

and by Taylor expansion,

$$g'^{\mu\nu}(x') = g'^{\mu\nu}(x^\alpha - \xi^\alpha(x)) = g'^{\mu\nu}(x) - g^{\mu\nu}_{,\alpha} \xi^\alpha + \mathcal{O}(\xi^2).$$

So comparing at the same point,

$$g'^{\mu\nu}(x) - g^{\mu\nu}(x) = -\xi^\nu_{,\alpha} g^{\mu\alpha}(x) - \xi^\mu_{,\alpha} g^{\alpha\nu}(x) + g^{\mu\nu}_{,\alpha} \xi^\alpha;$$

indeed the induced change

$$g'^{\mu\nu}(x) - g^{\mu\nu}(x) = \delta_{4-diffeo.} g^{\mu\nu}(x) \tag{1.11}$$

is a four-diffeomorphism associated with the vector field $\xi^\mu(x)$. However, the distinction between "internal" and "external" is not merely a "point of view"; since one may ask whether the "gauge group" of Einstein's theory is $GL(4, R)$ as generally $\frac{\partial x'^\mu}{\partial x^\alpha} \in GL(4, R)$; or are the gauge transformations only *infinitesimal* translations from the "active" point of view? Or perhaps the Poincaré or de Sitter group? Note these Lie groups, as fibers, do not even have the same number of generators! Four-diffeomorphisms are symmetries of the Einstein-Hilbert action, but are they true gauge symmetries of classical gravitation? Of quantum gravity?

The surest and soundest path to extract and reveal the true gauge symmetry and the d.o.f. of a theory is the canonical treatment developed by Dirac [30]. In the book there is a *reductio ad absurdum* example of "1 = 0", a caution that Lagrangian methods should not be devoutly trusted. On the same note, many of the theories of gravitation surveyed in Ref. [44] lack proper canonical analyses, this also complicates determination of the d.o.f. in the theories. As will be discussed and espoused, in gravitation the gauge transformations are spatial diffeomorphisms which compare changes of the fields *at the same point*, just as in Yang-Mills theory. A true distinction that is advocated and emphasized in ITG is that gravity is geometrodynamics with the spatial (not spacetime) metric being the basic configuration field variable; whereas in the case of Yang-Mills it is the connection that is fundamental. What about alternative formulations of gravitation with connection as fundamental configuration variables? We are of the opinion geometrodynamics bequeathed with well-defined concept of space-like hypersurfaces is requisite for causality; the Cotton-York tensor which is crucial in the Horava ultraviolet completion of Einstein's theory is also available only to geometrodynamics.

1.7.1 *Canonical analysis of Einstein-Hilbert action*

The fundamental symmetries of Einstein's theory can be revealed by carrying out its full canonical analysis. The Einstein-Hilbert action for pure GR is

$$S = \frac{c^3}{16\pi G} \int R\sqrt{-g}d^4x. \tag{1.12}$$

(In the presence of non-vanishing cosmological constant Λ, the four-dimensional Ricci scalar curvature term should be modified from $R \to R - 2\Lambda$.) As space-time is not inert in gravitation, the canonical analysis has to be carried out consistently without the benefit of a fixed background; this had to wait until the work of Dirac and ADM [10,15]. The four-dimensional metric can be expressed, essentially without loss of generality, in the ADM form as

$$ds^2 = g_{\mu\nu}dx^\mu dx^\nu = -N^2(dt)^2 + q_{ij}(dx^i + N^i dt)(dx^j + N^j dt). \tag{1.13}$$

Equivalently,

$$g_{\mu\nu} = \begin{bmatrix} q_{ij}N^iN^j - N^2 & q_{ij}N^i \\ q_{ij}N^j & q_{ij} \end{bmatrix}. \qquad (1.14)$$

Straightforward substitution of this decomposition into the action yields,

$$S = \int dt d^3x \, (\tilde{\pi}^{ij} \frac{dq_{ij}}{dt} - N^iH_i - NH) + \text{boundary term}; \qquad (1.15)$$

wherein several important features are revealed: only the spatial or three-metric, q_{ij}, and its conjugate canonical momentum, $\tilde{\pi}^{ij}$, are dynamical variables; and the shift and lapse functions, respectively N^i and N, play the role of Lagrange multipliers which are associated, respectively, with the momentum and Hamiltonian constraints,

$$H_i = -2q_{ik}\nabla_j\tilde{\pi}^{jk} = 0, \qquad (1.16)$$

$$H = \frac{8\pi G}{c^3\sqrt{q}}(q_{ik}q_{jl} + q_{il}q_{jk} - q_{ij}q_{kl})\tilde{\pi}^{ij}\tilde{\pi}^{kl} - \frac{c^3\sqrt{q}}{16\pi G}\,^3R = 0; \qquad (1.17)$$

wherein ^3R is the Ricci scalar curvature of the spatial metric.[13]

1.7.2 *Counting the d.o.f. in canonical theory of gravitation*

There are 6 field d.o.f. associated with the canonical pairs $(q_{ij}, \tilde{\pi}^{ij})$ for GR in four dimensions. These are symmetric in the indices ij, and they are subject to four constraints which are first class. The constraints obey the Dirac algebra,

$$\{H_i[N^i], H_j[M^j]\}_{P.B.} = H_i[(\mathcal{L}_{\vec{N}}\vec{M})^i], \qquad (1.18)$$

$$\{H_i[N^i], H[M]\}_{P.B.} = H[\mathcal{L}_{\vec{N}}M], \qquad (1.19)$$

$$\{H[N], H[M]\}_{P.B.} = H_i[q^{ij}(N\partial_j M - M\partial_j N)]; \qquad (1.20)$$

wherein the smeared constraints are defined as

$$H[N] := \int N(x)H(x)d^3x; \qquad H_i[N^i] := \int N^i(x)H_i(x)d^3x. \qquad (1.21)$$

By the usual canonical method of counting d.o.f. in a first class constrained system, there are $(6-4) = 2$ net field d.o.f. in the theory (in Chapter 8, GR and its extensions are discussed within the context of second class Dirac

---

[13]In later discussions, when there is no fear of confusion, for simplicity R will be used to denote the same spatial entity ^3R.

constraint systems; and there are still 2 d.o.f. in these theories). H_i do form a closed algebra of the generators of spatial diffeomorphisms, and H does transform as a spatial scalar density. However, the commutator between the Hamiltonian constraint at different points depends on the dynamical variable q_{ij}; so the Dirac algebra comes with structure functions instead of structure constants. Taken together, the four constraints do not generate the algebra of four-dimensional diffeomorphisms. The Hamiltonian constraint is also quadratic in the canonical momentum; its interpretation as a generator of local gauge symmetry should be treated with caution.

1.7.3 *Spatial diffeomorphisms as local gauge symmetries*

The momentum constraint (also called diffeomorphism constraint) are associated with spatial diffeomorphism symmetry. Its action on the phase space variables produces changes which are Lie derivatives,

$$\{f(q_{ij}, \tilde{\pi}^{ij}), H_k[N^k]\}_{\text{P.B.}} = \mathcal{L}_{\vec{N}} f. \tag{1.22}$$

The Lie derivative of a tensor is defined by

$$\mathcal{L}_{\vec{\xi}} T^{i_1 \ldots i_m}_{j_i \ldots j_n}(x) = \xi^k \partial_k T^{i_1 \ldots i_m}_{j_1 \ldots j_n}(x) - T^{k i_2 \ldots i_m}_{j_1 \ldots j_n}(x) \partial_k \xi^{i_1} - \ldots$$
$$- T^{i_1 \ldots k}_{j_1 \ldots j_n}(x) \partial_k \xi^{i_m} + T^{i_1 \ldots i_m}_{k j_2 \ldots j_n}(x) \partial_{j_1} \xi^k + \ldots$$
$$+ T^{i_1 \ldots i_m}_{j_1 \ldots k} \partial_{j_n} \xi^k.$$

Lie derivatives obey the Leibniz rule for derivations. Under general coordinate transformations, a tensor behaves as,

$$T'^{i_1 \ldots i_m}_{j_1 \ldots j_n}(x') = \frac{\partial x'^{i_1}}{\partial x^{k_1}} \ldots \frac{\partial x'^{i_m}}{\partial x^{k_m}} \frac{\partial x_{l_1}}{\partial x'^{j_1}} \ldots \frac{\partial x_{l_n}}{\partial x'^{j_n}} T^{k_1 \ldots k_m}_{l_1 \ldots l_n}(x), \tag{1.23}$$

and an infinitesimal coordinate transformation $x'^i = x^i - \xi^i(x)$ induces the change

$$T'^{i_1 \ldots i_m}_{j_1 \ldots j_n}(x) - T^{i_1 \ldots i_m}_{j_1 \ldots j_n}(x) = \mathcal{L}_{\vec{\xi}} T^{i_1 \ldots i_m}_{j_i \ldots j_n}(x), \tag{1.24}$$

at the same point.

Although it is possible to interpret H_i as the causing the same effect of an infinitesimal general coordinate transformation, it should be emphasised that

- the true symmetry of the theory is dictated by the form of the constraints and the *precise transformations they generate on the dynamical variables*; not by changes in the coordinates which are just dummy integration variables in both the action and in the Hamiltonian.

- $H_i(x)$ generates (spatial) diffeomorphisms which are changes in the dynamical fields evaluated at the *same* coordinate point,

$$\delta_{\vec{N}} q_{ij}(x) = \{q_{ij}(x), H_k[N^k]\}_{P.B.}$$
$$= \mathcal{L}_{\vec{N}} q_{ij}(x), \tag{1.25}$$
$$\delta_{\vec{N}} \tilde{\pi}^{ij}(x) = \{\tilde{\pi}_{ij}(x), H_k[N^k]\}_{P.B.}$$
$$= \mathcal{L}_{\vec{N}} \tilde{\pi}^{ij}(x). \tag{1.26}$$

In this aspect there is no distinction between usual Yang-Mills gauge transformations which are deemed "internal" symmetries, while GR is often naively associated with "external" symmetry due to the confusion with *coordinate* transformations.

The canonical analysis clearly reveals coordinate labels are inert and "transformation of coordinates" play no role in the description of symmetries of the dynamical variables. Thus as far as spatial diffeomorphisms are concerned, they are as valid and genuine *gauge symmetries* of Einstein's theory on equal footing with the usual Yang-Mills "internal" gauge symmetries.

1.7.4 *Changes associated with the Hamiltonian constraint*

The interpretation of the changes associated with the Hamiltonian or Wheeler-DeWitt constraint [9, 12], $H(x) = 0$, is not as straightforward, due primarily to its quadratic dependence on momentum. It can be shown that

$$\delta_N q_{ij} = \{q_{ij}, H[N]\}_{P.B.}$$
$$= \frac{16\pi G N}{c^3 \sqrt{q}} (q_{ik} q_{jl} + q_{il} q_{jk} - q_{ij} q_{kl}) \tilde{\pi}^{kl}$$
$$\cong 2N K_{ij} = \mathcal{L}_{Nn^\mu} q_{ij}, \tag{1.27}$$
$$\delta_N \tilde{\pi}^{ij} = \{\tilde{\pi}^{ij}, H[N]\}_{P.B.}$$
$$\cong \mathcal{L}_{Nn^\mu} \tilde{\pi}^{ij}, \tag{1.28}$$

wherein the congruence symbol $\cong$ denotes equality modulo EOM and constraints.

Thus the Hamiltonian constraint generates a diffeomorphism or Lie derivative in the normal direction to the Cauchy surface only on the constraint surface and *on-shell* (i.e. when the EOM are imposed). Off-shell quantum fluctuations will spoil all this. More importantly, and as explained and emphasized in ITG, compared to the arbitrariness of its *a priori* value, the value of the lapse function N is fixed *a posteriori* and it is physical.

This is akin, in simple mechanical systems, to Lagrange multipliers taking on physical significance when they are solved *a posteriori* from the EOM (for the case of GR, see Chapters 2 and 3; and also the relativistic point particle analogy at the end of Chapter 2). This is in contradistinction with true gauge parameters which are arbitrary and cannot be solved (in a first class system without imposition of supplementary conditions) in terms of other variables, even when the EOM and constraints are imposed. The conclusion is H does generate a diffeomorphism, but with $N_{a\,posteriori}$; this is true physical evolution and not an arbitrary *gauge* transformation. It offers a resolution on the reason true *physical evolution* (this is not a boundary effect) occurs in Einstein's theory, despite a total Hamiltonian consisting entirely of constraints.

1.7.5 *Local gauge symmetry generators are first-order in canonical momentum*

In quantum field theories, true local gauge symmetries have clear physical meaning in the context of first class Dirac quantization [30]. Physical quantum states are to be annihilated by the constraints; this has the direct interpretation that the state is invariant under symmetry transformations of the fundamental configuration field variables. For instance, in the case of Yang-Mills gauge theories, the Gauss Law constraint is

$$\mathcal{G}^b(x)\Psi[A_{ia}] = D_j\tilde{\pi}^{jb}(x)\Psi[A_{ia}] = 0. \tag{1.29}$$

With the conjugate momentum of A_{ia} realized in the connection representation by the field operator,

$$\tilde{\pi}^{ia} = \frac{\hbar}{i}\frac{\delta}{\delta A_{ia}}, \tag{1.30}$$

Taylor expansion leads to,

$$\begin{aligned}
\Psi[A_{ia} + \delta_{gauge}A_{ia}] &= \Psi[A_{ia}] + \int (\delta_{gauge}A_{jb}(x))\frac{\delta\Psi}{\delta A_{jb}(x)}d^3x \\
&= \Psi[A_{ia}] + \frac{i}{\hbar}\left(\int \eta_b\mathcal{G}^b d^3x\right)\Psi[A_{ia}] \\
&= \Psi[A_{ia}],
\end{aligned} \tag{1.31}$$

wherein for a Yang-Mills infinitesimal gauge transformation, $\delta_{\text{gauge}}A_{jb} = -D_j\eta_b$; and integration by parts over closed manifold (or with vanishing gauge parameter η_b on the boundary, if there is one) has been performed. This manifests the *invariance under local gauge transformations* of all physical states.

In precisely the same manner, with regard to spatial diffeomorphisms in GR,

$$\Psi[q_{ij} + \delta_{\text{diffeo.}} q_{ij}] = \Psi[q_{ij}] + \int (\delta_{\text{Diffeo.}} q_{ij}) \frac{\delta \Psi}{\delta q_{ij}} d^3 x$$

$$= \Psi[q_{ij}] + \frac{i}{\hbar} \left(\int N^i H_i d^3 x \right) \Psi[q_{ij}]$$

$$= \Psi[q_{ij}]. \tag{1.32}$$

This comes about because

$$\delta_{\text{diffeo.}} q_{ij} = \mathcal{L}_{\vec{N}} q_{ij} = \nabla_i N_j + \nabla_j N_i, \tag{1.33}$$

and the momentum operator in the metric representation is

$$\tilde{\pi}^{ij} = \frac{\hbar}{i} \frac{\delta}{\delta q_{ij}}, \tag{1.34}$$

and spatial diffeomorphism constraint annihilates quantum states i.e.

$$H_i \Psi[q_{ij}] = 0. \tag{1.35}$$

In the above, both the Yang-Mills and GR constraint generators are first order in the corresponding canonical momenta. For a constraint operator such as H which is quadratic in the momentum, the Taylor expansion and the above interpretation of invariance of quantum states under gauge transformation of the configuration variables fail. Instead, the quantum version of Hamiltonian constraint is the Wheeler-DeWitt equation [9, 12] which governs quantum dynamics; but "time must be determined intrinsically" [12]. The DeWitt supermetric of Einstein's theory, $G_{ijkl} = \frac{1}{2}(q_{ik}q_{jl} + q_{il}q_{jk} - q_{ij}q_{kl})$ in the Hamiltonian constraint has signature $(-, +, +, +, +, +)$ which singles out the sole negative eigenmode as a preeminent intrinsic time variable to deparameterize the theory. This intrinsic time interval corresponds to $\delta \ln q$, but many-fingered time is the immediate consequence; as discussed earlier in Sec. 1.5, it is spatial diffeomorphism invariance that comes to the rescue.

An oft-quoted rule of thumb attributed to Dirac is that first class constraints generate gauge transformations. By this criterion the first class Dirac algebra of GR would imply the Hamiltonian constraint is also responsible for a gauge symmetry as well. The exact statement on page 21 of Dirac's book, Ref. [30], is that these constraints "... lead to changes in the q's and p's that do not affect the physical state." The clause ending the sentence is crucial; on this count H fails the criterion both in the quantum and classical context. As demonstrated above, it certainly cannot be interpreted as leaving the quantum state invariant; and the *a posteriori* lapse function implies H generates *nontrivial physical evolution*.

1.7.6 General Relativity possesses only spatial diffeomorphism invariance

The analyses above and the decoupling of H as a generator of symmetry imply that GR does not possess the full gauge symmetry of four-dimensional diffeomorphisms, but spatial diffeomorphism invariance is genuine. This is corroborated by the fact that the Dirac algebra in GR is not the algebra of four-dimensional diffeomorphisms. Fundamental reality of time and four-covariance are not compatible with each other. Neither is perturbative renormalizability and unitarity of GR compatible with four-covariance. The fact that only the spatial metric is dynamical and the simplification of the Hamiltonian analysis of GR led Dirac to conclude and declare in no uncertain terms "four-dimensional symmetry is not a fundamental property of the physical world" [15]. In his seminal article, Wheeler emphasized that classical spacetime is a concept of "limited applicability", because quantum gravity states cannot generically be infinitely peaked at a classical configuration. If spacetime does not exists fundamentally, why should four-dimensional symmetry be an overriding concern? A three-geometry is the equivalence class of a metric modulo spatial diffeomorphisms; and it is three, rather than four, geometry, that is fundamental in quantum geometrodynamics [9]. Even classical geometrodynamics is the dynamics of three-geometry, and not of four-geometry, the apparent "four-covariance" of the Hilbert-Einstein action notwithstanding.

A consistent quantum theory of gravity should recognise the fundamental nature of spatial diffeomorphism as a genuine gauge symmetry which perfectly parallels Yang-Mills symmetry. But a theory of gravitation should be based, not on a Yang-Mills connection variable, but on the primacy of the spatial metric. From gravitation must come Wheeler's "notion of simultaneity", the well-defined hypersurface of *spacelike separated points*; on which fundamental quantum commutation (or anti-commutation) relations of all fields, gravitational or not, is defined; and on which microcausality hinges. With the paradigm shift from four-covariance, the fundamental reality of time (gauge invariant under spatial diffeomorphisms) is redeemed, together with the substantiation of its indispensable role in temporal causation and quantum dynamics. Gravitation no longer stands apart from the rest; its geometrical foundation on fundamental metric variable remains untouched by the shift, but its true symmetries can be understood within the same context, and its workings can be reconciled with the dynamics and triumphs of modern quantum gauge field theories.

Table 1.1   Gravitational and Yang-Mills gauge symmetries: similarities and differences.

	Spatial diffeomorphism gauge symmetry	Yang-Mills gauge symmetry
Fundamental variable	Spatial metric tensor q_{ij}	Gauge connection A_{ia}
Symmetry generators[♮]	$H_i(x) = -2q_{ik}\nabla_j \tilde{\pi}^{jk}(x)$	$\mathcal{G}^a(x) = D_i \tilde{\pi}^{ia}(x)$
Gauge transformation	$\frac{1}{i\hbar}\left[q_{ij}(x), H_k[N^k] \right]_- = \mathcal{L}_{\vec{N}}\, q_{ij}(x);$	$\frac{1}{i\hbar}\left[A_{ia}(x), \mathcal{G}^b[\eta_b] \right]_- = -D_i \eta_a(x);$
	$H_i[N^i] = \int N^i H_i d^3 x$	$\mathcal{G}^b[\eta_b] = \int \eta_b \mathcal{G}^b d^3 x$
Group algebra	$[H_i(x), H_j(y)]_- = (i\hbar)(H_j(x)$	$\left[\mathcal{G}^a(x), \mathcal{G}^b(y) \right]_-$
	$\partial_i \delta^3(x,y) - H_i(y)\partial_j \delta^3(x,y))^{\sharp}$	$= (i\hbar)f^{ab}{}_c \mathcal{G}^c(x)\delta^3(x,y)$
Potential	$V \propto \tilde{C}_j^i \tilde{C}_i^j, \quad \tilde{C}^{ij} = \frac{\delta(W_{CS})^b}{\delta q_{ij}}$	$V \propto q_{ij}\tilde{B}^{ia}\tilde{B}^{ja}, \quad \tilde{B}^{ia} = \frac{\delta CS}{\delta A_{ia}}$
	Not product of same group at each point of base manifold.	Infinite product $\prod_x G_x$ of same group G (finite dimensional
Group Properties	Not of principal fiber bundle structure (cannot identify a "fiber" at each point); but $H_i(x)$ generates the infinite-dimensional group of diffeomorphisms.	Lie group with same structure) constants $f^{ab}{}_c$ usually called the "Yang-Mills gauge group". Principal fiber bundle structure with fiber G.

[♮]The constraints enforce transversality of the conjugate momentum.
[♯]Note the partial derivative acting on the delta function c.f. delta function for Yang-Mills.
[b]W_{CS} is the gravitational analog of the Yang-Mills Chern-Simons functional CS.

1.7.7   Physical understanding of spatial diffeomorphism invariance in gravitation

Focusing on the spatial diffeomorphisms as the genuine local gauge symmetries of GR on equal footing with Yang-Mills gauge symmetry, there are similarities and subtle but important differences. Table 1.1 provides a comparison of some of these attributes.

There are several salient features. By far the most important is *gauge invariance is the statement of (covariant) transversality of the conjugate momentum*. The corresponding Gauss Law constraints $H_i = 0, \mathcal{G}^a = 0$ are the mathematical actualization of this physical requirement. The gauge transformations generated on fundamental connection and metric variables may appear quite different (gauge covariant derivative $D_i\eta^a$ for Yang-Mills connection, and Lie derivative for the spatial metric), but they are really similar, since $\mathcal{L}_{\vec{N}}q_{ij} = \nabla_i N_j + \nabla_j N_i$ in covariant derivative form. The auxiliary (supplementary, or "gauge-fixing") conditions to turn the systems into Dirac second class are that the infinitesimal[14] physical excitations of the configuration variables must be transverse as well i.e. $D^i\delta_{phys}A_{ia} = 0; \nabla^j\delta_{phys}q_{ij} = 0$ (this is discussed in detail in Chapters 8

[14]Recall that Gauss Law constraints only generate infinitesimal gauge transformations.

and 9). Thus gauge symmetry of Gauss Law constraints can be understood physically as *enforcing the absence of "longitudinal modes"*, to guarantee that physical excitations are transverse. In weak field perturbative quantum field theory about flat background, this requirement means only *massless pure* spin 1 (respectively spin 2), transverse excitations of Yang-Mills gauge bosons (respectively transverse graviton) are physically allowed, without the complications of longitudinal polarizations and massive modes. Ergo the viability and existence of long-ranged interactions for Yang-Mills and gravitation — the *raison d'etre* of Gauss Law constraints. It is also well known that an anomalous quantum field theory in which local gauge invariance is broken is generally pathological. The overriding physical concern of eliminating undesirable modes is the commonality between Yang-Mills gauge theories and gravitation, and the basis of gauge invariance. This transcends the consideration and role of principal (frame) bundles on four-manifolds, generalized spacetime isometries, the gauging of tangent space symmetries and/or spacetime translations etc. in determining the local gauge invariance of gravitation. These constructs fail to be even well-defined when four-dimensional spacetime loses its validity in the domain of quantum gravity.

Continuing with the table, in the algebra of constraints, the commutator between two Gauss Law generators of field transformations in Yang-Mills theory yields an ultralocal delta function. The group of gauge transformations in Yang-Mills theory is an infinite-dimensional group; it is generated by $\mathcal{G}^{ia}(x)$ which has, at each point, the group action of a finite-dimensional Lie group G with the same structure constants i.e. the "equivalent action" of a product group $\Pi_x G_x$, each G_x isomoprhic to G. This confirms a principal fiber bundle geometrical structure, with G as the fiber which is often called the "Yang-Mills gauge group" even though the group of gauge transformations on fields is infinite-dimensional and its generator $\mathcal{G}^{ia}(x)$ is a composite *field operator*. This has an important consequence in particle physics model building; staying within the context of a principal fiber bundle, the group G can be changed and its effects can be studied, without leaving the paradigm and physical framework of Yang-Mills gauge theories. In contradistinction, the group of spatial diffeomorphisms does not exhibit this fiber structure, because of the extra partial derivative on the delta function in the commutator of two diffeomorphism generators. While the diffeomorphism group is itself an infinite-dimensional group, there is no avenue to identify a Lie group fiber as in the context of fiber bundle theory. The loss of ultralocality can be traced to the dependence of the covariant derivative ∇_i on derivatives of the spatial metric which is the fundamental configuration variable (on the other hand, $\mathcal{G}^a$ depends on the connection A_{ia} but not its derivatives). These are mathematical technicalities, whereas transversality is the primary physical concern; however, the revelation of the non-product nature of diffeomorphisms casts doubts on the wisdom and relevance of confining models of gravitation to the Yang-Mills fiber bundle framework and on the attempts to "gauge" and test various theories of GR by changing the fiber, be it $GL(3, R)$ or $ISO(3, 1)$, or any other Lie group for this matter. At the very least, beyond the agreement of classical solutions, the viability of perturbative massless transverse spin 2 physics and the suppression of pathological modes should be demonstrated in the proposed alternatives. Moreover, gravitational wave detections in recent years [45] have spectacularly confirmed the two transverse-traceless long-ranged modes which arise from Einstein's theory of gravitation.[15] Chapter 6 covers gravitational waves from the perspective of ITG; the effects of the Cotton-York addition to gravitational waves and its "$\frac{1}{k^3}$ signature" in the early universe are discussed therein and also in Chapter 10. ITG does not alter the 2 d.o.f. of Einstein's geometrodynamics, nor does it differ with regard to physical excitations being transverse-traceless changes of the spatial metric.

In Chapter 5, it is pointed out Klauder's momentric variables [39] $\bar{\pi}_j^i(x)$ generate an $sl(3, R)_x$ algebra at each point; and it is possible to embed the infinite-dimensional group of spatial diffeomorphisms in $\Pi_x SL(3, R)_x$, because the relevant momentum constraint therein can be expressed in terms of momentric as $\mathcal{H}_i = -2\nabla_j \bar{\pi}_i^j$, and when smeared, $\mathcal{H}_i[N^i] = \bar{\pi}_i^j[2\nabla_j N^i] := \int (2\nabla_j N^i)\bar{\pi}_i^j \, d^3x$. As the kinetic operator is a Casimir invariant constructed out of the momentric operators, in the limit of vanishing potential, the "free" Hamiltonian of ITG possesses $\Pi_x SL(3, R)_x$ symmetry.

1.7.8 *Geometrodynamics, and gauging the Lorentz group*

In the absence of ideal classical four-dimensional spacetimes, it is imperative to define and quantize gravitation and matter based only upon fundamental spatial three-geometry and not four-geometry. Hitherto, the incorporation of fermions compatible with generic curved spacetime manifolds (these need

---

[15]If the *local gauge symmetry* of GR were the $GL(4, R)$ (which has 16 generators) of Eq. (1.9), it would be hard to explain why, modulo these symmetries, there are two observed metric d.o.f. in gravitational waves; or for that matter, why the metric or any of its components is an observable. This applies to all those proposed groups which have more generators than there are metric components.

not have any isometry, much less Lorentz isometry) is achieved by gauging the symmetry group of the tangent space which is isomorphic to Minkowski spacetime. This is the gauging of local Lorentz invariance as $SO(3,1)$. Quantum gravity spells the demise of four manifolds; we must turn instead to the spatial metric or dreibein to provide the necessary and sufficient framework.

1.7.9 $SL(2,C)$ Weyl fermions and the Lorentz group as $SO(3,C)$

The dreibein, e_{ai}, is related to the spatial metric through $q_{ij} = e_{ai}e^a{}_j$; with the metric being invariant under local full $SO(3,C)$ rotations of the dreibein. One of the miracles in Lie group theory is complex $SO(3,C)$ is actually isomorphic to $SO(3,1)$. Thus full Lorentz symmetry, including boosts and not just real spatial $SO(3)$ rotations, is already captured in this complex rotational freedom of the dreibein fields (they are the inverse of the (anti)self-dual triads [46]). The gauging of the Lorentz group as $SO(3,C)$ and its double covering group as $SL(2,C)$ allows the natural introduction of fundamental Weyl fermions compatible with the chiral nature of the Standard Model of particle physics. Under local Lorentz transformations, the left and right-handed chiral Weyl fermions transform as $\psi'_{L,R} = u_{L,R}\psi_{L,R}$, with

$$u_{L,R} = e^{(\mp\alpha_a + i\theta_a)\frac{\tau^a}{2}} \in SL(2,C)_{L,R} \qquad (1.36)$$

wherein τ^a are the Pauli matrices, boost and rotation parameters denoted by α_a and θ_a respectively, and $u_R^\dagger = u_L^{-1}$. On spatial hypersurfaces, right and left-handed spin connections, $A_{ia}^\pm$, in the adjoint representation transform as

$$A'^\pm_{ia}\frac{\tau^a}{2} = u_{R,L}A^\pm_{ia}\frac{\tau^a}{2}u_{R,L}^{-1} + u_{R,L}i\partial_i u_{R,L}^{-1}; \qquad A_{ia}^{\dagger\pm} = A_{ia}^{\mp}. \qquad (1.37)$$

The corresponding gauge-covariant derivative, for instance $D_i^{A^-}$, acting on the Weyl fermion fields (fermion fields are scalars under spatial diffeomorphisms) can be introduced. Under local Lorentz transformations,

$$(D_i^{A^-}\psi)' = u_L D_i^{A^-}\psi, \qquad iD_i^{A^-} := i\partial_i + A_{ia}^-\frac{\tau^a}{2}. \qquad (1.38)$$

Construction of the Lorentz-invariant fermionic Hamiltonian is carried out in detail in Chapter 9 which also reveals how the spin connections are built from the dreibein and its conjugate momentum. It should be noted only fundamental spatial dreibein and its conjugate variable, but not the full spacetime metric, are involved; no appeal or recourse to four-dimensional entities is needed to gauge the full local Lorentz symmetry. In this sense, *fermionic elementary particles can remain as fundamental entities even in the full regime of quantum gravity when four-dimensional spacetimes and their tangent spaces lose validity.* Furthermore, the chiral nature of fundamental fermionic particles transforming according to $SL(2, C)_{L,R}$ in Eq. (1.36) is naturally and consistently actualized by this gauging of the Lorentz symmetry. It must also be pointed out that unlike the gauging of only unitary groups to preserve norms of quantum mechanical wave functions; in QFT, the fermionic fields, though fundamental, are themselves not physical states, and there is no objection to the gauging of a non-unitary representation. This is what happens with regard to finite-dimensional (hence non-unitary) fermionic representations of the Lorentz group which is non-compact. Besides many other related topics, Chapter 9 also addresses the solution of all constraints in the presence of matter, the need and realization of generalised spin-structure with Yang-Mills gauge fields, and the question of chiral anomalies in quantum field theories and ITG.

1.7.10   *Physical Hamiltonian for Einstein's theory and its extensions*

In the previous sections, the incompatibility of four-covariance with renormalizability plus unitarity and with the fundamental existence of time has been elucidated and highlighted, together with the revelation that four-covariance is not a true local gauge symmetry of Einstein's theory. In Mark Twain's piquant words, "It ain't what you don't know that gets you into trouble. It's what you know for sure that just ain't so." Even though many have long-believed it is responsible for dynamics rather than a generator of symmetry, the Hamiltonian constraint has a perplexing role as one of the constraints making up the total ADM Hamiltonian which consists solely of constraints. In a four-covariant theory, apart from a spacetime constant, no combination of the fundamental variables can be can be invariant under four-diffeomorphisms; so it is difficult to believe a d.o.f. can play the role of time. But existence of a fundamental time variable does not conflict with just spatial diffeomorphim invariance.

On the nature of an inherent time, Einstein's theory does yield a clue, one that DeWitt already noticed [12], and one that is present in the decomposition of the phase space variables of geometrodynamics that is carried out in Chapter 2. In commuting with the rest of the variables, $(\tilde{\pi}, \ln q^{\frac{1}{3}})$ can be cleanly separated from them. $\delta \ln q^{\frac{1}{3}}$ is also precisely the sole negative eigenmode (thus the suggested intrinsic time interval) in the DeWitt superspace interval, $G^{ijkl}\delta q_{ij}\delta q_{kl}$, which is of signature $(-,+,+,+,+,+)$. In Einstein's theory, $G^{ijkl} = \frac{1}{2}(q^{ik}q^{jl} + q^{il}q^{jk}) - q^{ij}q^{kl}$ is the inverse of the De-Witt supermetric G_{ijkl} in the Hamiltonian constraint. Had this been the only upshot, we would still end up with a second-order-in-intrinsic-time Wheeler-DeWitt equation that had been the preoccupation of canonical quantum gravity for decades. There is more. First, by substituting the decomposition of the variables into the Hamiltonian constraint it is noted that a simple algebraic factorization is possible. The Hamiltonian constraint is equivalently,

$$0 = H = -\beta^2 \tilde{\pi}^2 + \bar{\pi}^j_i \bar{\pi}^i_j + \mathcal{V}(\bar{q}_{ij}, q) \qquad (1.39)$$

$$= -(\beta \pi - \bar{H})(\beta \pi + \bar{H}); \qquad (1.40)$$

$$\text{with} \quad \bar{H}(\bar{\pi}^i_j, \bar{q}_{ij}, q) = \sqrt{\bar{\pi}^i_j \bar{\pi}^j_i + \mathcal{V}(\bar{q}_{ij}, q)}; \qquad (1.41)$$

wherein $\beta^2 := \frac{1}{3(3\lambda-1)}$ plays the role of a deformation parameter in a generalized DeWitt supermetric compatible with spatial diffeomorphism invariance. Einstein's GR, with $\lambda = 1$ or $(\beta^2 = \frac{1}{6})$ and

$$\mathcal{V}(\bar{q}_{ij}, q) = -\frac{q}{(2\kappa)^2}[R(\bar{q}_{ij}, q) - 2\Lambda_{eff}], \qquad (1.42)$$

is a particular realization of a wider class of Horava gravity theories. $\bar{H}$ is independent of $\tilde{\pi}$, the trace of the canonical momenta;[16] thus the factorization implies the solution of the Hamiltonian constraint is $\tilde{\pi} = \pm\frac{1}{\beta}\bar{H}$. As will be discussed further in later chapters the chosen sign is related to direction of time and the physical Hamiltonian of ITG, $H_{phys} = \frac{1}{\beta} \int \bar{H} d^3x$ that is embraced. The gauge-invariant part of $\delta \ln q^{\frac{1}{3}}$ is $dT = \frac{2}{3} d\ln V$; equivalently,

$$T - T_{ref.} = \frac{2}{3}\ln(V/V_{ref}). \qquad (1.43)$$

For an ever-expanding closed universe and a Hamiltonian bounded from below, the branch

$$(\beta\tilde{\pi}(x) + \bar{H}(x)) = 0 \qquad (1.44)$$

should be adopted. The analogy of this equation with Dirac's $H(x) - W(x) = 0$ in the extended phase space proposal[17] has been pointed out earlier.

---

[16]In classical GR, $\tilde{\pi} = -\frac{\sqrt{q}}{\kappa} K$, with K being the trace of the extrinsic curvature.

[17]In Chapter 2, the expression of the *a posteriori* lapse function shows that when $\bar{H}$ becomes imaginary, the geometry can be analytically continued to Euclidean signature manifold.

The question of the reality of $\bar{H}$ which has a square root springs to mind; this is a concern which is addressed in Chapter 7. But it must be pointed out that at this juncture the situation is the same as, thus no worse than, Einstein's theory in which $H = 0$ or equivalently $\beta^2 \tilde{\pi}^2 = \bar{\pi}_i^j \bar{\pi}_j^i - \frac{q}{(2\kappa)^2}(R - 2\Lambda_{eff})$ requires, for real variables, the L.H.S. of this latter equation to be positive semidefinite. Some practitioners treat this as a restriction on the configurations that are allowed on the R.H.S.; in Chapter 7, the positivity of the Hamiltonian of ITG is discussed. The emergence of the Einstein Ricci scalar potential from a quantum commutator and regularization is elucidated; a simple harmonic oscillator analogy at the end of the chapter serves as a useful device to explain the essential reasonings.

Dirac quantization of Eq. (1.44) in the metric representation with[18] $\tilde{\pi} = \frac{\hbar}{i}\frac{\delta}{\delta \ln q^{\frac{1}{3}}}$ yields not a second order Wheeler-DeWitt equation, but a Schrödinger equation,

$$-\beta \frac{\hbar}{i} \frac{\delta}{\delta \ln q^{\frac{1}{3}}} \Psi = \hat{\bar{H}} \Psi. \tag{1.45}$$

Alas, it is an equation with Schwinger-Tomonaga time $\delta \ln q^{\frac{1}{3}}(x)$; as elucidated in earlier sections, consistent dynamical many-fingered evolution is problematic. The correct physical question to ask is not how the physical state of the universe changes with the local gauge-dependent intrinsic time, but how it evolves with the gauge-invariant global changes dT. Here, the use of $\delta \ln q$ (rather than q to some power) is crucial in that for the logarithmic function, the spatial dependence of its local variation can be exactly cancelled by a Lie derivative. The upshot is global dT, rather than many-fingered time interval, is precisely the gauge invariant part of $\delta \ln q^{\frac{1}{3}}(x)$; and w.r.t. the global time the physical dynamical quantum equation is

$$i\hbar \frac{d\Psi}{dT} = \frac{1}{\beta} \left(\int \hat{\bar{H}} \, d^3x \right) \Psi. \tag{1.46}$$

Chapter 2 recounts the transition from Einstein's theory with Hamiltonian *constraint* to the final Schrödinger equation and the emergence of classical spacetime from constructive interference. The *a posteriori* lapse function in classical solutions of GR theory is given by Eq. (2.43). In it the remarkable relation $N(x)dt \propto \frac{[\delta \ln q^{\frac{1}{3}}]_{equiv}}{H}$ emerges, wherein $[...]_{equiv}$ denotes equivalence up to spatial diffeomorphisms. This means that redshifts and effects of time dilation associated with the lapse function can in fact be understood in the physical terms of intrinsic time intervals dT and energy

---

[18]This satisfies the commutation relation Eq. (2.8) of Chapter 2.

density of the Hamiltonian (in ITG $H_{phys} \propto \int \bar{H} \, d^3x$). No resolution of the problem of time can be deemed complete before it can account for and clarify relativistic time dilation effects encountered in moving test particles and in the rate of test clocks placed at different locations in a non-constant external gravitational field. These signature effects of Einstein's theories of relativity (Special and General) can be encompassed and elucidated from a dynamical perspective in a single framework for a generic ADM space-time (see Sec. 2.9 of Chapter 2); and there is no conflict between these phenomena and the existence of a global intrinsic time.

1.7.11 *Preferred clock and the foliation of classical spacetimes*

In canonical ADM formalism, Einstein's theory comes with the Hamiltonian *constraint*, but instead of its *a priori* arbitrariness, the lapse function N takes on an *a posteriori* value on solving the EOM and Hamiltonian constraint. When the Hamiltonian constraint is dispensed with in ITG, as detailed in Chapters 3, 4, and 8, a physical Hamiltonian generating intrinsic time translation reproduces an effective value of N which equals the aforementioned *a posteriori* value. The nature and explicit form of either the *a posteriori* or effective lapse function carries a further significance. The hitherto assumed arbitrariness of the lapse and shift functions, N and N^i, in the foliation of spacetime is taken to be a vindication of the absence of preferred slicings and preferred frames in Einstein's theory of the universe. From Newton's bucket experiment[19] of 1689 to the debates, and the philosophical and physical reasonings of George Berkeley, Leibniz, Mach, Einstein (who coined the phrase "Mach's principle" [47]), Weyl, Wheeler, and many others, the existence or non-existence of preferred frames in the universe has long been a subtle topic of contention (see for instance Ref. [48], and references therein).

With the fall of four-covariance; and with its preferred cosmic clock, and the dependence of either *a posteriori* or effective lapse function on $\bar{H}$ (ergo energy density), ITG implies the foliation of spacetime is in fact determined by the distribution of local energy density[20] associated with the Hamiltonian generating cosmic time translation.

---

[19]Newton wrote, "Absolute space by its own nature, without reference to anything external, always remains similar and unmovable." In the bucket experiment Newton concluded it was the spinning of the bucket with respect to absolute space that curved the water surface.

[20]The same effective value and interpretation of N appears in the motion and accelerations of test particles (see Sec. 2.9 of Chapter 2).

1.7.12 *Consistent Horava completion of Einstein's theory*

The transition to the final global Hamiltonian of Chapter 2 and the fall of four-covariance naturally prompts the extension of Einstein's theory by modifying the potential without adding time derivatives. Horava's emphasis on the crucial role of the Cotton-York tensor [31], and considerations discussed in later chapters lead the ITG framework to settle on

$$H_{phys} = \frac{1}{\beta} \int \bar{H} d^3 x$$

$$= \frac{1}{\beta} \int \sqrt{\bar{\pi}_i^j \bar{\pi}_j^i - \frac{q}{(2\kappa)^2}(R - 2\Lambda_{eff}) + g^2 \hbar^2 \tilde{C}_j^i \tilde{C}_i^j} \; d^3 x. \quad (1.47)$$

In ITG, intrinsic time is dimensionless, and H_{phy} is of dimension $\hbar$, while the conjugate momentum to the spatial metric and the momentric variable are of dimension $\hbar/L^3$ and are density weight one tensors. The Cotton-York tensor density is of weight one and dimension $1/L^3$, it is third order in spatial derivatives of the metric. H_{phys} is the integral (over closed spatial manifold) of a density weight one spatial scalar; it is thus invariant under spatial diffeomorphisms. It commutes with the momentum or Gauss Law constraint $H_i = -2q_{ik}\nabla_j \tilde{\pi}^{jk} = 0$.

The Cotton-York tensor (density), $\tilde{C}^{ij} = \tilde{\epsilon}^{imn}\nabla_m(R^j{}_n - \frac{1}{4}\delta^j{}_n R)$, is special to three dimensions.[21] Its vanishing is the necessary and sufficient condition for conformal flatness (see Chapter 10, which discusses its properties and physical ramifications). In four-dimensions this role is played by the Weyl tensor which is identically zero in three dimensions. An action incorporating the square of the Weyl tensor associated with dimensionless coupling is renormalizable; alas, loss of unitarity, due to the presence of higher time derivatives, comes with this Weyl extension of GR. With only spatial diffeomorphism invariance, the (trace of the) square of the Cotton-York tensor density has just the right scalar weight, dimensionless coupling g, and higher spatial derivatives to be incorporated into the Hamiltonian of ITG. It is conformal invariant, and thus a function of the unimodular spatial metric $\bar{q}_{ij}$, instead of the full metric. The independence[22] of $\tilde{C}_j^i$ from q (hence the intrinsic time parameter) means that it dominates the early universe at small spatial volumes when the usual GR terms in the Hamiltonian are negligible, whereas at late times and large volumes

---

[21] Three-dimensional Levi-Civita $\tilde{\epsilon}^{imn}$ appears in its definition; and it is the functional derivative w.r.t. the spatial metric of the three-dimensional gravitational Chern-Simons invariant of the affine Christoffel connection.

[22] It is the mixed tensor density $\tilde{C}_j^i$, with one covariant and one contravariant index, that is conformally invariant (see Chapter 10).

the theory becomes Einsteinian. It is further discussed in Chapters 7 and 8, that the Cotton-York tensor is precisely the gravitational counterpart of the Yang-Mills magnetic field. Both are functional derivatives of their respective Chern-Simons invariant; the crucial difference is the spatial metric is fundamental to GR, whereas in Yang-Mills it is the gauge connection; consequently the Cotton-York tensor is third order in spatial derivatives of the metric. The incorporation of the Cotton-York tensor into the GR potential is as natural as the appearance of the magnetic potential in renormalizable Yang-Mills theory. These parallels are totally missed in a four-covariant paradigm obsessed with Lagrangians, Yang-Mills $F_{\mu\nu a}F^{\mu\nu a}$, and gravitational four-curvature invariants.

Although the Hamiltonian constraint is dispensed with, the same effective *a posteriori* value of lapse function of GR can be reproduced by the square root form of the Hamiltonian. This is demonstrated in Chapters 3 and 4. The square root form, which is not in Horava's original proposal [31], is essential to reproduce the correct correspondence with Einstein's theory. A square root Hamiltonian for Einstein's GR was first noted in Ref. [49], but with $\sqrt{q}$, rather than $\ln q$ and its consequent benefits, as intrinsic time. In DeWitt's seminal work [12] the intrinsic time variable is $q^{\frac{1}{4}}$. Misner, in Ref. [50], also deployed $\ln q$ as intrinsic time, but in a minisuperspace context.

The dispersion relation, or the relation between energy (Hamiltonian) and momentum is given by the explicit Hamiltonian of ITG above. In Special Relativity, it can be argued that, from the dynamical point of view, the "spirit of relativity" lies in the square root dispersion relation $E = \sqrt{p^2c^2 + m_0^2c^4}$ which supersedes the Newtonian $p^2/2m$; the square root form and Hamilton's equation yield the upper bound of c for all particle velocities (see Sec. 2.11 of Chapter 2); and the dispersion relation for massless particles is $E = pc$. In the Hamiltonian of ITG, the dispersion relation retains this "relativistic form" of a square root. In the limit of vanishing potential, the free theory Hamiltonian does yield $E \propto \int \sqrt{\bar{\pi}^i_j \bar{\pi}^j_i}\, d^3x$, but in general there are nontrivial potential terms. Perturbations about the de Sitter background with $k = +1$ (or S^3) closed spatial slicings are considered in Chapter 6; it is shown that classical gravitational waves are produced; with the dispersion relation of Eq. (6.78) for the perturbations. These excitations are "pure spin 2" i.e. transverse-traceless modes (they can be expanded in terms of S^3 tensor harmonics [51,52]), without the complication of longitudinal polarizations. Gravitational waves in the Cotton-York dominated era have a characteristic frequency dependence different from Einstein's theory (see Chapter 6 for the details); detection of these primordial gravitational waves will be a smoking gun that confirms the presence of the Cotton-York term in the Hamiltonian.

In the full Hamiltonian, the kinetic operator can be expressed totally in terms of Klauder's momentric variables [39], $\bar{\pi}^i_j(x)$; these generate an $sl(3, R)$ algebra at each point. The kinetic operator is moreover a Casimir invariant, which makes its action well-defined in the group theoretic sense. In ITG, the conjugate momentum is superseded by the self-adjoint momentric operator. The fundamental dynamical variables are the unimodular spatial metric and the momentric. The resultant commutation relations, as well as other features, are discussed in Chapter 5. Unitarity (which follows from self-adjointness of the Hamiltonian) and a Hamiltonian bounded from below are also desirable to make the quantum theory well-defined; Chapter 7 demonstrates that, with appropriate regularization, the Hamiltonian of ITG can be expressed as $\frac{1}{\beta} \int \sqrt{\tilde{Q}^{\dagger i}_j \tilde{Q}^j_i + q\mathcal{K}}$, with positive constant $\mathcal{K}$. In the Schrödinger equation β can be absorbed into dT by redefinition; or into $\bar{H}$ by redefining the couplings to the R, Λ_{eff} and Cotton-York terms and rescaling, in the kinetic term, the momentric variables which are generators of $sl(3, R)$.

1.8   ITG and quantum gravity as Schrödinger-Heisenberg quantum mechanics

1.8.1   *Circumventing the difficulties of the Wheeler-DeWitt equation*

The Wheeler DeWitt equation [12] which emerges from Dirac first class canonical quantization of the Hamiltonian constraint, $H(x) = 0$, of Einstein's theory is

$$\left[-\frac{\delta^2}{\delta\zeta^2} + \frac{32}{3\zeta^2} \bar{G}^{AB} \frac{\delta^2}{\delta\zeta^A \delta\zeta^B} + \frac{3}{32}\,^3R \right] \Psi = 0, \qquad \zeta = q^{\frac{1}{4}}. \qquad (1.48)$$

This singles out ζ as the intrinsic time variable ($\zeta^{A=1,\cdots,5}$ denotes the rest of the variables with $\bar{G}^{AB}$ as the positive-definite supermetric of this complementary set). It is the first quantum field theoretical equation of gravitation. Despite its intuitive beauty and emotive appeal that one is finally seeing the face of quantum gravity, there are at least three main formidable problems:

1) The equation has to hold at each spatial point because the intrinsic time, $\zeta(x)$, is a field variable. Time is many-fingered.

2) The solution of an infinite-dimensional second order functional differential equation is an intimidating task — one that may not even be mathematically well-defined.

3) It is second order in intrinsic time (i.e. of the "Klein-Gordon" type) and with a nontrivial potential; even if a well-defined solution can be found, the norm of the solution does not yield conserved positive probabilities. Probabilistic interpretation of the norm, first advocated by Max Born, is a key postulate of quantum mechanics.

All these three daunting problems are, remarkably, nimbly and cleanly resolved in ITG. The Hamiltonian of ITG, $H_{phys} = \frac{1}{\beta} \int \bar{H}(x) \, d^3x$, is a global entity integrated over all spatial points; cosmic time is global, and not many-fingered; the resultant quantum equation is the familiar first-order-in-time Schrödinger equation which is a equation for cosmic time-development, not an equation at each spatial point (unitarity or conservation of positive definite norm holds provided the Hamiltonian of ITG is self-adjoint); and there is a guaranteed formal solution to the Schrödinger equation no matter how complicated the theory is.

1.8.2 *The cosmic clock and Schrödinger-Heisenberg quantum mechanics*

In the long journey and progression from Newton, to Euler-Lagrange, to Hamilton, and to Schrödinger-Heisenberg quantum mechanics, the search for the fundamental equation(s) of mechanics had taken almost three centuries. But will the demise of spacetime in quantum gravity be science's ultimate assault on the "immutable ordering of happenings", and on cause and consequence?

Quantum mechanics, the last word (so far), is surprisingly, or rather, befittingly, the simplest description. In the quantum wave function lies all that can be known about the state of the system. That is a postulate. Its evolution in time is determined by the rate of change $\frac{d\Psi}{dT}$; quantum superposition is a fundamental principle of quantum mechanics, ergo the dynamical equation is $\frac{d\Psi}{dT} = \hat{L}\Psi$, wherein $\hat{L}$ is a linear operator ensuring any linear combination of solutions will also be a solution. Unitarity or preservation of norm under evolution implies $\hat{L}$ is $i\hat{O}$, with self-adjoint $\hat{O}$.

With these considerations, the fundamental equation is thus of the form

$$i\hbar\frac{d\Psi}{dT} = H_{phys}\Psi, \qquad (1.49)$$

with self-adjoint H_{phys}; and $\hbar$ is a conversion factor which must be of physical dimension times×energy. There is stupendous power and staggering simplicity in this equation. Due to its first order dependence on time, there is always a solution, no matter how complicated H_{phys} may be (this includes time-dependent Hamiltonians); and the solution is unique given an initial state $\Psi(T_0)$. The formal solution can be derived by integrating, yielding $\Psi(T) = U(T, T_0)\Psi(T_0)$, with

$$U(T, T_0) := \mathcal{T}\left(\exp\left[-\frac{i}{\hbar}\int_{T_0}^{T}\hat{H}_{phys}(T')dT'\right]\right). \qquad (1.50)$$

This is a T-ordered operator, due to, in general, non-commutation of the Hamiltonian operator at different times. A temporal ordering of the state emerges. Without the requirement and availability of a c-number time T, the solution would not have come about. This time-development operator can be expanded explicitly as a time-ordered Dyson series (see the expression following Eq. (3.17) in Chapter 3). In ITG the intrinsic time parameter is dimensionless and H_{phys} is of dimension $\hbar$. What is paramount to causality is not the physical dimension of time, but the existence of an ordering. Quantum gravity spells the demise of spacetime, but not of temporal ordering. *The fundamental gauge-invariant entity that can be, and is in fact time-ordered, is the quantum state of the entire physical universe.* Therein lies the substantiation of what remains of "cause and effect". No more, but no less, is allowed by quantum physics.

The conjunction of the paradigm shift from four-covariance and solution of the Hamiltonian constraint as $\tilde{\pi} = -\frac{1}{\beta}\bar{H}$ allows ITG to reconcile the fundamental dynamics of quantum gravity with Schrödinger-Heisenberg quantum mechanics. Unitary and diffeomorphism-invariant $U(T, T_0)$ follows if H_{phys} is self-adjoint and spatial diffeomorphim invariant (this holds if $\bar{H}$ is a spatial scalar density of weight one. The quantum condition is that H_{phys} commutes with the generator of spatial diffeomorphisms; this is true of Eq. (1.47), and of the Hamiltonians discussed in later chapters). The intrinsic time interval, $dT = \frac{2}{3}(dV/V) = \frac{2}{3}d\ln V$, is a measure of the fractional change of the volume of the universe (this is gauge invariant under spatial diffeomorphisms). The whole description is gauge invariant if $\Psi(T_0)$ is also invariant under spatial diffeomorphisms (this is true of the explicit initial Chern-Simons Hartle-Hawking state advocated in Chapter 10).

All evolution is with respect to the cosmic clock; not only does gravity no longer stand apart from the rest, all are at one with the universe. It is the quantization of gravity that leads to universal Schrödinger-Heisenberg mechanics and fundamental temporal-ordering. A closed universe yields a finite volume; and an ever-expanding universe yields a direction of cosmic time. Four-dimensional classical spacetime does not exists in quantum gravity, but even quantum gravity has no dominion over the cosmic clock. It is not subject to quantum fuzziness (see Sec. 1.3); it is infinitely robust and all-encompassing. Time is not many-fingered but global (it is everywhere but nowhere in particular). In its own quantum evolution, the universe "needs no and has no external clock".

1.8.3 *Time-dependent Hamiltonian of ITG*

Time dependence of the theory comes through the q-dependence of $q(R - 2\Lambda_{eff})$ in the Hamiltonian, while conformally-invariant $\tilde{C}_j^i$ is independent of q. The Hamiltonian can be arrived at from different complementary perspectives. In ITG, $\tilde{\pi}$ is eliminated in terms of the other variables, the corresponding q which has nontrivial commutation relation with $\tilde{\pi}$ is a field degree of freedom which should also be fixed. This consideration can be put on sounder footing; Chapter 8 espouses the theory from the perspectives of a Dirac second class constrained system and of the canonical functional integral. The auxiliary condition, $\chi = \frac{1}{3}\ln[\frac{q}{q_0}] - (T - T_0) = 0$, yields a generally non-vanishing Poisson bracket with the Hamiltonian constraint $H = 0$ (see Chapter 8). Thus these two conditions can be used to eliminate one full field degree of freedom from the theory, specifically $\tilde{\pi}$ and q. The condition $\chi = 0$, or $q(x, T) = q_0(x)e^{3(T-T_0)}$, is the same as $\delta \ln q^{\frac{1}{3}} = dT$ i.e. *the fixing of the local intrinsic time interval to its gauge-invariant value.* With $q_{ij} = q^{\frac{1}{3}}\bar{q}_{ij}$ and

$$R = q^{-1/3}\left[\bar{R} - 2\bar{q}^{ij}\bar{\nabla}_i\partial_j \ln q^{1/3} - \frac{1}{2}\bar{q}^{ij}(\partial_i \ln q^{1/3})(\partial_j \ln q^{1/3})\right], \quad (1.51)$$

the potential, $q(R - 2\Lambda_{eff})$ scales as $q^{2/3}$ for the qR term and as q for the cosmological constant term. So these terms scale exponentially with T.

From another perspective, as H_{phys} contains no $\tilde{\pi}$, it commutes with $\ln q^{\frac{1}{3}}$; and $\frac{\partial \ln q^{\frac{1}{3}}}{\partial T} = 1$ obtains from the Hodge decomposition $\delta \ln q^{\frac{1}{3}} = dT + \nabla_i \delta Y^i$ into orthogonal complements. Consequently, the Heisenberg EOM for operators (see Eq. (9.48) in Chapter 9) yields

$$\frac{d \ln q_H^{\frac{1}{3}}}{dT} = U^\dagger(T, T_0)\left(\frac{\partial \ln q^{\frac{1}{3}}}{\partial T} + \frac{1}{i\hbar}[\ln q^{\frac{1}{3}}, H_{phys}]\right)U(T, T_0) = 1; \quad (1.52)$$

and $q_H(x,T) = q_0(x)e^{3(T-T_0)}$ is the solution; thus this can be substituted in Eq. (1.47) in quantum Heisenberg evolution. No ad hoc "fixing" of q in H_{phys} is necessary in the Heisenberg picture.

In the quantum context, if the physical Hamiltonian commutes an anti-unitary time reversal operator T, then whenever $\Psi(T)$ is a solution of the Schrödinger equation the time-reversed wave function, $T\Psi(T)$, satisfies $i\hbar\frac{d}{d(-T)}T\Psi(T) = H_{phys}T\Psi(T)$.[23] The physical Hamiltonian of ITG, $H_{phys}(T)$, has explicit time-dependence through q, and it is not an even function of the intrinsic time T; the theory is time-irreversible. Time dependence of a Hamiltonian is allowed in mechanics; it does not spoil unitarity which is ensured by the self-adjointness of the Hamiltonian.

A time-dependent theory has an added feature: it can allow different physics to come into dominance at different times; without having to postulate time-dependent, or ever-changing, fundamental laws. It allows the universe to have a beginning very different from how it will end. An in-built time asymmetry in theory carries an arrow of time. In ITG, the Hamiltonian is Cotton-York dominated at early times of small spatial volumes and Einsteinian at late times and large volumes. This agrees with the dominance and correctness of GR in the current epoch. As the universe continues to expand, the cosmological constant term will become more and more dominant.

Penrose has pointed out that there is indeed an asymmetry in our universe [23, 25, 53, 54]. Observations support the picture that our universe has a rather smooth Robertson-Walker type early stage, and as it ages structures from gravitational clumping turn into stars, galaxies, and black holes. Penrose further argues that when gravitational d.o.f. are included, black hole entropy (according to the Bekenstein-Hawking [55] formula $S_{bh} = \frac{1}{4}(\frac{A_{horizon}}{L^2_{Planck}})k_B$) by far outweighs all other forms of entropy; so overall entropy of a universe which starts smoothly increases as the universe ages and black holes appear. According to Boltzmann's famed formula,

---

[23]In Minkowski spacetime, time reversal is an improper Lorentz transformation; in a generic curved spacetime it can be generalized to an orientation reversal which maps the vierbein one-form $e^0 \to -e^0$ while preserving the spatial dreibein one-forms $e^{a=1,2,3}$ (the metric $g_{\mu\nu}dx^\mu dx^\nu = -(e^0)^2 + e^a e^a$). Even though there is no recourse to four-dimensional classical spacetime in quantum gravity, Schrödinger-Heisenberg mechanics in ITG allows time reversal invariance or violation to be properly phrased in Wigner's formulation [40] discussed above. In ITG, the unimodular spatial metric $\bar{q}_{ij}$ should be invariant while the momentric is odd (this preserves the fundamental commutation relations in Chapter 5) under time reversal; q on the other hand scales exponentially with intrinsic time T.

$S = k_B \ln \Omega$, entropy is essentially the logarithm of the number of compatible microstates Ω; since black hole entropy is by far the most dominant, black hole states would swamp all other; and extreme fine tuning (with odds worse than one in a googleplex ($10^{10^{100}}$)) would be necessary for the universe to steer away from being born a black hole! Strict thermodynamic adherence is not really needed (as emphasized by Penrose); what can be argued on general physical grounds is a phase space that includes gravitational d.o.f. would quite certainly be predominantly black hole initial data. To pick out a smooth Robertson-Walker low entropy Big Bang would require imposing an extreme form of selection rule. Penrose identifies this as the vanishing of the four-dimensional Weyl curvature tensor [23,25,53,54]. A requirement on the Weyl tensor is essentially extraneous to Einstein's field equations; they directly relate the Ricci tensor, rather than the Weyl curvature, to energy-momentum. But Cotton-York dominance brings new physical perspectives, on the nature of the early universe and on the very origin of the universe.

1.9 ITG and the origin of the universe

The question whether time is finite or infinite has been asked and debated even before Aristotle and Plato. If time is finite, is it meaningful to ask what happened before time came about? On the other hand, an immediate tension arises from the more common assumption that time should run from infinite past to infinite future (this is certainly true of Minkowski spacetime). If time were infinite, would it not, paradoxically, take forever before the universe gets to where is it now? CMB, WMAP, and Planck data along with the Λ_{CDM} model, are consistently showing the lookback time of the universe is about 13.7 billion years [5–7]. GN-z11, the galaxy with the highest redshift (11.1) thus far, is calculated to have a light travel time of 13.4 billion years. But did the universe start infinitely far back in time?

In ITG, $dT = \frac{2}{3}d\ln V$, or $T - T_{now} = \frac{2}{3}\ln(V/V_{now})$, does imply, for $0 < V/V_{now} < \infty$, that $-\infty < T < +\infty$. Intrinsic time should run from infinite past to infinite future. But quantum gravity has yet to be taken into account. As a first order equation in time, the Schrödinger equation needs an additional physical input: $\Psi(T_0)$. The nature of the Hamiltonian in the very early universe may yield hints, but the initial condition of the universe demands further insights and input.

Hawking in Ref. [54] had proposed a semiclassical picture of how our universe came into being. It arose out of Euclidean tunneling from literally almost nothing, from a single point. This is intuitive and appealing to many in its simplicity. In Ref. [54], there is an image of this Euclidean-Lorentzian tunneling for the de Sitter four-manifold (see Fig. 1.5); of an Euclidean origin from a single point to the size of the Lorentzian de Sitter throat and then expanding exponentially onwards as a Lorentzian de Sitter spacetime. This is an exact solution of Einstein's equations with cosmological constant (the continuation to imaginary time for the Euclidean manifold remains a solution (see Chapter 10); due to the conformally flat spatial slicings, it is also a solution of ITG).

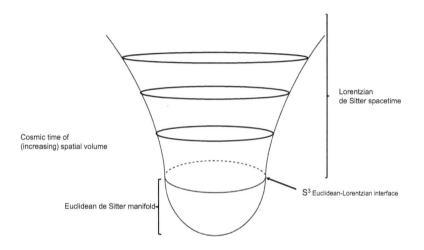

Fig. 1.5   Origin of the universe from Euclidean tunneling.

More importantly, Hartle and Hawking [22] advocated the no-boundary proposal (sometimes referred to, perhaps more appropriately, as the "one-boundary proposal") for $\Psi(T_0)$, the wave function and quantum origin of the universe. The proposal states that the wave function or probability amplitude on a closed spatial hypersurface Σ is given by e^{-S_E} summed over all Euclidean manifolds with Σ as their sole boundary, and S_E is the Euclidean gravitational action (see Eq. (10.36)).

ITG takes the Hartle-Hawking proposal seriously and finds it amazingly in concord with the nature and clues of the Hamiltonian in the early

universe that is Cotton-York dominated. The details of the origin of the
universe that ITG pieces together are recounted in the final chapter of this
book which advocates the quantum Chern-Simons Hartle-Hawking initial
state. In ITG, time, which corresponds to the spatial volume, is common to
both Euclidean and Lorentzian manifolds; and Σ can serve as the common
interface, provided the junction condition of vanishing extrinsic curvature
(or momentum) [56] is satisfied. This implies the vanishing of the kinetic
term (a Casimir of the traceless momentric $\bar{\pi}^i_j$) in the Hamiltonian of ITG.
If, in addition, the Cotton-York tensor vanishes, $\tilde{C}^i_j = 0$, together with
$(R - 2\Lambda_{eff})$, then the total Hamiltonian density which is proportional to
trace of the momentum, $\tilde{\pi}$, is zero. The junction condition is then fully
satisfied. Remarkably, the Chern-Simons Hartle-Hawking state yields vanishing expectation values of the momentric and Cotton-York tensor; and
the condition for the vanishing of $\tilde{C}^{ij} = \frac{\delta W_{CS}}{\delta q_{ij}}$, which is the statement of
conformal flatness in there-dimensions, is just the behavior or characteristic
of a critical point of the gravitational Chern-Simons functional W_{CS} which
is gauge invariant under spatial diffeomorphisms (large gauge transformations and relative Chern-Simons functionals are addressed in Chapter 10).
Since standard S^3 is the only simply-connected conformally flat manifold,
this produces a homogeneous Robertson-Walker $R = 6/a^2$ which can can-
cel the effective cosmological constant then.[24] This semiclassical picture is
explained further in Chapter 10; Chapter 4 arrives at the same conclusions
from the perspective of the Lichnerowicz-York equation. What is more significant is there is an exact quantum state that exhibits this semiclassical
behavior: it is the Chern-Simons state, which is essentially $e^{-gW_{CS}}$. Moreover, a relation in Chern-Weil theory allows this Chern-Simons state to be
interpreted as an exact Hartle-Hawking wave function, with S_E being the
Euclidean Pontryagin action of instanton tunneling. Furthermore, beyond
the semiclassical picture, expectation values and multi-point correlations
of operators w.r.t. to this Chern-Simons Hartle-Hawking quantum gravity
state can be explicitly calculated. In particular, at the lowest approximation, two-point transverse-traceless physical metric excitations are found to
exhibit scale-invariant power spectrum. Nothing less that quantum gravity fluctuations of an exact state and their physical signatures are being
manifested.

---

[24]The large effective value of the cosmological constant which determines the initial
temperature and size of the universe (see Chapter 10) is very different from its current observational value of $\sim 10^{-120} L_{Planck}^{-2}$; the "cosmological constant problem" [57]
remains an important topic and challenge for ITG.

A Euclidean-Lorentzian interface at a small throat implies a hot beginning through Euclidean periodicity and the Kubo-Martin-Schwinger condition citeKMS. The quantum universe did not begin from infinitely small size (at $T \to -\infty$), but semiclassically, at a small Euclidean-Lorentzian spatial interface. The Chern-Simons Hartle-Hawking state yields a semiclassical smooth Robertson-Walker beginning, an origin in accord with Penrose's Weyl Curvature Hypothesis [23]. As black hole entropy by far outweighs all other forms of entropy, such a beginning implies from the initial low entropy to formation of structure into stars, black holes, galaxies and gigantic black holes, the overall entropy increases as the universe ages. This entropy arrow of time is in accordance with the second law of thermodynamics. In ITG, the cosmological arrow of time points in the direction of increasing volume in an ever-expanding universe. Thus the gravitational-cosmological arrow of time concurs with the second law of thermodynamics.

"If we do discover a complete theory, it should in time be understandable in broad principle by everyone, not just a few scientists. Then we shall all, philosophers, scientists, and just ordinary people, be able to take part in the discussion of the question of why it is that we and the universe exist. If we find the answer to that, it would be the ultimate triumph of human reason — for then we would know the mind of God."— Stephen Hawking

Chapter 2

General Relativity without paradigm of space-time covariance; and resolution of the problem of time

"I am inclined to believe from this that four-dimensional symmetry is not a fundamental symmetry of the physical world."
—P.A.M. Dirac

2.1 Conceptual and technical challenges of the paradigm of four-covariance

Four-covariance of spacetime has been a steadfast paradigm of Einstein's theories of relativity, of both Special Relativity and the General Theory of Relativity. In the former, Lorentz isometry peculiar Minskowski spacetime holds sway, even though a generic spacetime may not have any isometry at all. The assumption of full four-dimensional diffeomorphism symmetry in GR entails a number of technical and conceptual difficulties. On the technical front, Einstein's theory fails to be renormalizable as a perturbative quantum field theory; and GR remains incomplete as a quantum theory, despite some recent advances [46]. At the conceptual level, the Hamiltonian constraint which is believed to dictate dynamics appears to have a "dual role", also as a generator of temporal translation "symmetry", with the lapse function as Lagrange multiplier. It is difficult to reconcile the physical reality of time and time evolution with full four-dimensional invariance and the entailing seemingly arbitrary lapse function and "gauged histories" generated by first class Hamiltonian and momentum constraints.

Canonical quantum gravity has the guise of a formalism "frozen in time". Quantum states which are required to be annihilated by the constraints in first class Dirac quantization do not evolve in ADM coordinate "time"; and dx^0 is merely one component of space-time displacement, with no invariant physical meaning in a theory with full space-time covariance.

Yet a notion of "time" is needed to make sense of quantum mechanical interpretations: both for the discussion of dynamics and evolution, and for the interpretation of probabilities which are normalized at particular instants of time. More importantly, causation and temporal ordering of quantum states require such a notion. Even if a suitable degree of freedom is chosen as the intrinsic time (as many have advocated), a successful quantum theory will still need to reconcile a Klein-Gordon type Wheeler-DeWitt (WDW) equation [9,12] quadratic in momenta with the positivity of "probabilities". A resolution of the "problem of time" [11] cannot be deemed complete if it fails to account for the intuitive physical reality of time and does not provide a satisfactory correlation between intrinsic time development in quantum dynamics and the passage of time in classical space-times.

At the deeper level of symmetries, the constraints of GR satisfy the Dirac algebra (Eqs. (1.18)–(1.20) in Chapter 1). However, closer inspection reveals it is not the algebra of the generators of four-dimensional diffeomorphisms. The constraint algebra has structure functions dependent upon dynamical variables, rather than structure constants. As explained in Sec. 1.7.5, the Hamiltonian constraint is quadratic in its dependence on the canonical momentum, and unlike the momentum constraints, it cannot be interpreted as generating the desired diffeomorphisms. Off-shell, full four-dimensional Lie derivatives involve time derivatives; these cannot be generated by constraints[1] which depend only on the spatial metric q_{ij} and its conjugate momentum $\tilde{\pi}^{ij}$. In Einstein's theory, the EOM, among other things, relate $\dot{q}_{ij}$ to $\tilde{\pi}^{ij}$ (as shown in Eqs. (1.27)–(1.28)), and the constraints then generate four-dimensional diffeomorphisms only on-shell [37] (but as we shall see, with an *a posteriori* value for the lapse function). This is a clue that, without the help of EOM, a *quantum* theory of GR cannot, and need not, enforce full four-dimensional space-time covariance *off-shell*. Furthermore, spacetime is a classical ideal of "limited applicabilty" [9], ergo the rationale of imposing, and holding on, to a symmetry of classical entities which do not even exist in quantum gravity is dubious, if not bewildering.

2.1.1 *Paradigm shift from four-covariance*

The call to abandon four-covariance is not new. On encountering the simplification of the Hamiltonian analysis of GR, and the fact that only the spatial metric is physical, Dirac concluded and declared in no uncertain terms "four-dimensional symmetry is not a fundamental property of the

---

[1]See Eqs. (1.18)–(1.20) in Chapter 1.

physical world" [15]. In his seminal article, Wheeler emphasised that space-time is a concept of 'limited applicability', and it is three-geometry, rather than four-geometry, which is fundamental in quantum geometrodynamics. That semiclassical space-time is emergent begs the question what, if anything at all, takes the place of "time" in quantum gravity? Wheeler went as far as to claim we have to forgo time-ordering, and to declare "there is no spacetime, there is no time, there is no before, there is no after" [9]. But without "time-ordering", how can "causality", which is requisite in any "sensible physical theory", be ensured and enforced in quantum gravity?

A key obstacle to the viability of GR as a perturbative quantum field theory lies in the conflict between unitarity and space-time general covariance: improved ultraviolet convergence and even renormalizability can be attained with the introduction of higher-order curvature terms into the theory; but space-time covariance requires time as well as spatial derivatives of the same (higher) order, thus compromising unitarity. As explained in Sec. 1.6.2, Horava relinquished full four-dimensional symmetry and achieved power-counting renormalizable modifications of GR with higher-order spatial curvature terms [31].

GR as geometrodynamics leaves almost no room for any modification within the context of a first class Dirac algebra [33]; this has also played a role in restricting canonical GR to its present form of a Ricci scalar potential plus at most a cosmological constant. This appears to eliminate, almost entirely, the prospects of modifying Einstein's theory with higher spatial but not time derivatives within the context of a first class constrained system. In loop quantum gravity, the non-perturbative master constraint program [38] seeks representations not of the Dirac algebra, but of the master constraint algebra which has the advantages of having structure *constants* (rather than functions), and of decoupling the equivalent quantum Hamiltonian constraint from spatial diffeomorphism generators H_i. Reference [36] consistently realized Horava gravity theories as canonical theories with first class master constraint algebra. This removes not only the canonical inconsistencies of "projectable Horava theories" which lack of a local $H(x) = 0$ Hamiltonian constraint, but also captures the essence of the theory in retaining spatial diffeomorphisms as the *only* local gauge symmetries. With full four-covariance, Dirac observables, O, of Einstein's theory are required to commute both with H_i *and* H, leading to the demand (above and beyond usual canonical rules) of $\{O, \{O, M\}\}|_{M=0} = 0$ [38]. In contradistinction this restriction is not required in ITG which marks a real paradigm shift in the symmetry, from full four-covariance to invariance only with respect to spatial diffeomorphisms. Enforcing $H(x) = 0$ through the master constraint effectively removes it from one of its "dual roles" since M generates no extra symmetry, and determines only dynamics. It allows us to disentangle the Hamiltonian constraint from its "dual role", and be cast solely as a generator of dynamics in a theory which clearly possesses only spatial diffeomorphism symmetry. In conjunction with the decomposition of the symplectic potential and the identification of $\ln q^{\frac{1}{3}}$ as intrinsic time interval,[2] the physical content of Einstein's theory from this paradigm shift is revealed. With the identification of the *a posteriori* lapse function (instead of its *a priori* arbitrariness) dynamical evolution in Einstein's theory is not only manifestly physical but also correlated to the cosmic clock! Although q is a scalar density (a criticism which has been used to disqualify it from the role of time variable), $\delta \ln q = \frac{\delta q}{q}$ is a spatial diffeomorphism scalar. Even in Galilean-Newtonian physics, it is *time interval*, and not absolute time, which is physical. Remarkably, it is spatial diffeomorphism symmetry which collapses multi-fingered (spatially-dependent) intrinsic time interval $\delta \ln q^{\frac{1}{3}}$ to a global (one-parameter spatially-independent) gauge-invariant time interval dT. The effective Hamiltonian generating this corresponding global cosmic time translation is ascertained and derived, in this, and in later chapters from different complementary perspectives, including from the point of view of a second class Dirac system with reduced phase space variables all identically Dirac commuting with every constraint and subsidiary condition. ITG affirms the reality and physicality of time, and of microcausality in the temporal-ordering of the quantum state of our universe. The resolution of "problem of time" arises from the synergy between the theory of gravitation and the cosmic clock that is essentially the logarithm of expanding spatial volume of our universe.

The goals of ITG are to explicitly demonstrate the fundamental reality of time in Einstein's theory and its extensions; to reveal that causal temporal ordering is dictated by the first order Schrödinger evolution with respect to the cosmic clock of our expanding closed universe and its equivalent Heisenberg time-development operator; to advocate that a spatial diffeomorphism-invariant theory of gravitation, instead of a theory obfuscated by four-covariant paradigm, prompts modification of Einstein's theory to include a Cotton-York term for power-counting ultraviolet renormalizability in addition to infrared convergence from spatial compactness; and to study the ramifications the paradigm shift brings to question of the

---

[2] In a mini-superspace context, Misner had also advocated $\ln q$ as a time variable and obtained the corresponding Hamiltonian [50].

origin of the universe and of the arrow of time. The physical contents of Einstein's GR are captured at low energies and low curvatures; and the time-dependent physical Hamiltonian ensures the primacy of the Einstein's theory and its observational consequences, except in the very early Cotton-York dominated era.

It is the universe itself that plays the role of the all-encompassing and infinitely robust clock that makes "time" in gravitation concrete, universal, and comprehensible; while its own dynamics, origin and destiny are dictated by the laws of gravitation and quantum mechanics.

2.2   Theory of gravity without paradigm of four-covariance

With the preceding concerns, observations and goals in mind, a framework for a theory of gravity, from quantum to classical regimes, will be presented wherein several factors collude to result in a minimal and compelling framework to realize the paradigm shift.

The master constraint formulation is put into action in Sec. 2.4; and it will be demonstrated the classical content of GR in an expanding universe may be captured by $\tilde{\pi} = -\frac{1}{\beta}\bar{H}$. That $\tilde{\pi}$, the trace of the momentum, is essentially conjugate to $\ln q^{\frac{1}{3}}$ leads to a quantum theory described by a Schrödinger equation *first order in intrinsic time*. The resultant semi-classical Hamilton-Jacobi (HJ) equation is also first order in intrinsic time, with the implication of completeness [59] (its integral solution provides a complete set of gauge-invariant integration constants of motion). Classical space-time emerges from constructive interference of quantum wave functions [60]. The physical content of GR is regained from a theory with $\bar{H}/\beta$ generating $\delta \ln q^{\frac{1}{3}}$ time translations, subject to only spatial diffeomorphism invariance. Furthermore, the emergent classical space-times have ADM metrics with proper times directly correlated to intrinsic time intervals by the explicit relation Eq. (2.53). The distinction between the *a priori* and *a posteriori* value of the lapse function can be found in Sec. 2.5; and the resolution of the "problem of time" is presented in Sec. 2.6. There is the promise of a viable unitary quantum theory of gravity if the Hamiltonian generating intrinsic time development can be rendered real, bounded from below, and also power-counting renormalizable by including suitable higher-order spatial curvature terms. These modifications are discussed in Sec. 2.7. Emergence of a *global* intrinsic time interval due to diffeomorphism invariance, and the physical Hamiltonian governing quantum geometrodynamics in superspace are covered in Sec. 2.8, together with *gauge-invariant temporal ordering* in quantum gravity.

A resolution of the problem of time cannot be deemed complete if it fails to show consistency with relativistic time effects, such as time dilation and gravitational redshifts, in Einstein's theories of relativity. The framework must also be able to accommodate the known fermionic matter content of the universe without recourse to four covariance. Section 2.9 introduces the Lorentz group as $SO(3, C)$ to pave the way for the inclusion of chiral fermions which will be discussed in detail in Chapter 9. Sections 2.10 and 2.11 address the motion of test point particles in generic ADM curved space-times and the relativistic time effects that are encountered. Without recourse to Lorentz isometry in Special Relativity or to four-covariance in GR, these time dilation effects are shown to arise dynamically from the change in the Hamiltonian of a test particle when it is in motion in a gravitational field; and the existence of such effects are completely compatible with the presence of the intrinsic cosmic clock. Section 2.12 on the relativistic particle in flat spacetime and the description of its dynamics in three different ways serves as an apposite analogy to simplify and grasp the main essence of this long chapter.

2.3   Geometrodynamics, symplectic potential, and the Hamiltonian constraint

A theory of geometrodynamics, with $(q_{ij}, \tilde{\pi}^{ij})$ as fundamental variables, is bequeathed with a number of remarkable features: among this positivity of the spatial metric guarantees space-like separation between any two points on the initial value hypersurface. This allows, to quote Wheeler in Ref. [9], "a notion of 'simultaneity' and a common moment of a rudimentary 'time'". Quantum states however do not depend on x^0; the link between intrinsic and coordinate times which will be revealed is more subtle.

In the ADM $(3 + 1)$-decomposition, the metric is,

$$ds^2 = g_{\mu\nu}dx^\mu dx^\nu = -N^2(dx^0)^2 + q_{ij}(dx^i + N^i dx^0)(dx^j + N^j dx^0); \quad (2.1)$$

equivalently the metric components are explicitly,

$$g_{\mu\nu} = \begin{pmatrix} q_{ij}N^i N^j - N^2 & q_{ij}N^i \\ q_{ij}N^j & q_{ij} \end{pmatrix}, \quad (2.2)$$

$$g^{\mu\nu} = \begin{pmatrix} -1/N^2 & N^i/N^2 \\ N^i/N^2 & q_{ij} - N^i N^j/N^2 \end{pmatrix}. \quad (2.3)$$

2.3.1   Symplectic potential, and decomposition of variables

In the initial value problem, York [17,61] explored a conformal decomposition of the spatial metric $q_{ij} = \phi \bar{q}_{ij}$, of which $\phi = q^{\frac{1}{3}}$ with unimodular $\bar{q}_{ij}$ is a special case ($q := \det(q_{ij})$). However, in focusing on the generic conformal factor ϕ, the "miracle" of $\phi = q^{\frac{1}{3}}$ was not fully revealed. Among other advantages, it results in clean separation of $(\ln q^{\frac{1}{3}}, \tilde{\pi})$ from other canonical pairs, with the symplectic potential,[3]

$$\int \tilde{\pi}^{ij} \delta q_{ij} = \int \bar{\pi}^{ij} \delta \bar{q}_{ij} + \tilde{\pi} \delta \ln q^{\frac{1}{3}}, \qquad (2.4)$$

wherein $\tilde{\pi} := q_{ij} \tilde{\pi}^{ij}$ and

$$\bar{\pi}^{ij} := q^{\frac{1}{3}} \left[\tilde{\pi}^{ij} - \frac{q^{ij}}{3} \tilde{\pi} \right]. \qquad (2.5)$$

Thus the only non-trivial Poisson brackets are[4]

$$\{\bar{q}_{kl}(x), \bar{\pi}^{ij}(x')\}_{P.B.} = P^{ij}_{kl} \delta^3(x, x'), \qquad (2.6)$$

$$\{\bar{\pi}^{ij}(x), \bar{\pi}^{kl}(x')\}_{P.B.} = \frac{1}{3}(\bar{q}^{kl}\bar{\pi}^{ij} - \bar{q}^{ij}\bar{\pi}^{kl})\delta^3(x, x'), \qquad (2.7)$$

and[5]

$$\{q(x), \tilde{\pi}(x')\}_{P.B.} = 3q\delta^3(x, x'); \qquad (2.8)$$

with the trace-free projector,

$$P^{ij}_{kl} := \frac{1}{2}(\delta^i_k \delta^j_l + \delta^i_l \delta^j_k) - \frac{1}{3}\bar{q}^{ij}\bar{q}_{kl}. \qquad (2.9)$$

This separation carries over to the quantum theory, and permits a d.o.f., namely $\ln q$, separate from the others, to be identified as the carrier of temporal information.

An intrinsic clock is in fact not tied to four covariance, since the generic ultra-local DeWitt supermetric [12] compatible with spatial diffeomorphism is

$$G_{ijkl} = \frac{1}{2}(q_{ik}q_{jl} + q_{il}q_{jk}) - \frac{\lambda}{3\lambda - 1}q_{ij}q_{kl}, \qquad (2.10)$$

with deformation parameter λ. The supermetric which has signature

$$\left(\text{sgn} \left[\frac{1}{3} - \lambda \right], +, +, +, +, + \right), \qquad (2.11)$$

---

[3]Another noteworthy feature of this decomposition is unimodular $\bar{q}_{ij}$ which contains the d.o.f. in ITG is non-degenerate by definition, and invertible; whereas the degeneracy in $q = 0$ or $\ln q^{\frac{1}{3}} \to -\infty$ is identified as the beginning of intrinsic time.

[4]These follow from $\{A, B\}_{P.B.} = \frac{\partial A}{\partial q_{ij}} \frac{\partial B}{\partial \tilde{\pi}^{ij}} - \frac{\partial B}{\partial q_{ij}} \frac{\partial A}{\partial \tilde{\pi}^{ij}}$.

[5]The inverse of $\bar{q}_{ij}$ is denoted by $\bar{q}^{ij}$; also $\int \tilde{\pi}\delta \ln q^{\frac{1}{3}} = \int \frac{1}{3}\frac{\tilde{\pi}}{q}\delta q$.

comes equipped with $(-, +, +, +, +, +)$ as long as $\lambda > \frac{1}{3}$; and the single negative eigenvalue for $\lambda > \frac{1}{3}$ corresponds to the $\delta \ln q$ mode in $G^{ijkl}\delta q_{ij}\delta q_{kl}$. Crucially, it is also this same $\ln q^{\frac{1}{3}}$ which has a nontrivial commutation relation with $\bar{\pi}$, and which can be so neatly separated from the rest of the variables.

2.3.2 *Hamiltonian constraint; and its factorization*

The Hamiltonian constraint of a theory of geometrodynamics quadratic in momenta and with ultralocal supermetric is generically,

$$0 = \frac{\sqrt{q}}{2\kappa}H$$
$$= G_{ijkl}\tilde{\pi}^{ij}\tilde{\pi}^{kl} + \mathcal{V}(q_{ij})$$
$$= -(\beta\tilde{\pi} - \bar{H})(\beta\tilde{\pi} + \bar{H}); \tag{2.12}$$

with

$$\bar{H}(\bar{\pi}^{ij}, \bar{q}_{ij}, q) = \sqrt{\bar{G}_{ijkl}\bar{\pi}^{ij}\bar{\pi}^{kl} + \mathcal{V}(\bar{q}_{ij}, q)}$$
$$= \sqrt{\frac{1}{2}[\bar{q}_{ik}\bar{q}_{jl} + \bar{q}_{il}\bar{q}_{jk}]\bar{\pi}^{ij}\bar{\pi}^{kl} + \mathcal{V}(q_{ij})}; \tag{2.13}$$

and

$$\beta^2 := \frac{1}{3(3\lambda - 1)}. \tag{2.14}$$

Einstein's GR which corresponds to $\lambda = 1$ or $\beta^2 = \frac{1}{6}$, and

$$\mathcal{V}(\bar{q}_{ij}, q) = -\frac{q}{(2\kappa)^2}[R - 2\Lambda_{eff}], \tag{2.15}$$

is a particular realization of a wider class of theories.

As written, Eq. (2.12) is a local constraint, in addition to those of spatial diffeomorphism invariance $H_i = 0$. This leads to a quandary: as the constraint algebra is Dirac first class, $\int N(x)H(x)d^3x$ is apparently consistently generating multi-fingered time translation *symmetry*, whereas it is also supposed to *generate dynamical physical evolution rather than gauge transformations* of the d.o.f. of the theory. The resolution for GR, as will be discussed in detail in this and later chapters, is true physical evolution can be discerned in the *a posteriori* value of N as opposed to its *a priori* arbitrariness. This implies full four-covariance is not needed to capture the physical content of Einstein's theory, and modifications which will reduce to GR in appropriate limits can be considered.

2.4 Master constraint as a provisional formalism in disentangling the dual roles of the Hamiltonian constraint

In this section, within the restrictive context of a first class Dirac algebra and a Hamiltonian consistently wholly of constraints, the method of master constraint [38] is brought in to decouple the dual role of the Hamiltonian constraint. It can equivalently enforce the local content of Eq. (2.12), while allowing the Hamiltonian constraint of GR to be treated as purely dynamical in content in a framework without the paradigm of four-covariance. The method has the added feature of consistently allowing deformations of GR (as in Horava theories of gravitation [31]) while remaining within the context of a Dirac first class constrained system.

The comparison and analogy to simple relativistic point particle mechanics is recounted in Sec. 2.12. The commencing action is

$$S = \int [\tilde{\pi}^{ij} \dot{q}_{ij} - N^i H_i] d^3x dt - \int m(t) \mathrm{M} dt; \tag{2.16}$$

$$\mathrm{M} := \int (\beta \tilde{\pi} + \bar{H})^2 / \sqrt{q} d^3x = 0; \tag{2.17}$$

$$H_i := -2 q_{ik} \nabla_j \tilde{\pi}^{jk}(x). \tag{2.18}$$

Master constraint, M, is a spatial integral; and it is easy to see that the resultant constraint algebra,

$$\{\mathrm{M}, \mathrm{M}\}_{P.B.} = 0, \tag{2.19}$$

$$\{H_i[N^i], \mathrm{M}\}_{P.B.} = 0, \tag{2.20}$$

$$\{H_i[N^i], H_j[N'^j]\}_{P.B.} = H_i[\mathcal{L}_{\vec{N}} N'^i], \tag{2.21}$$

is first class and exhibits only spatial diffeomorphism gauge symmetry, both on- and off-shell. M decouples from H_i in the constraint algebra; and the result is a theory with *only spatial diffeomorphism invariance*; with physical dynamics dictated by H, but encoded in M.

Modulo M = 0, the total constraints of the theory generate only spatial diffeomorphisms since

$$\{f(q_{ij}, \tilde{\pi}^{ij}), m(t)\mathrm{M} + H_k[N^k]\}|_{\mathrm{M}=0 \Leftrightarrow H=0}$$
$$= \{f, H_k[N^k]\}_{P.B.} = \mathcal{L}_{\vec{N}} f. \tag{2.22}$$

Thus $m(t)$ plays no role in the dynamics. Instead, true physical evolution can only be with respect to an intrinsic time extracted from the WDW or M constraint. As detailed above, $\ln q^{\frac{1}{3}}$ is the preeminent choice.[6]

---

[6]It should be obvious the power $\frac{1}{3}$ in $\ln q^{\frac{1}{3}} = \frac{1}{3} \ln q$ is not physically significant; it is retained since $\hat{\bar{\pi}} = \frac{\hbar}{i} \frac{\delta}{\delta \ln q^{\frac{1}{3}}}$, in accordance with the symplectic potential in Eq. (2.4).

2.4.1   First-order in intrinsic time Wheeler-DeWitt equation

The factorization in Eq. (2.12) reveals that,

$$(\beta\tilde{\pi} \pm \bar{H}) = 0$$

is necessary and sufficient to recover the classical content of the Hamiltonian constraint in GR. The positive sign is singled out to render the final physical Hamiltonian

$$H_{phys} = \frac{1}{\beta} \int \bar{H}(x) d^3 x \qquad (2.23)$$

(which shall be elaborated on) positive semidefinite. In closed FLRW cosmological models this corresponds to an expanding universe. Chapter 8 will address the phase space, cosmic time, and reduced Hamiltonian in detail.

Despite its apparent simplicity, Eq. (2.23) is a breakthrough: a semiclassical HJ equation first order in (intrinsic) time (since $\tilde{\pi}$ is conjugate to $\ln q^{\frac{1}{3}}$) comes with the consequence and upshot of completeness [59]; and quantum gravity will now be dictated by a corresponding Schrödinger equation first order in intrinsic time (with consequent positive semidefinite probability density at any instant of intrinsic time). This resolves the deep divide between a quantum mechanical interpretation (in which *both* the notion of time and positive semidefinite probabilities are needed) and the usual Klein-Gordon type WDW equations second-order in intrinsic time and its difficulties with positive definite "probabilities". A spatial diffeomorphism-invariant quantum theory with, $M := \int (\beta\tilde{\pi}+\bar{H})^2/\sqrt{q}d^3x = 0$, will translate into

$$[\beta\hat{\tilde{\pi}} + \bar{H}(\hat{\tilde{\pi}}^{ij}, \hat{\bar{q}}_{ij}, \hat{q})]|\Psi\rangle = 0, \qquad (2.24)$$

and

$$\hat{H}_i|\Psi\rangle = 0. \qquad (2.25)$$

In the metric representation, the momentum operators are realized by

$$\hat{\pi} = \frac{3\hbar}{i} \frac{\delta}{\delta \ln q}, \qquad (2.26)$$

$$\hat{\tilde{\pi}}^{ij} = \frac{\hbar}{i} P^{ij}_{lk} \frac{\delta}{\delta \bar{q}_{lk}} \qquad (2.27)$$

operating on $\Psi[\bar{q}_{ij}, q]$. The Schrödinger equation and HJ equation for semiclassical states $Ce^{\frac{iS}{\hbar}}$ are respectively,

$$i\hbar \frac{\delta}{\delta \ln q} \Psi = \frac{\bar{H}(\hat{\tilde{\pi}}^{ij}, q_{ij})}{3\beta} \Psi, \qquad (2.28)$$

$$\frac{\delta S}{\delta \ln q} = -\frac{\bar{H}(\tilde{\pi}^{ij} = P^{ij}_{kl} \frac{\delta S}{\delta \bar{q}_{kl}}; q_{ij})}{3\beta}; \qquad (2.29)$$

and

$$\nabla_j \frac{\delta \Psi}{\delta q_{ij}} = 0 \qquad (2.30)$$

enforces spatial diffeomorphism symmetry. Behold the appearance of a true Hamiltonian density $\bar{H}(x)/\beta$ generating evolution with respect to intrinsic, albeit still multi-fingered, time $\delta \ln q^{\frac{1}{3}}(x)$. The physical Hamiltonian generating global cosmic time translations will be explained and introduced later.

The method of master constraint is only one of the complementary approaches, and the same physics can be deduced from the Schrödinger equation above, and the Heisenberg formulation and generalized Baierlein-Sharp-Wheeler action [62] which shall all be discussed later. What is important is the paradigm shift from four-covariance to spatial diffeomorphism invariance which reveals the primacy of dynamics, with respect to intrinsic time, dictated by $\bar{H}$. This fundamental change in perspective also permits a satisfactory resolution of "the problem of time" and consistent extensions and improvements to Einstein's theory.

2.5 Emergence of classical spacetime

2.5.1 *Constructive interference*

Many years ago Gerlach [60] demonstrated that classical spacetime and its EOM can be recovered from the quantum theory through HJ theory and constructive interference. The first order HJ equation, which bridges quantum and classical regimes, has complete [59] solution

$$S = S(^{(3)}\mathcal{G}; \alpha) \qquad (2.31)$$

which depends on 3-geometry, $^{(3)}\mathcal{G}$, and integration constants (denoted generically here by α). Constructive interference with

$$\mathcal{S}(^{(3)}\mathcal{G}; \alpha + \delta\alpha) = \mathcal{S}(^{(3)}\mathcal{G}; \alpha); \qquad (2.32)$$

$$\mathcal{S}(^{(3)}\mathcal{G} + \delta^{(3)}\mathcal{G}; \alpha + \delta\alpha) = \mathcal{S}(^{(3)}\mathcal{G} + \delta^{(3)}\mathcal{G}; \alpha) \qquad (2.33)$$

leads to

$$\frac{\delta}{\delta\alpha} \left[\int \frac{\delta\mathcal{S}(^{(3)}\mathcal{G}; \alpha)}{\delta q_{ij}} \delta q_{ij} \right] = 0; \qquad (2.34)$$

subject to constraints

$$M = H_i = 0.$$

2.5.2   *Recovery of the classical EOM from constructive interference*

With the momenta identified with

$$\tilde{\pi}^{ij}(\alpha) := \frac{\delta \mathcal{S}(^{(3)}\mathcal{G}; \alpha)}{\delta q_{ij}}, \tag{2.35}$$

and Lagrange multipliers δm and $\delta \mathrm{N}^i$, the requirement of constructive interference is equivalent to

$$0 = \frac{\delta}{\delta \alpha} \left[\int (\tilde{\pi}^{ij} \delta q_{ij} - \delta \mathrm{N}^i H_i) - \delta m \mathrm{M} \right] \tag{2.36}$$

$$= \frac{\delta}{\delta \alpha} \left[\int \tilde{\pi} \delta \ln q^{\frac{1}{3}} + \bar{\pi}^{ij} \delta \bar{q}_{ij} + \frac{2q^{ij}}{3} \delta \mathrm{N}_i \nabla_j \tilde{\pi} + 2q^{-\frac{1}{3}} \delta \mathrm{N}_i \nabla_j \tilde{\pi}^{ij} \right]. \tag{2.37}$$

Happily, for master constraint theories, there is no δm contribution (since

$$\mathrm{M} = 0 \Leftrightarrow H = 0 \tag{2.38}$$

and M is quadratic in H). Imposing $\tilde{\pi} = -\bar{H}/\beta$, integrating by parts, and bearing in mind $\bar{H}(\tilde{\pi}^{ij}(\alpha), q_{ij})$, the resultant EOM is,

$$\frac{\delta \bar{q}_{ij}(x) - \mathcal{L}_{\vec{N}dt} \bar{q}_{ij}(x)}{\delta \ln q^{\frac{1}{3}}(y) - \mathcal{L}_{\vec{N}dt} \ln q^{\frac{1}{3}}(y)} = P^{kl}_{ij} \frac{\delta[\bar{H}(y)]}{\beta \delta \bar{\pi}^{kl}(x)} \tag{2.39}$$

$$= \frac{\bar{G}_{ijmn} \bar{\pi}^{mn}}{\beta \bar{H}} \delta(x, y), \tag{2.40}$$

wherein $\mathcal{L}_{\vec{N}}$ denotes Lie derivative and $\delta \vec{\mathrm{N}} =: \vec{N}dt$. Proceeding as in Ref. [60], the other half of Hamilton's equations,

$$\frac{\delta \bar{\pi}^{ij}(x) - \mathcal{L}_{\vec{N}dt} \bar{\pi}^{ij}(x)}{\delta \ln q^{\frac{1}{3}}(y) - \mathcal{L}_{\vec{N}dt} \ln q^{\frac{1}{3}}(y)} = -\frac{\delta[\bar{H}(y)/\beta]}{\delta \bar{q}_{ij}(x)}, \tag{2.41}$$

can be recovered. As predicted by Eq. (2.28), $\bar{H}/\beta$ is the Hamiltonian for evolution of $(\bar{q}_{ij}, \bar{\pi}^{ij})$ with respect to $\ln q^{\frac{1}{3}}$.

Although the derivation above bear similarities to Gerlach's work, fundamental differences must be noted. In Ref. [60], $\delta \mathrm{N} =: Ndt$ (associated with local constraint $H = 0$) will always contribute to the final EOM resulting in multi-fingered time with arbitrary lapse function. In contradistinction, δm contribution does not arise for a master constraint theory. This is part and parcel of the paradigm shift. Not only is unphysical time development with arbitrary lapse function now evaded, the "shortcoming" that M does not generate dynamical evolution with respect to coordinate time is redeemed at a much deeper level through physical evolution with respect to intrinsic time.

2.5.3   *Emergent classical ADM spacetime; time evolution in Einstein's theory is physical*

Through Eq. (2.40) and $\tilde{\pi}(x) = -\bar{H}(x)/\beta$, the emergent ADM classical space-time has momentum

$$G_{ijkl}\tilde{\pi}^{kl}(x) = \frac{\sqrt{q}}{4N\kappa}\left(\frac{dq_{ij}}{dt} - \mathcal{L}_{\vec{N}}q_{ij}\right), \tag{2.42}$$

$$N(x)dt := \frac{\delta\ln q^{\frac{1}{3}} - \mathcal{L}_{\vec{N}dt}\ln q^{\frac{1}{3}}}{(4\beta\kappa\bar{H}/\sqrt{q})}. \tag{2.43}$$

In conventional canonical formulation of Einstein's GR, the EOM with *arbitrary a priori* lapse function is,

$$\frac{dq_{ij}(x)}{dt} = \left\{q_{ij}, \int (NH + N_i H^i)d^3x'\right\}_{P.B.} \tag{2.44}$$

$$= \frac{4N\kappa}{\sqrt{q}}G_{ijkl}\tilde{\pi}^{kl} + \mathcal{L}_{\vec{N}}q_{ij}. \tag{2.45}$$

The extrinsic curvature is related to $\tilde{\pi}^{ij}$ by

$$K_{ij}(x) := \frac{1}{2N}\left(\frac{dq_{ij}(x)}{dt} - \mathcal{L}_{\vec{N}}q_{ij}(x)\right) \tag{2.46}$$

$$= \frac{2\kappa}{\sqrt{q}}G_{ijkl}(x)\tilde{\pi}^{kl}(x). \tag{2.47}$$

Taking the trace yields

$$\frac{1}{3}Tr(K(x)) = \frac{1}{2N}\left(\frac{\partial\ln q^{\frac{1}{3}}}{\partial t} - \mathcal{L}_{\vec{N}}\ln q^{\frac{1}{3}}\right) \tag{2.48}$$

$$= \frac{2\kappa\beta}{\sqrt{q}}\bar{H}(x), \tag{2.49}$$

wherein the constraint

$$(\beta\tilde{\pi}(x) + \bar{H}(x)) = 0 \tag{2.50}$$

has been used to arrive at the last step. Eq. (2.49) demonstrates that the lapse function and intrinsic time are precisely related *a posteriori*, by the EOM and constraints, by the same formula as in Eq. (2.43). For a theory with apparent full four-covariance (such as Einstein's GR with $\beta^2 = 1/6$ and consistent Dirac algebra of constraints), this relation is an *a posteriori* identity which does not compromise the *a priori* arbitrariness of N. However, it reveals, even in Einstein's GR, the physical meaning of the lapse function and its relation to the intrinsic time. It also demonstrates *time evolution in Einstein's theory is physical*, with precise *a posteriori* Ndt as in Eq. (2.43); and not seemingly arbitrary because N has been introduced as a Lagrange multiplier of the Hamiltonian constraint in a "four-covariant" formulation. The situation is similar to Lagrange multipliers in simple mechanics which can take on physical significance upon solving all the EOM and constraints. A relativistic particle analogy apposite to GR is discussed in Sec. 2.12.

2.6   Paradigm shift and resolution of the problem of time

2.6.1   *Only spatial diffeomorphism invariance is needed and realized*

Starting with only spatial diffeomorphism invariance and through constructive interference, Eq. (2.40) with physical evolution in intrinsic time is obtained. This relates the momentum to coordinate time derivative of the metric precisely as in Eq. (2.46). It is thus possible to interpret the emergent classical space-time (which can generically be described with ADM metric) to possess extrinsic curvature which corresponds precisely to the derived *a posteriori* lapse function displayed in Eq. (2.43). However, *only* the freedom of spatial diffeomorphism invariance is realized, as the *emergent* lapse is now completely described by the intrinsic time $\ln q^{\frac{1}{3}}$ and N^i. The EOM w.r.t. ADM coordinate time t generated by

$$H_{ADM} = \int (NH + N^i H_i) d^3 x, \qquad (2.51)$$

in Einstein's GR can be recovered from evolution with respect to $\ln q^{\frac{1}{3}}(x)$ and generated by $\bar{H}(x)/\beta$ iff $N(x)$ assumes the form of Eq. (2.43). Dispensing with ADM Hamiltonian constraint, the effective Hamiltonian for an ever-expanding universe which produces the same *a posteriori* lapse function is elucidated in Chapters 3, 4, and 8.

2.6.2   *Four-dimensional space-time covariance in GR is a red herring*

All the previous observations lead to the central revelation: full four-dimensional space-time covariance is a red herring which obfuscates the physical reality of time; and all that is necessary to consistently capture the classical physical content of Einstein's GR is a theory invariant only with respect to spatial diffeomorphisms accompanied by a true Hamiltonian which will produce the same *a posteriori* lapse function. Chapters 3,

4 and 8 will elaborate on this, and exhibit from many perspectives the synergy between the resultant Hamiltonian and the cosmic clock of our expanding universe. The paradigm shift points to a *complete resolution of the problem of time*, from quantum to classical GR: classical spacetime, with consistent lapse function and ADM metric,

$$ds^2 = -\left[\frac{(\partial_t \ln q^{\frac{1}{3}} - \mathcal{L}_{\vec{N}} \ln q^{\frac{1}{3}})dt}{4\beta\kappa(\bar{H}/\sqrt{q})}\right]^2 + q_{ij}[dx^i + N^i dt][dx^j + N^j dt], \quad (2.52)$$

emerges from constructive interference of a spatial diffeomorphism invariant quantum theory with Schrödinger and Hamilton-Jacobi equations which are first-order in intrinsic time development.

2.6.3   Classical proper time and its relation to intrinsic time interval

It is also gratifying to note the correlations, via Eq. (2.43), of classical proper time $d\tau$ and intrinsic time $\ln q^{\frac{1}{3}}(x)$,

$$d\tau^2 = \left[\frac{\delta \ln q^{\frac{1}{3}}}{(4\beta\kappa\bar{H}/\sqrt{q})}\right]^2 \quad (2.53)$$

(for vanishing shifts); and of Wheeler's notion of ADM "simultaneity" (a constant-t spacelike ADM hypersurface) to simultaneity in intrinsic time,

$$dt = 0 \Rightarrow \delta \ln q = 0. \quad (2.54)$$

In particular, by Eq. (2.44) and Eq. (2.49), proper time intervals measured by physical clocks in space-times which are solutions of Einstein's equations always agree with the result of Eq. (2.52).

In conventional formulations, four-covariance is the underlying paradigm; and it has been argued that the chosen time variable should be a space-time scalar. It follows that no function of the intrinsic quantity to just three-geometry can fulfill this criterion. "Scalar field time" has thus been advocated by many, notably by Kuchar; despite the fact that such a choice would not be available in the context of pure gravity. Chapter 8 compares scalar field time, York extrinsic time and our advocated intrinsic time. The novel and crucial feature is that the paradigm shift to only spatial diffeomorphism invariance is just what is needed to permit a degree of freedom constructed from the intrinsic spatial geometry to act as the time variable. Indeed $\delta \ln q^{\frac{1}{3}} = \frac{1}{3}(\delta q)/q$ which is a scalar under spatial diffeomorphisms is the apposite choice of intrinsic time interval; and its (spatial) diffeomorphism-invariant part is directly proportional to the change in the logarithm of the spatial volume of the universe. With just spatial diffeomorphism, rather than four-covariance, this cosmic intrinsic clock of our expanding closed universe becomes meaningful and well defined.

2.7   Modifications of the Hamiltonian, and improvements to the quantum theory

The framework of the theory also prompts improvements to $\bar{H}$. Requirement of a real physical Hamiltonian density $\bar{H}$ compatible with spatial diffeomorphism symmetry suggests supplementing the kinetic term in the square root with a positive semidefinite quadratic form, i.e.

$$
\begin{aligned}
\bar{H} &= \sqrt{\bar{G}_{ijkl}\bar{\pi}^{ij}\bar{\pi}^{kl} + \left[\frac{1}{2}(q_{ik}q_{jl} + q_{jk}q_{il}) + \omega q_{ij}q_{kl}\right]\frac{\delta W}{\delta q_{ij}}\frac{\delta W}{\delta q_{kl}}} \\
&= \sqrt{[\bar{q}_{ik}\bar{q}_{jl} + \omega\bar{q}_{ij}\bar{q}_{kl}]\left(\bar{\pi}^{ij}\bar{\pi}^{kl} + q^{\frac{2}{3}}\frac{\delta W}{\delta q_{ij}}\frac{\delta W}{\delta q_{kl}}\right)}.
\end{aligned} \tag{2.55}
$$

$\bar{H}$ is then real if $\omega > -\frac{1}{3}$.

A slight generalization of Eq. (2.55) is to replace $\frac{\delta W}{\delta q_{ij}}$ in the positive semidefinite quadratic form with

$$
\frac{\delta W}{\delta q_{ij}} \to \sqrt{q}(\Lambda' q^{ij} + a' R q^{ij} + b R^{ij} + g' C^{ij}), \tag{2.56}
$$

which is the most general symmetric second rank tensor (density) containing up to third derivatives of the spatial metric. The potential of the Ricci scalar and cosmological constant of Einstein's theory is recovered at low curvatures.

2.7.1   *Chern-Simons term, power-counting renormalizability; and positive-definite Hamiltonian*

Reference [31] proposes, for perturbative power-counting renormalizabilty, to include a gravitational Chern-Simons term which has dimensionless coupling and is third-order in spatial derivatives of the metric. To this order,

$$
W = \int \left[\sqrt{q}(aR - \Lambda) - \frac{g}{2}\tilde{\epsilon}^{ikj}\left(\Gamma^l_{im}\partial_j\Gamma^m_{kl} + \frac{2}{3}\Gamma^l_{im}\Gamma^m_{jn}\Gamma^n_{kl}\right)\right]d^3x \tag{2.57}
$$

i.e. it is of the form of a three-dimensional Einstein-Hilbert action with cosmological constant supplemented by a Chern-Simons action with dimensionless coupling constant g; and the Cotton-York tensor density is the functional derivative of the Chern-Simons action w.r.t. the spatial metric.

The effective value of κ and cosmological constant from Eq. (2.55) can thus be determined as

$$\kappa = \frac{8\pi G}{c^3} = \sqrt{\frac{1}{2a\Lambda(1 + 3\omega)}} \qquad (2.58)$$

and

$$\Lambda_{eff} = \frac{3}{2}\kappa^2\Lambda^2(1 + 3\omega) = \frac{3\Lambda}{4a} \qquad (2.59)$$

respectively. The possibility of having a new parameter ω in the potential (different from the deformation parameter λ in the supermetric) has been overlooked in previous works. Furthermore, positivity of $\bar{H}^2$ (with $\omega > -\frac{1}{3}$) is correlated with *real* κ and *positive* Λ_{eff}. There is also the intriguing feature that the lowest classical energy of the physical Hamiltonian $\bar{H}$ occurs when zero modes are present, i.e. $\omega \to -\frac{1}{3}$, leading, in this limit and for fixed κ, to $\Lambda_{eff} \to 0$. This, however, requires a thorough investigation of the renormalization group flow of ω and other parameters to deduce the exact behavior of Λ_{eff} with physical energy scale, especially when matter and other forces are also taken into account.

Equation (2.55) is in the spirit of the "detailed balance" scheme of [31], but the departure here is the square root form of the Hamiltonian in ITG which is necessary to agree with the EOM in GR in the low curvature limit (see, for instance, Chapter 3 on the *a posteriori* lapse function). It must be pointed out that the Hamiltonian density above involves a square root and its rigorous definition through spectral decomposition remains a most formidable challenge. While this is not necessarily a defect of the theory, issues in quantum gravity which involve the rigorous non-perturbative definition of the Hamiltonian operator are thus not yet addressed in the current work. There are some intriguing and encouraging signs. Out of the many possible alternatives which respect three-covariance (rather than the more restrictive four-covariance paradigm for Einstein's theory), Horava's power-counting renormalizable "detailed balance" form of the Hamiltonian constraint [31] (or its slight generalization discussed above which also yields a dimensionless coupling constant in the highest order term C^{ij}) is essentially pinned down by the square root form and positivity of the Hamiltonian density. In Ref. [31], the Hamiltonian density with "detailed balance" is proportional to

$$\frac{G_{ijkl}}{\sqrt{q}}\left(\tilde{\pi}^{ij}\tilde{\pi}^{kl} + \frac{\delta W}{\delta q_{ij}}\frac{\delta W}{\delta q_{kl}}\right). \qquad (2.60)$$

Without factoring out $\tilde{\pi}$, the negative mode in the full supermetric, G_{ijkl}, compromises the positivity of the kinetic term, $\frac{G_{ijkl}}{\sqrt{\bar{q}}}\tilde{\pi}^{ij}\tilde{\pi}^{kl}$, in the theory. Formulations which use an extra scalar field [11], or variables other than $\ln q^{\frac{1}{3}}$ as time to attain deparametrization will also be afflicted with the same problem. In contradistinction, with $\ln q^{\frac{1}{3}}$ as intrinsic time, $\tilde{\pi}$ is singled out and isolated as the conjugate variable, with the upshot that in the Schrödinger equation, $\bar{H}$ (which does not contain $\tilde{\pi}$) always comes with positive-semidefinite kinetic term.

In Chapter 7 self-adjoint and positive-definite requirements on the Hamiltonian achieved by applying a similarity transformation on the momentric variables will be elucidated; and its analogy to Yang-Mills theory will be clarified. Its decisive role in the initial state of our universe will be explored in Chapters 4 and 10. To wit, instead of Eq. (2.55), the Hamiltonian advocated in ITG is with the form of

$$\bar{H}(x) = \sqrt{Q_j^{\dagger i}(x)Q_i^j(x) + q(x)\mathcal{K}}; \qquad (2.61)$$

wherein

$$Q_j^i(x) = e^W \bar{\pi}_j^i(x) e^{-W} = \frac{\hbar}{i}\bar{E}^i_{j(lm)}\left[\frac{\delta}{\delta\bar{q}_{mn}} - \frac{\delta W}{\delta\bar{q}_{mn}}\right] \qquad (2.62)$$

$$Q_j^{\dagger i}(x) = e^{-W} \bar{\pi}_j^i(x) e^{W} = \frac{\hbar}{i}\bar{E}^i_{j(lm)}\left[\frac{\delta}{\delta\bar{q}_{mn}} + \frac{\delta W}{\delta\bar{q}_{mn}}\right], \qquad (2.63)$$

and

$$\bar{E}^k_{l(ij)} = \frac{1}{2}(\delta^k_i\bar{q}_{lj} + \delta^k_j\bar{q}_{li}) - \frac{1}{3}\delta^k_l\bar{q}_{ij}, \qquad (2.64)$$

being a trace-free projector. In comparison with the "detailed balance" form of Eq. (2.55), this Hamiltonian density is explicitly self-adjoint and positive-definite. The emergence of Einstein's theory with Cotton-York addition and the Hamiltonian of Eq. (1.47) from this shall be discussed in Chapter 7. An interesting feature is this Hamiltonian obtained by applying a similarity transformation on the momentric variables implies, to third order in spatial derivatives, W, as in Eq. (2.57), contains only the cosmological, Einstein Ricci scalar and Chern-Simons terms, with the latter associated with dimensionless coupling.[7] For the Hamiltonian of Eq. (1.47), every $1/k^2$ Ricci contribution in the ultraviolet behavior of the propagator is tempered and damped by a $1/k^6$ Cotton-York $\tilde{C}^i_j\tilde{C}^j_i$ contribution.

---

[7]Higher curvature integrals in W will come with couplings of negative mass dimensions.

2.8 Gauge-invariant global time, superspace dynamics, and temporal order

2.8.1 *Many-fingered time, Hodge decomposition, and invariant global time interval*

In our intrinsic time formulation, the would be quantum Wheeler-DeWitt equation, which is a constraint which must be satisfied at each spatial point x, is replaced by a Schrödinger equation with $\frac{\bar{H}(x)}{\beta}$ generating translations in $\ln q^{\frac{1}{3}}(x)$ which is a Tomonaga-Schwinger [13] many-fingered time variable. However, consistent time-ordering can be realized only with a global time parameter. The transcription from, apparently, many-fingered dynamics to evolution with respect to a gauge-invariant global time is, remarkably, unequivocal for three-hypersurfaces which are compact Riemannian manifolds without boundary. For $\delta \ln q^{\frac{1}{3}}$, spatial diffeomorphism invariance collapses multi-fingered time interval into a global invariant change. Hodge decomposition[8] of the 0-form $\delta \ln q^{\frac{1}{3}}$ uniquely yields

$$\delta \ln q^{\frac{1}{3}}(x) = \delta h + \nabla_i \delta Y^i(x), \qquad (2.65)$$

wherein δh is harmonic, *independent* of x and gauge-invariant, whereas δY^i can be gauged away because, fortuitously,

$$\mathcal{L}_{\delta N^i} \ln q^{\frac{1}{3}}(x) = \frac{2}{3}\nabla_i \delta N^i(x). \qquad (2.66)$$

This leads, on identification of the global time interval $\delta T := \delta h$, and bearing in mind Eq. (2.28), to the transcription,

$$i\hbar \frac{\delta \Psi}{\delta T} = \int i\hbar \frac{\delta \Psi}{\delta \ln q^{\frac{1}{3}}(x)} \frac{\delta \ln q^{\frac{1}{3}}(x)}{\delta T} d^3 x$$

$$= \left[\int \frac{\bar{H}(x)}{\beta} d^3 x \right] \Psi, \qquad (2.67)$$

which describes evolution with respect to intrinsic superspace time interval δT. The corresponding physical Hamiltonian

$$H_{phys.} := \int \frac{\bar{H}(x)}{\beta} d^3 x \qquad (2.68)$$

is moreover spatial diffeomorphism invariant as it is the integral of a tensor density of weight one. This remarkable Schrödinger equation,

$$i\hbar \frac{d\Psi}{dT} = H_{phys.} \Psi, \qquad (2.69)$$

---

[8]Please also see Sec. 1.8.3 for a brief discussion.

dictates quantum geometrodynamics in *explicit* superspace $^{(3)}\mathcal{G}$ entities $(\Psi[[q_{ij}] \in {}^{(3)}\mathcal{G}], H_{phys.}(T))$. For a closed three-manifold of spatial volume V, this intrinsic global time interval can be identified (see, for instance Sec. 3.3.1) as $dT = \frac{2}{3}d\ln V$.

2.8.2   *Emergent classical spacetime metric, and its relation to intrinsic time and the Hamiltonian density of ITG*

On equating $\delta t = \mathcal{K}\delta T$ in Eq. (2.52), the emergent classical space-time from constructive interference under superspace intrinsic time evolution will, as demonstrated in earlier subsections, be described by the ADM metric,

$$ds^2 = -\left[\frac{\delta T - \mathcal{L}_{\vec{\mathcal{N}}\delta T}\ln q^{\frac{1}{3}}}{4\beta\kappa(\bar{H}/\sqrt{q})}\right]^2 + q_{ij}[dx^i + \mathcal{N}^i\delta T][dx^j + \mathcal{N}^j\delta T], \quad (2.70)$$

wherein

$$\mathcal{N}^i := N^i/\mathcal{K}. \quad (2.71)$$

Since it can always be absorbed into the gauge parameter N^i, the constant $\mathcal{K}$ has no physical implication. Rather, it is the (scalar function) $\kappa\bar{H}/\sqrt{q}$ which provides the physical conversion between dimensionless intrinsic time interval δT and the proper time of ds^2. Gravitational redshifts and other physical effects are thus determined by the Hamiltonian density $\bar{H}$. The proper time (with vanishing shifts and $dx^i = 0$),

$$d\tau = \frac{dT\sqrt{q}}{4\beta\kappa\bar{H}} \quad (2.72)$$

exhibits the physically intuitive property of varying directly with intrinsic superspace time interval dT and reciprocally with energy density $\bar{H}$. Section 2.10 is devoted to the analysis of test particles dynamics and relativistic time effects in the framework of ITG.

2.8.3   *Diffeomorphism-invariant time-ordered evolution of physical quantum state of the universe*

The crucial time development operator can be derived by integrating the Schrödinger equation. This is now feasible without ambiguity because δT is "one-dimensional", more precisely, x-independent, *rather than many-fingered*. Moreover, the necessity of "time"-ordering, which underpins the notion of causality, emerges because quantum fields do not commute at different "times". Equation (2.67) implies the change

$$\delta\Psi = \left[-\frac{i}{\hbar}H_{phys.}\right]\delta T\Psi; \quad (2.73)$$

thus yielding

$$\Psi[[q_{ij}(T)] \in {}^{(3)}\mathcal{G}] = U(T, T_0)\Psi[[q_{ij}(T_0)] \in {}^{(3)}\mathcal{G}], \qquad (2.74)$$

with T-ordered evolution operator

$$U(T, T_0) := \mathcal{T}\left(\exp\left[-\frac{i}{\hbar}\int_{T_0}^{T} H_{phys}(T')dT'\right]\right). \qquad (2.75)$$

As $H_{phys.}$ is classically real and gauge-invariant, a unitary and diffeomorphism-invariant $U(T, T_0)$ is viable. Since dT is also unchanged under spatial diffeomorphisms, the temporal ordering in $U(T, T_0)$ is reassuringly gauge invariant. In quantum gravity, the physical state of the universe is the fundamental entity that is causally time-ordered and the embodiment of our shared past, present and future.

2.9   Further discussions

2.9.1   *Observables, two d.o.f., and the Hamilton-Jacobi equation*

That there are, for pure gravity, two physical d.o.f. can be ascertained. Chapter 8 addresses this in detail, with the reduced symplectic potential as displayed in Sec. 8.4.5. This decomposition yields 2 reduced physical canonical degrees of freedom $(\bar{q}_{ij}^{\text{phys.}}, \bar{\pi}_{TT}^{ij})$, and an extra pair $(\ln q^{\frac{1}{3}}, \tilde{\pi})$ to play the role of clock and Hamiltonian density. The momentum constraint generates infinitesimal, rather than finite, gauge transformations which are spatial diffeomorphisms. Physical gauge invariant infinitesimal excitations, $\delta\bar{q}_{ij}^{TT}$, are transverse traceless (TT) changes of the unimodular metric $\bar{q}_{ij}$. A finite change leading to the final $\bar{q}_{ij}^{\text{phys.}}$ can be obtained by accumulating these physical changes, each successive infinitesimal addition transverse ($\nabla^i\delta\bar{q}_{ij}^{TT} = 0$) w.r.t. the preceding metric defining the ∇_i.

Identification of a complete set of observables in theories with diffeomorphism invariance is often thought to be a formidable task; but due to the fact that only spatial diffeomorphism is realized, all Kuchar observables commuting with the momentum constraints are now Dirac observables. A theory with HJ equation which is first-order-in-time offers also offers a resolution from another perspective. The solution of the HJ equation is deemed complete [59], in that it has as many integration constants (denoted earlier by α) as the number of degrees of freedom in the theory, plus an overall additive constant. These are all gauge invariant, and together with

$$\phi := \frac{\delta S}{\delta \alpha} \qquad (2.76)$$

(which express the coordinates in terms of time and the constants (α, ϕ)), they provide general integrals of equations of motion which are well-suited to play the role of physical observables of diffeomorphism-invariant theories. Alternatively, emergent space-time manifolds obtained by integrating Hamilton's equations are characterized by $4 \times \infty^3$ freely specifiable initial data $(\bar{q}_{ij}^{\text{phys.}}, \bar{\pi}_{TT}^{ij})$.

With global cosmic time and first order evolution, for semiclassical states the HJ equation from the Schrödinger equation Eq. (2.69) will now be of the familiar form,

$$\frac{\partial S}{\partial T} + H_{phys.} = 0; \tag{2.77}$$

instead of the second-order "timeless" apparition [63],

$$G_{ijkl}\frac{\partial S}{\partial q_{ij}}\frac{\partial S}{\partial q_{kl}} + \mathcal{V} = 0. \tag{2.78}$$

Hamilton's equations for the dynamical variables,

$$\frac{d\bar{q}_{ij}}{dT} = \{\bar{q}_{ij}, H_{phys.}\}_{P.B.}, \qquad \frac{d\bar{\pi}^{ij}}{dT} = \{\bar{\pi}^{ij}(x), H_{phys.}\}_{P.B.}, \tag{2.79}$$

follow from a complete solution of the HJ equation [59]. It is noteworthy that *it is the same* global intrinsic time variable T that appears in Schrödinger, HJ, and Hamilton's equations; so quantum, semiclassical, and classical "time" are in harmony, even though it is fundamentally the quantum state that is temporally ordered. Microcausality is enforced at the quantum level.

Even within the context of a first class constrained system, a theory with only spatial diffeomorphism gauge symmetry requires physical observables to commute only with H_i (and not H); thus all Kuchar observables [64] *become* physical. Within the context of the master constraint framework, these observables will also consistently have weakly vanishing Poisson bracket with M, since

$$\{f(q_{ij}, \tilde{\pi}^{ij}), M\}|_{M=0 \Leftrightarrow H=0} \approx 0. \tag{2.80}$$

A Dirac second class system offers a more satisfactory and complete resolution. Therein Dirac, rather than Poisson, brackets rule; and by construction *all the variables* have vanishing Dirac brackets with all the constraints and auxiliary conditions. The route to quantum theory [30] lies not in promoting Poisson, but fundamental Dirac, brackets to commutators. In Chapters 8 and 9 the route of promoting GR and its extensions from a Dirac first class to a second class constrained system with resultant TT d.o.f. will be elucidated, and the auxiliary conditions and their role in "gauge-fixing" will be addressed in detail.

2.9.2   Dreibein and the Lorentz group

It should be pointed out that, for any classical ADM space-time, local $SO(3,1)$ Lorentz symmetry on the tangent space is intact, since the ADM metric,

$$ds^2 = \eta_{AB} e^A{}_\mu e^B{}_\nu dx^\mu dx^\nu, \tag{2.81}$$

is invariant under local Lorentz transformations of the vierbein fields

$$e'^A{}_\mu = \Lambda^A{}_B(x) e^B{}_\mu$$

which do not affect metric components

$$g_{\mu\nu} = \eta_{AB} e^A{}_\mu e^B{}_\nu.$$

This local gauge symmetry is associated with the freedom to choose (up to Lorentz transformations) different tangent spaces at different space-time points, rather than with the isometry of the metric under space-time coordinate transformations. In GR, a generic space-time may have no isometry at all. At the more fundamental level of spatial metric and without recourse to four-dimensional spacetime, Lorentz symmetry can be fully realized as $SO(3,C)$, rather than the isomorphic $SO(3,1)$, by defining [46]

$$q^{ij} := \frac{1}{(\det \tilde{E})} \tilde{E}^{ia} \tilde{E}^j{}_a, \tag{2.82}$$

wherein the densitized triad is

$$\tilde{E}^{ia} = \frac{1}{2} \tilde{\epsilon}^{ijk} \epsilon^{abc} e_{bj} e_{ck} = (\det e) [e_{ai}]^{-1}. \tag{2.83}$$

Both q^{ij} and $q_{ij} := e_{ai} e^a{}_j$, are invariant under local (anti)self-dual $SO(3,C)$ rotations of the triad, E^{ia}, and its inverse e_{ai}. Moreover, expressing the spatial metric as $q_{ij} = e_{ai} e^a{}_j$ leads to the decomposition of the symplectic potential as

$$\int \tilde{\pi}^{ij} \delta q_{ij} = \int \tilde{\pi}^{ai} \delta e_{ai} \tag{2.84}$$

$$= \int \left(\tilde{\pi} \delta \ln q^{\frac{1}{3}} + \bar{\pi}^{ai} \delta \bar{e}_{ai} \right); \tag{2.85}$$

with

$$\tilde{\pi}^{ai} := 2\tilde{\pi}^{ij} e^a{}_j, \tag{2.86}$$

$$\tilde{\pi} := q_{ij} \tilde{\pi}^{ij} = \frac{1}{2} (\tilde{\pi}^{ai} e_{ai}), \tag{2.87}$$

$$\bar{\pi}^{ai} := e^{\frac{1}{3}} \left[\tilde{\pi}^{ai} - E^{ai} \frac{(\tilde{\pi}^{bk} e_{bk})}{3} \right]; \tag{2.88}$$

and unimodular

$$\bar{e}_{ai} := e^{-\frac{1}{3}} e_{ai}. \tag{2.89}$$

Symmetry of $\tilde{\pi}^{ij}$ in Eq. (2.86) leads to the Gauss Law constraint,

$$\frac{1}{2} [\bar{e}^a{}_i \bar{\pi}^{bi} - \bar{e}^b{}_i \bar{\pi}^{ai}] = 0, \tag{2.90}$$

which generates $SO(3, C)$ Lorentz transformations of the variables. The pair

$$\left(\ln q^{\frac{1}{3}} := \ln e^{\frac{2}{3}}, \tilde{\pi} := \frac{1}{2} (\tilde{\pi}^{ai} e_{ai}) \right) \tag{2.91}$$

is Lorentz invariant, and commutes with the remaining variables $(\bar{\pi}^{ai}, \bar{e}_{ai})$.

Chapter 9 will address in detail the gauging of the full Lorentz group as $SO(3, C)$, together with the incorporation and coupling of $SL(2, C)$ chiral fermions to gravity. The inclusion of matter and other forces is rather straightforward as Standard Model fields do not couple to π; and the corresponding Hamiltonian density of these fields, H'_{non-GR}, can be appended to gravitational kinetic and potential energy terms. Consequently,

$$\bar{H}_T = \sqrt{\bar{H}^2 + H'_{non-GR}} \tag{2.92}$$

replaces $\bar{H}$ of Eq. (2.55); and the extension does not affect the basic form of the Hamiltonian constraint which is now

$$\beta \tilde{\pi} + \bar{H}_T = 0. \tag{2.93}$$

In H_T, Weyl fermions couple to $SL(2, C)$ spin connections as will be addressed in Chapter 9. For any classical ADM space-time, these spin connections which couple to Standard Model fermions are pullbacks of the self and anti-self-dual Lorentz spin connections to three-dimensional spatial slices [65].

2.9.3   *Generalized Baierlein-Sharp-Wheeler action of the theory*

The Schrödinger and HJ equations affirm that pure GR has the physical content of $(\bar{q}_{ij}, \tilde{\pi}^{ij})$ subject diffeomorphism spatial symmetry, and evolving with respect to $\ln q^{\frac{1}{3}}$ with effective Hamiltonian density $\bar{H}/\beta$. To wit, the action can be taken to be of the form

$$S = \int \left[\tilde{\pi}^{ij} \frac{\partial \bar{q}_{ij}}{\partial T} - \frac{\bar{H}}{\beta} \frac{\partial \ln q^{\frac{1}{3}}}{\partial T} - \int N^i H_i \right] dT d^3 x. \tag{2.94}$$

Inverting for $\bar{\pi}^{ij}$ in terms of $\frac{\partial \bar{q}_{ij}}{\partial T}$ from the EOM yields[9] the action functional as

$$S = -\int \sqrt{\mathcal{V}} \sqrt{\frac{1}{\beta^2} \left[\frac{\partial \ln q^{\frac{1}{3}}}{\partial T}\right]^2_{eq.} - \bar{G}^{ijkl} \left[\frac{\partial \bar{q}_{ij}}{\partial T}\right]_{eq.} \left[\frac{\partial \bar{q}_{kl}}{\partial T}\right]_{eq.}} \, dT d^3x; \quad (2.95)$$

$$\left[\frac{\partial \ln q^{\frac{1}{3}}}{\partial T}\right]_{eq.} := \frac{\partial \ln q^{\frac{1}{3}}}{\partial T} - \mathcal{L}_{\vec{N}} \ln q^{\frac{1}{3}}, \quad \left[\frac{\partial \bar{q}_{ij}}{\partial T}\right]_{eq.} := \frac{\partial \bar{q}_{ij}}{\partial T} - \mathcal{L}_{\vec{N}} \bar{q}_{ij};$$

and the potential

$$\mathcal{V} = \bar{H}^2 - \bar{G}_{ijkl} \bar{\pi}^{ij} \bar{\pi}^{kl}. \quad (2.96)$$

The entities $[\ ]_{eq.}$ denote equivalence classes modulo spatial diffeomorphisms or Lie derivatives.

The resultant action is thus proportional to the *"superspace proper time"* (analogous to $S = -m_0 c \int \sqrt{-g_{\mu\nu} \frac{dx^\mu}{dt} \frac{dx^\nu}{dt}} \, dt$ for a relativistic point particle), with $\mathcal{V}$ playing the role of "mass" if it were constant (in Einstein's theory it is as displayed in Eq. (2.15)). This regains the generalized Baierlein-Sharp-Wheeler action [62] which has also been studied in Ref. [66] in a different situation.

2.9.4 *York extrinsic time*

The symplectic 1-form

$$\int \tilde{\pi} \delta \ln q^{\frac{1}{3}} = \int \frac{2}{3} \left(\frac{\tilde{\pi}}{\sqrt{q}}\right) \delta \sqrt{q} \quad (2.97)$$

allows a different perspective. With the further restriction of

$$\nabla_i \tilde{\pi} = 0 \Leftrightarrow \frac{\tilde{\pi}}{\sqrt{q}} = p, \quad (2.98)$$

is spatially constant. York [61] interprets and deploys this as the "extrinsic time" variable (more on this in Chapter 8). It follows that the Hamiltonian in York's formulation is then proportional to $\sqrt{q}$, and the total energy to the volume. Note also that although the extrinsic time variable is then invariant under spatial diffeomorphisms, it is however not invariant under four-dimensional diffeomorpghisms which are supposed to be "symmetries" of Einstein's theory. In ITG, with the paradigm shift to just spatial diffeomorphism invariance, $\delta \ln q^{\frac{1}{3}}$ is well-suited to the role of physical time interval: it is a spatial diffeomorphism scalar with a gauge-invariant part δh which is spatially constant.

---

[9]This result can also be deduced from Eq. (2.39) and Eq. (2.41).

2.10   ITG and the dynamics of test particles

This section and the one which follows will be devoted to the description and elucidation of test particle dynamics and its understanding within the context of ITG.

The motion of a test point particle of mass, m_0, described canonically by (x_P^i, p_i) in the generic ADM space-time can be derived (see, for instance, Ref. [67]) from the particle Hamiltonian,

$$H_{\mathrm{P}} = \int \left[Nc\sqrt{q^{ij}p_i p_j + m_0^2 c^2} - N^i p_i c \right] \delta^3(\vec{x}, \vec{x}_{\mathrm{P}}) d^3 x \qquad (2.99)$$

$$= N(\vec{x}_{\mathrm{P}}, t) c \sqrt{q^{ij}(\vec{x}_{\mathrm{P}}, t) p_i p_j + m_0^2 c^2} - N^i(\vec{x}_{\mathrm{P}}, t) p_i c, \qquad (2.100)$$

wherein N and N^i are respectively the lapse and shift functions. Hamilton's equation,

$$\frac{dx_{\mathrm{P}}^i}{dt} = \{x_{\mathrm{P}}^i, H_{\mathrm{P}}\}_{\mathrm{P.B.}} \qquad (2.101)$$

relates the velocity and momentum by

$$\frac{dx_{\mathrm{P}}^i}{dt} + N^i c = \frac{Ncq^{ij}p_j}{\sqrt{q^{ij}p_i p_j + m_0^2 c^2}}. \qquad (2.102)$$

Inverting for p_i in terms of $\frac{dx_{\mathrm{P}}^i}{dt} + N^i c$ results in the action,

$$S = \int \left[p_i \frac{dx_{\mathrm{P}}^i}{dt} - H_{\mathrm{P}} \right] dt \qquad (2.103)$$

$$= -m_0 \int c \, dt \sqrt{N^2 c^2 - q_{ij} \left(\frac{dx_{\mathrm{P}}^i}{dt} + N^i c \right) \left(\frac{dx_{\mathrm{P}}^j}{dt} + N^j c \right)}$$

$$= -m_0 c \int \sqrt{-g_{\mu\nu}(\vec{x}_{\mathrm{P}}, t) dx^\mu dx^\nu}.$$

This is just the usual proper time action for test particle on identifying the ADM metric as in Eq. (2.81). Conversely, starting with the Lagrangian in the final step of Eq. (2.103), geodesic motion for the particle with the Hamiltonian of Eq. (2.100) is obtained. To wit, starting with

$$S = \int L \, dt = -m_0 c \int \sqrt{-g_{\mu\nu} dx^\mu dx^\nu} \qquad (2.104)$$

$$= -m_0 c \int \sqrt{-g_{\mu\nu} \dot{x}^\mu \dot{x}^\nu} \, dt \qquad (2.105)$$

in generic ADM spacetime with the metric,

$$ds^2 = -N^2 c^2 dt^2 + q^{\frac{1}{3}} \bar{q}_{ij}(x, t)(dx^i + N^i c dt)(dx^j + N^j c dt), \qquad (2.106)$$

the Lagrangian

$$L = -m_0 c \sqrt{N^2 c^2 - q_{ij} \left(N^i c + \dot{x}^i\right)\left(N^j c + \dot{x}^j\right)} \qquad (2.107)$$

leads to conjugate momentum p_i which is

$$p_i = \frac{\partial L}{\partial \dot{x}^i} = \frac{m_0 c q_{ij} \left(N^j c + \dot{x}^j\right)}{\sqrt{N^2 c^2 - q_{jk} \left(N^j c + \dot{x}^j\right)\left(N^k c + \dot{x}^k\right)}}. \qquad (2.108)$$

The Hamiltonian for the test particle will be

$$H_{\mathrm{P}} = p_i \dot{x}^i - L$$

$$= \frac{m_0 c^2}{\sqrt{N^2 - q_{jk}\left(N^j + \frac{v^j}{c}\right)\left(N^k + \frac{v^k}{c}\right)}}$$

$$\cdot \left[N^2 + q_{ij}\left(N^i + \frac{v^i}{c}\right)\frac{v^k}{c} - q_{jk}\left(N^j + \frac{v^j}{c}\right)\left(N^k + \frac{v^k}{c}\right)\right]$$

$$= Nc\sqrt{m_0^2 c^2 + p_i p^i} - N^i p_i c, \qquad (2.109)$$

upon using Eq. (2.108) and the consequent

$$p_i p^i = m_0^2 c^2 \left[\frac{N^2}{N^2 - q_{jk}\left(N^j + \frac{v^j}{c}\right)\left(N^k + \frac{v^k}{c}\right)} - 1\right]. \qquad (2.110)$$

This resultant Hamiltonian is precisely same as in Eq. (2.100).

Dynamical evolution of the test particle follows from Hamilton's equations which are first order in time; but the demonstrated correspondence of the Hamiltonian and Lagrangian implies the equations must be the same as the usual geodesic equation in generic curved spacetime,

$$\frac{d^2 x^\mu}{d\tau^2} + \Gamma^\mu_{\alpha\beta} \frac{dx^\alpha}{d\tau}\frac{dx^\beta}{d\tau} = 0. \qquad (2.111)$$

In flat spacetime, $N = 1$, $N^i = 0$ and $q_{ij} = \delta_{ij}$ and the resultant Hamiltonian is the old acquaintance, $H = \sqrt{p^2 c^2 + m_0^2 c^4}$. Note that the proper time action vanishes for massless particle; it is however well defined in term of the Hamiltonian. In the conventional Lagrangian formalism, one assumes null geodesics for massless particles; however, it can be readily shown to be indeed the case in the Hamiltonian formalism. As m_0 is an overall multiplicative constant in the proper time action, all particles have the same trajectories, independent of mass and in accordance with the Equivalence Principle. This is true despite the apparent mass dependence in the square root Hamiltonian of Eq. (2.100).

Another noteworthy feature is that the derivation of the geodesic motion is not sensitive to the particular form of the background lapse function and will remain true, in particular, for

$$N = \frac{\sqrt{q}(\partial_t \ln q^{1/3} - \frac{2}{3}\nabla_i N^i)}{4\beta\kappa\bar{H}} \tag{2.112}$$

in the intrinsic time formulation.

The particle Hamiltonian of Eq. (2.100) is motivated by the fact that in the presence of test particle of energy characterized by E_P; including the gravitational field and test particle, the total Hamiltonian constraint,

$$H_T = H_{\text{pure GR}} + \sqrt{q}E_P = 0, \tag{2.113}$$

with

$$\begin{aligned}
H_T &= \frac{2\kappa}{\sqrt{q}}(-\beta\tilde{\pi} + \bar{H})(\beta\tilde{\pi} + \bar{H}) + E_P \\
&= \frac{2\kappa}{\sqrt{q}}\left(-\beta\tilde{\pi} + \sqrt{\bar{H}^2 + \frac{\sqrt{q}E_P}{2\kappa}}\right)\left(\beta\tilde{\pi} + \sqrt{\bar{H}^2 + \frac{\sqrt{q}E_P}{2\kappa}}\right).
\end{aligned} \tag{2.114}$$

This implies the effective Hamiltonian for evolution with respect to ADM coordinate time t is[10]

$$\begin{aligned}
H_{\text{ADM}} &= \int d^3x \frac{1}{\beta}\left(\partial_t \ln q^{1/3} - \frac{2}{3}\nabla_i N^i\right)\sqrt{\bar{H}^2 + \frac{\sqrt{q}E_P}{2\kappa}} + N^i\mathcal{H}_i \\
&= \int d^3x \left(\partial_t \ln q^{1/3} - \frac{2}{3}\nabla_i N^i\right)\frac{\bar{H}}{\beta}\left(1 + \frac{\sqrt{q}E_P}{4\kappa\bar{H}^2} + ...\right) + N^i\mathcal{H}_i \\
&\approx \int d^3x \left(\partial_t \ln q^{1/3} - \frac{2}{3}\nabla_i N^i\right)\frac{\bar{H}}{\beta} + NE_P + N^i\mathcal{H}_i,
\end{aligned} \tag{2.115}$$

with

$$Ndt = \frac{\sqrt{q}(\partial_t \ln q^{1/3} - \frac{2}{3}\nabla_i N^i)dt}{4\beta\kappa\bar{H}} \tag{2.116}$$

being the lapse function of the *background geometry*; and in the approximation in Eq. (2.115) only the first order term in $\frac{\sqrt{q}E_P}{\kappa\bar{H}^2}$ is retained (this ratio compares the particle's energy density to the Hamiltonian density of the rest of the universe; in the context of "test particle" it must by definition be small enough to not disturb the background geometry, and the expansion is justified).

---

[10]See, for instance, Eq. (4.15). In ITG, the Hamiltonian constraint eliminates $\tilde{\pi}$ in terms of the rest of the variables, resulting in a reduced effective Hamiltonian. This is elaborated in Chapters 4 and 8.

The EOM of the test particle are then obtained as,

$$\frac{dx_P^i}{dt} = \{x_P^i, H_{\text{ADM}}\}_{\text{P.B.}}$$

$$\approx \left\{x_P^i, \left(\int d^3x N E_P\right) - N^i c p_i\right\}_{\text{P.B.}} = \{x_P^i, H_P\}_{\text{P.B.}}, \quad (2.117)$$

$$\frac{dp_i}{dt} = \{p_i, H_{\text{ADM}}\}_{\text{P.B.}}$$

$$\approx \{p_i, H_P\}_{\text{P.B.}}; \quad (2.118)$$

with

$$E_P = c\sqrt{q^{ij}p_i p_j + m_0^2 c^2}\, \delta^3(\vec{x}, \vec{x}_P), \quad (2.119)$$

and, consistently, H_P is as in Eq. (2.99). Instead of the usual simple addition of energies for subsystems, effects of energy addition within the big square root in first line of Eq. (2.115) will be manifest only when a subsystem under consideration carries energy comparable to the whole universe.

2.11 Dynamical basis of relativistic time effects in generic ADM spacetimes

Einstein's theories of Relativity, both Special and GR, brought about the most profound and far-reaching shifts in our understanding of space and time. Among the most fascinating and significant developments are relativistic time effects which include the relativity of time intervals for observers in relative motion, and the relativity of periods, hence frequencies, of signals received and emitted by observers placed in gravitational fields of different strengths. The former, especially in Special Relativity, is often expediently attributed to Lorentz transformations between observers; and the latter often to the difference in the lapse functions at different locations in curved spacetime. These "explanations" are obfuscating, revealing neither the physical and dynamical basis of these relativistic time effects, nor their generality. In the former, a physical time dilation effect is attributed to Lorentz boost which is an isometry of Minkowski spacetime; in the latter, the lapse function in canonical GR is *a priori* a Lagrange multiplier of the Hamiltonian constraint. Moreover, relativistic time effects due both to relative motion and curvature differences can occur together, for instance, in satellite orbits in curved spacetimes.

Physical reality of relativistic time effects can conjure the implication "time" is fundamentally "not real", or that at best it is "relative". We seek to reveal and emphasize the basis of these relativistic time effects by casting the framework in canonical and dynamical terms; by deriving the relevant formulas in terms of momentum and Hamiltonian of relativistic test particles in ADM spacetimes, and by relating the *a posteriori* emergent lapse function to the energy density of the reduced Hamiltonian in ITG. The effects arise fundamentally from energy differences; and it is noteworthy when phrased in terms of particle momenta rather than velocities, the shift function does not appear at all in the final formula. The framework not only ties the relativistic time dilation effects to the reality of cosmic time in ITG, but also demonstrates there is no contradiction in both intrinsic cosmic time being real and the physicality of relativistic time effects experienced by test particles in curved spacetime. Time interval correction due to both relativistic motion and gravitational redshift is also combined a single formula in a natural way; this is directly applicable to Global Positioning System and similar satellite navigational systems. The new formula also contains corrections which are not taken into account in the usual naive addition of separate contributions from relativistic motion and gravitational redshift. Future high precision measurements may be able to test and confirm these corrections.

2.11.1   *A general formula for relativistic time effects*

This subsection is based upon the collaborative work of Ref. [68] which contains further details. The main goal is the elucidation of the dynamical basis of relativistic time dilation effects, whether due to pure motion or to the gravitational field, manifested in test particles. The effects are derived in generic ADM spacetimes with no recourse to Lorentz isometry in Special Relativity, or to four covariance in GR. These time dilation effects are shown to arise dynamically from the change in the Hamiltonian of a test particle when it is in motion in a gravitational field.

2.11.1.1   *Proper time interval*

The relation between the proper time interval $d\tau$, coordinate time interval dt and the spacetime interval ds^2 is rather revealing when it is expressed in canonical rather than four-covariant terms. It is

$$d\tau = \frac{1}{c}\sqrt{-ds^2} = \sqrt{N^2 - q_{ij}\left(N^i + \frac{v^i}{c}\right)\left(N^j + \frac{v^j}{c}\right)}\,dt$$

$$= \frac{Nm_0 c}{\sqrt{m_0^2 c^2 + q^{ij}p_i p_j}}dt, \tag{2.120}$$

wherein p_i is the momentum of the particle and the general relation between momentum and velocity, Eq. (2.102), has been used. Thus proper time interval $d\tau$ depends on both the lapse function N the 3-metric q_{ij} and the particle momentum p_i which are all physical quantities. In ITG the lapse function of a classical spacetime is physical and agrees with the *a posteriori* value of N in Einstein's theory (as in Eq. (2.116), in which, up to multiplicative constants, Ndt is just the gauge equivalence class of $d\ln q^{1/3}$ divided by the gravitational Hamiltonian density $\bar{H}$ of ITG). Note that the shift function N^i (which a true gauge parameter, of spatial diffeomorphism invariance) drops out from the proper time interval when it is expressed in terms the momentum (the canonical, and more fundamental variable) instead of velocity. In essence, Eq. (2.120) says $d\tau \propto \frac{dT}{H}\frac{E_0}{E_P}$, wherein dT is the gauge invariant part of $d\ln q^{\frac{1}{3}}$, and $\frac{E_0}{E_{\vec{p}}}$ is the ratio of the test particle's rest mass energy to its energy at momentum $\vec{p}$ (in Special Relativity this factor is $\frac{1}{\gamma}$). From the expression, it is clear that variations in $d\tau$ arise from dynamical effects due to different momenta values and/or difference in energy at different spacetime points. Lorentz isometry of flat spacetime plays no fundamental role; and the canonical, rather than four-covariant, description conveys a more fundamental and physical picture.

2.11.1.2 *Examples of time dilation effects*

- Time dilation in Special Relativity

In Special Relativity with Minkowski spacetime, $N = 1$, $N^i = 0$ and $q_{ij} = \delta_{ij}$. Thus Eq. (2.108) yields

$$p_i = \frac{m_0 \dot{x}_i}{\sqrt{1 - \frac{v^2}{c^2}}}$$

$$= m_0\gamma\frac{dx^i}{dt} = m_0\frac{dx^i}{d\tau}, \tag{2.121}$$

which is the same as the usual expression of relativistic momentum; and the factor, $\gamma = \frac{\sqrt{m_0^2 c^4 + \delta^{ij}p_i p_j c^2}}{m_0 c^2} = \frac{1}{\sqrt{1 - \frac{v^2}{c^2}}}$ is the dilation factor in $d\tau = \frac{1}{\gamma}dt$.

- Gravitational redshift effects

Consider the motion of a test particle in the spherically symmetric Schwarzschild background geometry,

$$ds^2 = -\left(1 - \frac{2GM}{rc^2}\right)c^2dt^2 + \left(1 - \frac{2GM}{rc^2}\right)^{-1}dr^2 + r^2d\Omega^2,$$

with lapse function

$$N(r) = \sqrt{1 - \frac{2GM}{rc^2}}, \qquad (2.122)$$

and $N^i = 0$. For particles with zero momentum, the proper time interval Eq. (2.120) is dependent only upon Ndt i.e.

$$d\tau = = N(r)dt$$
$$= \sqrt{1 - \frac{2GM}{rc^2}}dt, \qquad (2.123)$$

wherein the ADM coordinate time interval, dt coincides with the proper time interval at $r \to \infty$. The ratio of the proper time intervals at different values of r is,

$$\frac{d\tau_2}{d\tau_1} = \sqrt{\frac{N(\vec{r}_2)}{N(\vec{r}_1)}} = \frac{\sqrt{1 - \frac{2GM}{r_2c^2}}}{\sqrt{1 - \frac{2GM}{r_1c^2}}}, \qquad (2.124)$$

with resultant frequencies of emitted and received waves at different positions inversely proportional to the respective proper time intervals.

- Global Positioning System and other similar satellite navigational systems

A GPS receiver calculates its position and time based on signals received from other satellites which carry atomic clocks of very high accuracy. Application of relativistic and gravitational effects to realistic GPS system will lead to the following corrections because of differences in location and relative speed between observer (receiver) and the orbital satellite in space

(1) In Special Relativity, proper time intervals on the satellite and on the surface of the earth, respectively $d\tau_s$ and $d\tau_0$, differ.

The fractional time difference $f_S = \frac{d\tau_s - d\tau_0}{d\tau_0}$ is,

$$f_S = \sqrt{1 - \frac{v_s^2}{c^2}} - 1 \qquad (2.125)$$

$$\approx -8.3 \times 10^{-11},$$

for a typical satellite orbital speed of $v_s \approx 3.8704 \times 10^3 m/s$.

(2) In the pure gravitational redshift, one can approximate the Earth's gravitational field with the Schwarzschild metric to calculate the effect at different heights from the earth surface. At r_s away form the center of the Earth, the fractional difference is then

$$f_G = \frac{\sqrt{1 - \frac{2GM_E}{r_s c^2}}}{\sqrt{1 - \frac{2GM_E}{r_o c^2}}} - 1 \qquad (2.126)$$

$$\approx +5.3 \times 10^{-10},$$

with the satellite in the orbit at $r_s \approx 2.656275 \times 10^7 m$, radius of the Earth, $r_o \approx 6.378 \times 10^6 m$, and mass of the Earth taken as $5.97 \times 10^{24} kg$.

It is worth emphasizing there is a sign difference between corrections arising from relative motion ("from Special Relativity") and from gravitational redshift (or "from General Relativity"), with the latter being larger. Without the benefit of a formula such as Eq. (2.120) which is applicable to a test particle of arbitrary momentum in motion in a generic spacetime, it is customary to just add the effects to get a total fractional difference of,

$$f_S + f_G \approx 4.5 \times 10^{-10},$$

which leads to a correction of about $38\mu s$ per day.

(3) Exact calculation and correction to conventional formula: Assuming a satellite in circular motion with velocity, $\dot{x}^i \leftrightarrow (\dot{r} = 0, \dot{\theta} = 0, \dot{\phi})$ in Schwarzschild background. From Eq. (2.108),

$$p_i \leftrightarrow \left(0, 0, \frac{m_o c r^2 \dot{\phi}}{\sqrt{N^2 - r^2 \dot{\phi}^2}}\right).$$

Now with Eq. (2.120) incorporating the total time effect, the result is

$$d\tau = \frac{Nm_0}{\sqrt{p^i p_i + m_0^2 c^2}} dt$$

$$= \sqrt{\left(1 - \frac{2GM_E}{rc^2}\right) - \frac{v^2}{c^2}} dt,$$

wherein $v = r\dot{\phi}$ is the orbital velocity of the satellite. Hence, the actual total fractional change f_a is,

$$f_a = \frac{dt_s - dt_o}{dt_o}$$

$$= \frac{\sqrt{\left(1 - \frac{2GM_E}{r_s c^2}\right) - \frac{v_s^2}{c^2}}}{\sqrt{\left(1 - \frac{2GM}{r_o c^2}\right) - \frac{v_o^2}{c^2}}} - 1$$

$$\approx \sqrt{1 - \frac{v_s^2}{c^2}} \frac{\sqrt{1 - \frac{2GM_E}{r_s c^2}}}{\sqrt{1 - \frac{2GM_E}{r_o c^2}}} - 1$$

$$= f_S + f_G + f_S f_G.$$

The correction term $f_S \times f_G \approx -4.4 \times 10^{-20}$ is very small and below current detection sensitivity; it induces an error in distance of about $4cm$ per century. However, it is a definite departure from usual treatments and a prediction. The dynamical basis of relativistic time effects is not just a conceptual advance, but it is also of practical importance in applications.

(4) Atomic experiments of gravitational redshift

Precise experiments on gravitational redshift have been carried out in Ref. [69] recently. The experiments have deployed the conventional derivation of the redshift formula to measure redshifts of free falling particles under the earth's gravitational field. It is assuring to confirm the dynamical picture yields the same results from a different perspective.

In the conventional treatment, one approximates the earth's gravitational field by a Schwarzschild background metric; the resultant proper time interval for a radially free falling particle in this background is,

$$d\tau = \sqrt{-g_{00} - g_{rr}\frac{v^2}{c^2}}\,dt$$

$$= \sqrt{1 - \frac{2GM}{rc^2} - \left(1 - \frac{2GM}{rc^2}\right)^{-1}\frac{v^2}{c^2}}\,dt \quad (2.127)$$

$$\approx 1 + \frac{gz}{c^2} - \frac{v^2}{2c^2} - \frac{GM}{r_o c^2}, \quad (2.128)$$

wherein, $r = r_o + z$ with z being the height from which the particle falls freely, and $\frac{GM}{r_o^2} \approx g$, is the free fall acceleration at the earth's surface.

It can checked that Eq. (2.127) is equivalent to the canonical dynamical derivation in Eq. (2.120). To wit, Eq. (2.120) yields

$$p_i \leftrightarrow \left(\frac{m_0 c q_{rr} v}{\sqrt{N^2 - q_{rr}v^2}}, 0, 0\right), \quad (2.129)$$

in spherical coordinates (r, θ, ϕ). Thus

$$d\tau = \frac{N m_0}{\sqrt{p^i p_i + m_0^2 c^2}}\,dt$$

$$= \frac{1}{c}\sqrt{N^2 - q_{rr}v^2}\,dt$$

$$= \sqrt{1 - \frac{2GM}{rc^2} - \left(1 - \frac{2GM}{rc^2}\right)^{-1}\frac{v^2}{c^2}}\,dt; \quad (2.130)$$

in agreement with (2.127) for the special case of spherical symmetric Schwarzschild metric. It should however be noted that the final line Eq. (2.128) is an approximation, and the exact total result Eq. (2.130) will yield corrections.

2.12   Relativistic point particle dynamics, analogy to Einstein's theory, and master constraint formulation

The example of the relativistic point particle in Minkowski spacetime is quite instructive as an elementary analog of GR with its Hamiltonian constrained to vanish and its *a priori* and *a posteriori* lapse function. The particle action is simply,

$$S = -m_0 c \int \sqrt{-\eta_{\mu\nu} dx^\mu dx^\nu}$$

$$= -m_0 c^2 \int dt \sqrt{1 - \left(\frac{d\vec{x}}{cdt}\right)^2}$$

$$= \int \left(p_i \frac{dx^i}{dt} - c\sqrt{\vec{p}^2 + m_0^2 c^2}\right) dt, \qquad (2.131)$$

with canonical momentum $\vec{p} = m_0 \gamma \frac{d\vec{x}}{dt}$. The physical Hamiltonian,

$$\bar{H} = \sqrt{\vec{p}^2 c^2 + m_0^2 c^4} \qquad (2.132)$$

emerges when t is identified as the time variable.

On the other hand, in the "manifestly covariant" and "reparametrization invariant" approach, introduction of an extraneous "time" parameter λ results in,

$$S = -m_0 c \int \sqrt{-\eta_{\mu\nu} \frac{dx^\mu}{d\lambda} \frac{dx^\nu}{d\lambda}} d\lambda$$

$$= \int L\left[x^\mu, \frac{dx^\mu}{d\lambda}\right] d\lambda, \qquad (2.133)$$

$$\Rightarrow p_\mu = \frac{\partial L}{\partial(\frac{dx^\mu}{d\lambda})} = m_0 \eta_{\mu\nu} \frac{dx^\nu}{d\tau},$$

with,

$$p_\mu \frac{dx^\mu}{d\lambda} - L = 0, \qquad (2.134)$$

i.e. the exact vanishing of the corresponding Hamiltonian; but the momenta are constrained by,

$$H = p^\mu p_\mu + m_0^2 c^2 = -\left(p_0 - \frac{\bar{H}}{c}\right)\left(p_0 + \frac{\bar{H}}{c}\right) = 0. \qquad (2.135)$$

Formulating the constrained theory with

$$S = \int \left[p_\mu \frac{dx^\mu}{d\lambda} - N(p^\mu p_\mu + m_0^2 c^2)\right] d\lambda, \qquad (2.136)$$

results in the equations of motion

$$\frac{dx^0}{d\lambda} = \frac{\partial(NH)}{\partial p_0} = 2Np^0, \qquad (2.137)$$

and

$$\frac{dx^i}{d\lambda} = \frac{\partial(NH)}{\partial p_i} = N(2p_i), \qquad (2.138)$$

on the constraint surface, $H = 0$. Thus the *a posteriori* value of the lapse function is

$$N d\lambda = \frac{cdt}{2p^0} = \frac{cdt}{2\sqrt{\vec{p}^2 + m_0{}^2 c^2}}$$
$$= \frac{c^2 dt}{2\bar{H}}. \tag{2.139}$$

Moreover, Eq. (2.138) and the *a posteriori* value of N imply there is true evolution w.r.t. the inherent time t, which is

$$\frac{dx^i}{dt} = N \frac{d\lambda}{dt}(2p_i) = \frac{c^2}{2\bar{H}}(2p_i) = \frac{p_i c^2}{\sqrt{p^2 c^2 + m_0^2 c^4}}. \tag{2.140}$$

This result would have emerged directly if Eq. (2.131) and Eq. (2.132) had been used in the first place.

The formulation, with reparametrization "symmetry" which deploys λ (which is not physically inherent or "intrinsic" to the theory) and a Hamiltonian constrained to vanish, is obfuscating. Correctly isolating one of the degree x^0 as the "intrinsic" time, and forgoing the "reparametrization invariance" regains the much more transparent description of Eq. (2.131), with dynamical variables $\vec{x}$ evolving with respect to "intrinsic" time $t = \frac{x^0}{c}$ and physical Hamiltonian $\bar{H}$. Even without the appearance of λ, the action in Eq. (2.131) is already explicitly Lorentz invariant; and the "Spirit of Relativity" is embodied in the dispersion relation Eq. (2.132), in the square root Hamiltonian versus Newtonian $\frac{p^2}{2m_0}$. From it, Hamilton's equations produce $|\frac{d\vec{x}}{dt}| = |\frac{c\vec{p}}{\sqrt{p^2 + m_0^2 c^2}}| \leq c$. The time dilation factor $\gamma = \frac{1}{\sqrt{1 - \frac{v^2}{c^2}}} = \frac{\sqrt{p^2 c^2 + m_0^2 c^4}}{m_0 c^2}$ is best interpreted dynamically and physically as the ratio of energy at momentum $\vec{p}$ to the rest mass energy.

In GR, the quandary is analogous to the task of recovering the physical dynamics of Eq. (2.131) and Eq. (2.132) *given the starting point of Eq. (2.133), of a theory with Hamiltonian constrained to vanish*. Within the context of capturing the contents of the theory with constraints, this goal can be achieved by introducing the master constraint, $M = 0$, which can be used to eliminate fictitious symmetry. To wit,

$$M = \left(p_0 + \frac{\bar{H}}{c}\right)^2 = 0, \tag{2.141}$$

which makes $H = 0$ redundant and in effect replaces it. Specifically, with

Lagrange multiplier, l, of the constraint,

$$S = \int \left[p_\mu \frac{dx^\mu}{d\lambda} \right] d\lambda - \int l \left(p_0 + \frac{\bar{H}}{c} \right)^2 d\lambda \tag{2.142}$$

$$= \int \left[p_0 c + \vec{p} \cdot \frac{d\vec{x}}{dt} \right] dt - \int l \left(p_0 + \sqrt{\vec{p}^2 + m_0^2 c^2} \right)^2 d\lambda$$

$$= \int \left[-c\sqrt{\vec{p}^2 + m_0^2 c^2} + \vec{p} \cdot \frac{d\vec{x}}{dt} \right] dt + \int l \left(p_0 + \sqrt{\vec{p}^2 + m_0^2 c^2} \right)^2 d\lambda.$$

This implies

$$\bar{H} = c\sqrt{\vec{p}^2 + m_0^2 c^2} \tag{2.143}$$

is the effective Hamiltonian for the variables $(\vec{x}, \vec{p})$; and the master constraint equivalent to,

$$p_0 = -\frac{\bar{H}}{c}, \tag{2.144}$$

can consistently be interpreted classically as the HJ equation

$$\frac{\partial S}{\partial x^0} + \frac{\bar{H}}{c} = 0, \tag{2.145}$$

and quantum mechanically as a Schrödinger equation first order in "intrinsic" time x^0.

The sign ambiguity in Eq. (2.135) is settled in M to ensure that the reduced Hamiltonian, which generates forward t-evolution in Eq. (2.142), is positive definite. In GR, the elimination of $\tilde{\pi}$ in terms of other variables as the solution of the Hamiltonian constraint is the analog of Eq. (2.144); the physical Hamiltonian of Eq. (2.68) with Eq. (2.13) should be compared to Eq. (2.143); and Eq. (2.116) is the GR analog of the *a posteriori* value of the lapse function in Eq. (2.139). Last but not least, when confronted with a Hamiltonian which is constrained to vanish, putting complete faith on a boundary term to generate sole "physical evolution" is ludicrous, in this example, if not in GR.

Chapter 3

Quantum geometrodynamics with intrinsic time development

"Thunder is good, thunder is impressive,
but it is lightning that does the work."
—Mark Twain

3.1 Geometrodynamics and the riddle of time

Is time an illusion of our perception, an emergent semi-classical entity, or is it present at the fundamental level even in quantum gravity? Does "time" have a beginning, or does it flow unbounded from past infinite to the remote infinite future? Is "time evolution" meaningful in Einstein's Geometrodynamics? How is it manifested, given the assumption of four-covariance and of dynamics totally dictated by constraints alone? How can physical time evolution be generated by a Hamiltonian made up totally of constraints? It would seem that to achieve real physical time evolution a true Hamiltonian is requisite. To put the GR twist to Mark Twain's words, "Geometry is good, geometry is impressive, but it is the Hamiltonian that does the work!"

Geometrodynamics bequeathed with positive-definite spatial metric is the simplest consistent framework to implement fundamental commutation relations (CR) predicated on space-like initial data hypersurfaces. Yet, in quantum gravity, space-time is a concept of "limited applicability". That semi-classical space-time is emergent begs the question what, if anything at all, plays the role of "time" in quantum gravity? Wheeler [9] went as far as to claim we have to forgo time-ordering, and to declare "there is no spacetime, there is no time, there is no before, there is no after". But without "time-ordering", how is "causality", which is requisite in any "sensible physical theory", enforced in quantum gravity? Furthermore, a resolution of the "problem of time" in quantum gravity cannot be deemed complete if it fails to account for the intuitive physical reality of time and does not provide satisfactory correlation between time development in quantum dynamics and the passage of time in classical space-times.

A mental block to the resolution of the problem of time is the belief that four-covariance is sacrosanct to the description of GR and to the success of Einstein's theory. Yet, Wheeler has emphasized it is three-geometry, rather than four-geometry, which is fundamental in quantum geometrodynamics. Without the fundamental role of four-geometry, the insistence on four-covariance as a beacon can only be puzzling, if not misguided. The call to abandon four-covariance is not new. Simplifications in the Hamiltonian analysis of GR, and the fact that the physical d.o.f. involve only the spatial metric, led Dirac to conclude that "four-dimensional symmetry is not a fundamental property of the physical world." A key obstacle to the viability of GR as a perturbative QFT lies in the conflict between unitarity and space-time general covariance: renormalizability can be attained with higher-order curvature terms, but space-time covariance requires time as well as spatial derivatives of the same (higher) order, thus compromising unitarity. Relinquishing four-covariance to achieve power-counting renormalizability through modifications of GR with higher-order *spatial*, rather than space-time, curvature terms was Horava's bold proposal [31].

3.1.1   *Geometrodynamics reexplored*

In this chapter we review what Einstein's theory reveal about the *a posteriori* nature of the lapse function, N, despite its arbitrary *a priori* value. Ndt is physical, and related to the Hamiltonian density and change in intrinsic time interval $\delta \ln q^{\frac{1}{3}}(x)$. This bears the important message that despite having a Hamiltonian made up totally of constraints, time evolution in GR is meaningful after all the equations of motion and constraints are solved. The physical content of GR in the canonical ADM framework and the precise *a posteriori* form of the lapse function can moreover be reproduced by a true Hamiltonian in lieu of the Hamiltonian constraint. In the previous chapter, the master constraint (MC) method was deployed as a mathematical device to retain the first class nature of the constraints of GR, keep the Hamiltonian as totally made up of constraints, and yet allow deformations of the theory away from Einstein's. It is a device to achieve mathematical consistency in circumventing the no-go theorem of *Geometrodynamics Regained* [33] when we seek to tweak GR within the restrictive confines of a first class constrained system. However the MC framework does reveal that the Hamiltonian constraint of Einstein's GR can be deformed while keeping spatial diffeomorphism invariance. Indeed within the MC context it would seem that even the momentum constraint can be deformed; but spatial diffeomorphism invariance is needed[1] to collapse the many-fingered time interval of $\delta \ln q^{\frac{1}{3}}(x)$ to its physically meaningful gauge invariant part that is a global intrinsic and cosmic time interval $dT = \frac{2}{3}d \ln V$. The Hamiltonian constraint can be solved exactly by $\tilde{\pi}$ and a spatial diffeomorphism-invariant Hamiltonian which generates time evolution w.r.t. cosmic time is obtained. The resultant lapse function is in accord with the *a posteriori* value of N produced by GR. It resolves 'the problem of time' and bridges the deep divide between quantum mechanics and conventional canonical formulations of quantum gravity with either Schrödinger or Heisenberg picture that comes with first-order-in-time evolution and causal time-ordering in global cosmic time. Such a framework has been advocated in a series of works [16, 67, 70–72]. This book explores and elucidates its many facets from various perspectives.

3.2 General Relativity and the lapse function *a priori* and *a posteriori*

The Hamiltonian constraint factorizes in an interesting way as,

$$\frac{\sqrt{q}}{2\kappa} H = G_{ijkl}\tilde{\pi}^{ij}\tilde{\pi}^{kl} + \mathcal{V}(q_{ij})$$
$$= -\left(\beta\tilde{\pi} - \bar{H}\right)\left(\beta\tilde{\pi} + \bar{H}\right)$$
$$= 0,$$

wherein

$$\bar{H} = \sqrt{\bar{G}_{ijkl}\tilde{\pi}^{ij}\tilde{\pi}^{kl} + \mathcal{V}(q_{ij})}$$
$$= \sqrt{\frac{1}{2}\left[\bar{q}_{ik}\bar{q}_{jl} + \bar{q}_{il}\bar{q}_{jk}\right]\tilde{\pi}^{ij}\tilde{\pi}^{kl} + \mathcal{V}(q_{ij})}.$$

Solving the Hamiltonian constraint algebraically results in $\bar{H} = \pm\beta\tilde{\pi}$. In Einstein's theory, when

$$\mathcal{V} = -\frac{q}{(2\kappa)^2}[R - 2\Lambda_{\text{eff.}}]$$

---

[1] Transversality of the momentum with auxiliary requirement of transversality of infinitesimal transverse metric excitations (see Chapters 8 and 9) is also needed to ensure GR as a theory of long-ranged interaction will not be compromised by longitudinal modes.

the Hamiltonian and momentum constraints form a first class algebra. In conventional ADM formulation, the lapse function, N, which is a Lagrange multiplier, is *a priori* an arbitrary function. This "arbitrariness" on the one hand raises the question whether evolution is physically meaningful, on the other it is often mistakenly regarded as a true gauge parameter and the manifestation of four-covariance in Einstein's theory. Much can be learned from working out the *a posteriori*, versus the arbitrary *a priori*, value of lapse function explicitly.

3.2.1   A posteriori lapse function determined by constraints and EOM

The physical meaning of N is remarkably revealed *a posteriori* by the equations of motion and constraints. This is akin in simple mechanics to Lagrange multipliers being determined and taking on physical significance (such as being related to constrained forces) upon solving all the equations governing the theory. In particular, the lapse function is related to $\partial_t \ln q$ and $\bar{H}$ (which assumes the role of Hamiltonian density in intrinsic time geometrodynamics) through,

$$
\begin{aligned}
\frac{\partial \ln q^{\frac{1}{3}}(x)}{\partial t} &= \{\ln q^{\frac{1}{3}}(x), \int N(y)H(y)d^3y\}_{\text{P.B.}} \\
&= \int N(y)\{\ln q^{\frac{1}{3}}(x), H(y)\}_{\text{P.B.}} d^3y \\
&= -\frac{4}{\sqrt{q}}N(x)\kappa\beta^2\tilde{\pi}(x) \\
&= \mp 4\kappa\beta N(x)\underset{\sim}{\bar{H}}(x),
\end{aligned}
\tag{3.1}
$$

wherein the Poisson bracket of Eq. (2.8) has been used to arrive at the intermediate step, and the Hamiltonian constraint at the last step even though N was *a priori* arbitrary. Thus the precise form of the *emergent* ADM lapse function N is in accord with Eq. (2.112).

3.2.2   Reproducing General Relativity with an equivalent ADM Hamiltonian instead of a constraint

Dispensing with the Hamiltonian constraint, the equation of motion of the physical d.o.f. (which lie in the transverse traceless excitations of $(\bar{q}_{ij}, \bar{\pi}^{ij})$) can equivalently be captured by the non-trivial Hamiltonian,

$$
H_{\text{ADM}} := \frac{1}{\beta} \int \frac{\partial \ln q^{\frac{1}{3}}(x)}{\partial t} \bar{H}(x)d^3x,
\tag{3.2}
$$

which generates evolution with respect to ADM coordinate time t. To wit, and with Eq. (3.1) for the lapse function,

$$\dot{\bar{q}}_{ij}(x) = \left\{ \bar{q}_{ij}(x), \frac{1}{\beta} \int \frac{\partial \ln q^{\frac{1}{3}}(y)}{\partial t} \bar{H}(y) d^3y \right\}_{\text{P.B.}} \tag{3.3}$$

$$= \int \frac{1}{\beta} \frac{\partial \ln q^{\frac{1}{3}}(y)}{\partial t} \left\{ \bar{q}_{ij}(x), \bar{H}(y) \right\}_{\text{P.B.}} d^3y$$

$$= \int \mp \underset{\sim}{N}(y) 4\kappa \bar{H}(y) \left\{ \bar{q}_{ij}(x), \bar{H}(y) \right\}_{\text{P.B.}} d^3y$$

$$= \int \mp \underset{\sim}{N}(y) 2\kappa \left\{ \bar{q}_{ij}(x), \bar{H}^2(y) \right\}_{\text{P.B.}} d^3y$$

$$= \int \mp \underset{\sim}{N}(y) \left\{ \bar{q}_{ij}(x), H(y) \right\}_{\text{P.B.}} d^3y. \tag{3.4}$$

This final step can be reached bearing in mind

$$\frac{\sqrt{q}}{2\kappa} H = \bar{H}^2 - \beta^2 \tilde{\pi}^2$$

and $\tilde{\pi}$ commutes with $(\bar{q}_{ij}, \tilde{\pi}^{ij})$. This is thus equivalent to the evolution generated through conventional

$$\int \mp N(y) \left\{ \bar{q}_{ij}(x), H(y) \right\}_{\text{P.B.}} d^3y,$$

with resultant *a posteriori* relation Eq. (3.1) between $N, \partial_t \ln q$ and $\bar{H}$. The $\mp$ sign accompanying N does not affect the resultant classical ADM four-metric which depends on the square of N. The equation of motion for $\tilde{\pi}^{ij}$ can be similarly demonstrated. Adding

$$\int N^i(y) H_i(y) d^3y$$

to the Hamiltonian merely leads to modification of the equations of motion by Lie derivatives of the variables with respect to N^i, with the resultant lapse function Eq. (3.1), generalized to[2]

$$N = \frac{\sqrt{q}(\partial_t \ln q^{1/3} - \frac{2}{3}\nabla_i N^i)}{4\beta\kappa\bar{H}}. \tag{3.5}$$

---

[2]The determinant of the ADM metric is equal to Nq, so the lapse is non-vanishing unless there is degeneracy. In Minkowski space-time for example, both the numerator and denominator vanish, but we should consider instead de Sitter space-time with compact slicings, and N is well-defined in the limit of arbitrarily small positive cosmological constant. Master constraint is neither needed nor invoked in arriving at these results.

In fact this ensures the Hamiltonian constraint is satisfied classically in the form

$$\frac{K^2}{9} = \frac{4\kappa^2\beta^2}{q}\bar{H}^2. \tag{3.6}$$

Thus the Hamiltonian H_{ADM}, *which is no longer constrained to vanish,* equivalently captures the physical content and equation of motion of Einstein's theory. This framework however allows the potential $\mathcal{V}$ *to depart from that of Einstein's theory without leading to inconsistencies in the constraint algebra.*

3.3   Global intrinsic time and quantum evolution

3.3.1   *Expansion of the universe and global intrinsic time*

While q is a tensor density, the multi-fingered intrinsic time interval,

$$\delta \ln q^{\frac{1}{3}} = \frac{1}{3}\frac{\delta q}{q} = \frac{q^{ij}}{3}\delta q_{ij}, \tag{3.7}$$

is a scalar entity.[3] Hodge decomposition for any compact Riemannian manifold without boundary yields

$$\delta \ln q^{\frac{1}{3}} = \delta T + \nabla_i \delta Y^i, \tag{3.8}$$

wherein the gauge-invariant part of $\delta \ln q^{\frac{1}{3}}$ is $\delta T = \frac{2}{3}\delta \ln V$, which is proportional to the 3dDI fractional change in the spatial volume $d \ln V = \frac{1}{V}dV$. This can be seen from

$$\int (\delta \ln q^{\frac{1}{3}})\sqrt{q}d^3x = \int (\delta T + \nabla_i \delta Y^i)\sqrt{q}d^3x$$
$$= (\delta T)\int \sqrt{q}d^3x$$
$$= (\delta T)V \tag{3.9}$$

(which implies that δT is the average value of $\delta \ln q^{\frac{1}{3}}(x)$ over V), and also

$$\int (\delta \ln q^{\frac{1}{3}})\sqrt{q}d^3x = \frac{1}{3}\int \frac{(\delta\sqrt{q}^2)}{\sqrt{q}^2}\sqrt{q}d^3x$$
$$= \frac{2}{3}\delta\left(\int \sqrt{q}d^3x\right)$$
$$= \frac{2}{3}\delta V, \tag{3.10}$$

---

[3]In DeWitt's seminal work [12], the intrinsic time variable was $q^{\frac{1}{4}}$ instead of $\ln q^{\frac{1}{3}}$.

thus yielding

$$(\delta T)V = \frac{2}{3}\delta V. \tag{3.11}$$

As the change in the logarithm of a tensor density, $\delta \ln q^{\frac{1}{3}}$ also has the distinct advantage over, for instance, the change of a scalar field as time interval. The $\nabla_i \delta Y^i$ piece in its Hodge decomposition is in fact a Lie derivative of $\ln q^{\frac{1}{3}}$; in particular,

$$\mathcal{L}_{\delta \vec{N}} \ln q^{\frac{1}{3}} = \frac{2}{3}\nabla_i \delta N^i, \tag{3.12}$$

and this can thus be gauged away completely with a spatial diffeomorphism. So the gauge invariant time interval is $dT = \frac{2}{3}d\ln V$, and T increases monotonically with the expansion of a spatially closed universe.

3.3.2 *Temporal ordering and quantum evolution*

Instead of describing the evolution of quantum states with respect to gauge-dependent multi-fingered time $\ln q^{\frac{1}{3}}(x)$, it is eminently more meaningful to ask how the wave function of the universe, Ψ, changes w.r.t. to the 3dDI *global* intrinsic time variable T. With T displacing the physically less concrete ADM coordinate time t in H_{ADM}, this dynamics is determined by the Schrödinger equation,

$$i\hbar\frac{d\Psi}{dT} = H_{\text{Phys}}\Psi, \qquad H_{\text{Phys}} := \int \frac{\bar{H}(x)}{\beta}d^3x; \tag{3.13}$$

wherein H_{Phys} is the physical Hamiltonian generating evolution in global intrinsic time. This *replaces, and circumvents, the naive imposition of the Hamiltonian constraint as a Schwinger-Tomonaga equation with multi-fingered time* $\ln q^{\frac{1}{3}}(x)$, which would have looked like

$$i\hbar\frac{\delta\Psi}{\delta \ln q^{\frac{1}{3}}(x)}\Psi = \frac{\bar{H}(x)}{\beta}\Psi \tag{3.14}$$

in the metric representation.

The formalism is applicable to the full theory of quantum gravity, and assumptions of mini-superpsace models have not been invoked. $\beta > 0$ is required for the Hamiltonian to be bounded from below, and global intrinsic time increases monotonically with our ever expanding universe.

In this framework dictated by first-order evolution in global intrinsic time, the Hamiltonian H_{Phys} is 3dDI, provided

$$\bar{H} = \sqrt{\bar{G}_{ijkl}\bar{\pi}^{ij}\bar{\pi}^{kl} + \mathcal{V}} \tag{3.15}$$

is a scalar density of weight one; and Einstein's GR (with its particular value of β and of $\mathcal{V}$) is a particular realization of this wider class of theories. The crucial time development operator can be derived by integrating the Schrödinger equation, yielding

$$\Psi(T) = U(T, T_0)\Psi(T_0), \tag{3.16}$$

with T-ordered operator

$$U(T, T_0) := \mathcal{T} \exp\left[-\frac{i}{\hbar} \int_{T_0}^{T} H_{\text{Phys}}(T')dT'\right], \tag{3.17}$$

or in the form of a time-ordered Dyson series,

$$
\begin{aligned}
U(T, T_o) = I &- \frac{i}{\hbar} \int_{T_o}^{T} dT_1 H_{\text{Phys}}(T_1) \\
&+ \left(\frac{-i}{\hbar}\right)^2 \int_{T_o}^{T} dT_2 \int_{T_o}^{T_2} dT_1 H_{\text{Phys}}(T_2) H_{\text{Phys}}(T_1) \\
&+ ... + \left(\frac{-i}{\hbar}\right)^n \int_{T_o}^{T} dT_n ... \int_{T_o}^{T_2} dT_1 H_{\text{Phys}}(T_n)...H_{\text{Phys}}(T_1) \\
&+ ... = \sum_{n=0}^{\infty} \left(\frac{-i}{\hbar}\right)^n \int_{T_o}^{T} ... \int_{T_o}^{T_2} H_{\text{Phys}}(T_n)...H_{\text{Phys}}(T_1)dT_1...dT_n.
\end{aligned} \tag{3.18}
$$

A unitary and diffeomorphism-invariant $U(T, T_0)$ follows if the 3dDI H_{Phys} is self-adjoint. Since dT is also unchanged under spatial diffeomorphisms, the temporal ordering in $U(T, T_0)$ will be reassuringly gauge invariant. Time-ordered integrability of the Schrödinger equation is feasible without ambiguity because time is "one-dimensional", spatially-independent *rather than many-fingered*, and H_{Phys} is a spatial integral, rather than a local Hamiltonian density. Moreover, the necessity of "time"-ordering, which underpins the notion of causality, emerges, because quantum fields and the Hamiltonian do not commute at different "times".[4] What is paramount to causality is not the physical dimension of time, but the sequence, or time-ordered development, of the physical state.

Wheeler's provocative call to forgo all time-ordering may ring true intuitively, but it does not hold up to deeper scrutiny. While space-time events and their ordering have no primary role and space-time itself is an entity of "limited applicability" in quantum gravity, the fundamental gauge invariant entity that can be, and is in fact, time-ordered is the physical quantum state.

---

[4]The Hamiltonian density $\bar{H}$, in fact depends explicitly upon the intrinsic time parameter T.

Chapter 4

Intrinsic time gravity and the Lichnerowicz-York equation

Dedicated to the memory of Niall Ó Murchadha (1946–2021)

"Hold on to the now, the here,
through which all future plunges to the past."
—James Joyce

4.1 Introduction

Niall visited us in Taiwan about a year after the completion and submission for publication of Ref. [16]. During his stay we discussed the initial data formulation in GR in detail. Niall was intrigued by the scheme in Ref. [16], and as a leading expert on the formulation and solution of initial data and constraints in gravitation, he understood its advantages and promise. All three of us then decided to collaborate on a work to compare and contrast with York's formulation [61]. "You should call your scheme 'Intrinsic Time Gravity'", Niall suggested to one of us. It was apposite, in contradistinction to York's extrinsic time formalism. The resultant work, Ref. [72], on which this chapter is based, appeared in print earlier than Ref. [16], due in no small part to Niall's prestige among peers. We would like to dedicate this chapter to the memory of our collaborator and friend, Niall Ó Murchadha. Some of the earlier arguments and derivations have been tweaked and sharpened, and the notations for various physical entities have been made consistent with the rest of the book. The discussion on the Cotton-York addition to the Lichnerowicz-York equation has been expanded, and within this new context, a complete section at the end, on the Penrose Weyl Curvature Hypothesis and the initial state of the universe, has been added.

In canonical GR, a formalism in which the trace of the extrinsic curvature plays the role of clock was developed by York to solve the initial data problem. The Hamiltonian constraint translates into the Lichnerowicz-York

(LY) equation [73] in York's extrinsic time program which usually employs constant mean curvature (CMC) slicings [74, 75]. As has been laid out in the earlier chapters, the intrinsic time variable of the logarithm of the spatial volume as clock allows a physical reduced Hamiltonian to emerge from solving the Hamiltonian constraint and to evolve the initial data. Proper time in the ADM space-time manifold and intrinsic time can be correlated, as shown previously; even though the intrinsic time scheme does not really need to rely on the mathematical constructs of an ADM space-time foliation with lapse and shift functions. ITG with ADM space-time can however be used to discuss and reproduce the physical contents of Einstein's GR in conventional canonical language. The intrinsic time paradigm solves the Hamiltonian constraint by expressing the trace of the momentum in terms of the rest of the variables, so it does not require the Lichnerowicz-York equation for initial data evolution. However, in comparing and contrasting the intrinsic time paradigm with the extrinsic time program [72], certain clarifications and confirmation of the subtleties involved in determining the real symmetries in GR reveal themselves. Both schemes demonstrate that Einstein's theory can best be thought of, and solved, as a theory with a Hamiltonian constraint modulo spatial diffeomorphism invariance, rather than a theory of four-covariance. And with extrinsic or intrinsic time as clock, a reduced Hamiltonian emerges. In the York scheme, this reduced Hamiltonian is proportional to the volume which is dependent on the conformal factor determined from the Lichnerowicz-York equation, whereas with intrinsic time the reduced Hamiltonian is an explicit function of the remaining variables. The salient revelation from both schemes is that, unlike the first class Dirac algebra of Einstein's theory, the Lichnerowicz-York equation and the potential in the Hamiltonian of ITG both permit additions, moving away from Einstein's theory, with extra higher curvature terms, especially with the addition of the trace of the square of the Cotton-York tensor. With regard to the Cotton-York term, not only does it not present any inconsistency in both frameworks, it is a natural and preeminent entity to be included, once the fixation on four-covariance is overcome. The last section discusses the remarkable attendant effects this brings.

We investigate the effect on the Hamiltonian structure of general relativity of choosing an intrinsic time to fix the time slicing. Three-covariance with momentum constraint is maintained, but the Hamiltonian constraint is replaced by a algebraic equation for the trace of the momentum. This reveals a very simple structure with a local reduced Hamiltonian. The theory is easily generalized; in particular, the square of the Cotton-York tensor density can be added as an extra part of the potential while at the same time maintaining the classic $2 + 2$ degrees of freedom. Initial data construction is simple in the extended intrinsic time formulation; we get a generalized Lichnerowicz-York equation with nice existence and uniqueness properties. Adding standard matter fields is quite straightforward.

4.2   Einstein's theory and initial data

After the seminal work of Dirac [15], the Hamiltonian theory for GR was first clearly laid out by Arnowitt, Deser, and Misner [10] more than half a century ago. This important development and formulation of GR allow the initial data problem to be studied systematically. The phase space of the problem concerned, consists of the Riemannian spatial metric and its conjugate $(q_{ij}, \tilde{\pi}^{ij})$. However, they cannot be freely chosen because they must satisfy the constraints,

$$- qR + \tilde{\pi}^{ij}\tilde{\pi}_{ij} - \frac{1}{2}\tilde{\pi}^2 = 0; \tag{4.1}$$

and

$$\nabla_i \tilde{\pi}^{ij} = 0,$$

where $\tilde{\pi} = q_{ij}\tilde{\pi}^{ij}$ is the trace. These conditions are known respectively as the "Hamiltonian and momentum constraints".[1] To them are associated, respectively, a coordinate scalar and a vector (N, N^i). These are the "lapse" and "shift". The total Hamiltonian, in the closed (or compact without boundary) case, is

$$H_{ADM} = \int \left[q^{-1/2}N \left(-qR + \tilde{\pi}^{ij}\tilde{\pi}_{ij} - \frac{1}{2}\tilde{\pi}^2 \right) - 2N_j\nabla_i\tilde{\pi}^{ij} \right] dx. \tag{4.2}$$

Therefore (N, N^i) are the "Lagrange multipliers" of the constraints in the 4-covariance paradigm. In Chapter 3, it was pointed out that N is emergent and can actually be determined *a posteriori* when the equations of motion and constraints are imposed [16]. *This is akin, in elementary mechanics, to Lagrange multipliers taking on physical values (e.g. being related to constraint forces) upon solving the equations of the theory.* The resultant symmetry of the theory is in fact just three-dimensional diffeomorphisms, together with a Hamiltonian constraint to be solved. The first of Hamilton's

---

[1] For simplicity the coupling constant 2κ has been set to unity.

equations gives the relationship between $\tilde{\pi}^{ij}$ and the "time"[2] derivative of q_{ij}

$$\frac{\partial q_{ij}}{\partial t} = 2Nq^{-1/2}\left(\tilde{\pi}_{ij} - \frac{1}{2}q_{ij}\tilde{\pi}\right) + \nabla_i N_j + \nabla_j N_i. \qquad (4.3)$$

Each choice of N gives a different foliation of spacetime. A difficulty in describing and understanding the dynamics of GR is that, as formulated above, we lack a nontrivial true Hamiltonian and a physical time variable. We want to break this ambiguity, choose a natural time variable, and compute the emergent lapse. James York in [61] pointed out that the local volume, $\sqrt{q}$, and $\tilde{\pi}/\sqrt{q}$ are canonically conjugate, and suggested that $\tilde{\pi}/\sqrt{q}$ is a natural time, and that $\sqrt{q}$ is the local energy. This is called an "extrinsic" time because $\tilde{\pi}^{ij}$ is essentially the extrinsic curvature of the slice. We have

$$q^{-1/2}\left(\tilde{\pi}_{ij} - \frac{1}{2}q_{ij}\tilde{\pi}\right) = K_{ij}, \qquad (4.4)$$

where K_{ij} is the extrinsic curvature. With York's choice, $\sqrt{q}$ is *the associated energy density and the total volume of the slice is the Hamiltonian.* Two years earlier Charles Misner [50] made the opposite choice, albeit in a mini-superspace context. He picked the local volume as his time and $\tilde{\pi}$ is then the energy density. This is where he introduced the "mixmaster universe". The great advantage of Misner's choice is that a reduced Hamiltonian appears. Reference [16] realized the full theory of GR is amenable to such a description, and presented a theory of gravity passing from classical to quantum regimes with a paradigm shift from 4-covariance to 3-covariance. Its framework revealed the primacy of a local reduced Hamiltonian density and intrinsic time interval proportional to $\delta \ln q$.

4.3   Intrinsic time gravity made simple

We start by choosing the intrinsic time function as $\ln q^{1/3}$. Therefore $\delta \ln q^{1/3}$, for any variation, is a three-dimensional scalar entity. Further, the canonical conjugate is $\tilde{\pi}$, with coefficient of unity. We split the metric into unimodular and determinant parts irreducibly via

$$\bar{q}_{ij} = q^{-1/3}q_{ij}, \qquad (4.5)$$

and $\tilde{\pi}^{ij}$ into tracefree and trace parts via

$$\bar{\pi}^{ij} := q^{1/3}(\tilde{\pi}^{ij} - q^{ij}\tilde{\pi}/3). \qquad (4.6)$$

---

[2] As the ADM Hamiltonian is made up solely of constraints, the time t (or x^0) here is just a coordinate label in the ADM foliation of spacetime.

The symplectic one-form becomes

$$\int \tilde{\pi}^{ij} \delta q_{ij} = \int \bar{\pi}^{ij} \delta \bar{q}_{ij} + \tilde{\pi} \delta \ln q^{1/3}. \qquad (4.7)$$

We see immediately that $(\bar{q}_{ij}, \bar{\pi}^{ij})$ and $(\ln(q/q_0)^{1/3}, \tilde{\pi})$ form conjugate pairs. This is clearly the generalization of the $p_i dx^i + p_0 dt$ one gets in the case of the mechanics of a simple particle. With the choice of time as $\ln(q/q_0)^{1/3}$ (which varies from $-\infty$ to $+\infty$, instead of 0 to ∞), we can see that $\tilde{\pi}$ is analogous to

$$p_0 = -E.$$

We now substitute the decomposition of $(q_{ij}, \tilde{\pi}^{ij})$ into the Hamiltonian constraint, Eq. (4.1), to give

$$- qR + \bar{q}_{ik}\bar{q}_{jl}\bar{\pi}^{ij}\bar{\pi}^{kl} - \beta^2 \tilde{\pi}^2 = 0, \qquad (4.8)$$

where, $\beta^2 = \frac{1}{6}$ for GR. It is a free positive parameter for the extended theory which we discuss later. The important observation here is that we know that the Hamiltonian as the generator of time translation, is conjugate to time and should equal to the energy, and therefore the true Hamiltonian density with this choice of intrinsic time should be $-\tilde{\pi}$. Since there is only one $\tilde{\pi}^2$ term and no other $\tilde{\pi}$ factor buried inside Eq. (4.8), we can then just simply solve the Hamiltonian constraint algebraically, which is equivalent to solving

$$(\tilde{\pi} - \bar{H}/\beta)(\tilde{\pi} + \bar{H}/\beta) = 0, \qquad (4.9)$$

to obtain the reduced Hamiltonian,

$$- \tilde{\pi} = \bar{H}/\beta = \frac{1}{\beta} \sqrt{\bar{q}_{ik}\bar{q}_{jl}\bar{\pi}^{ij}\bar{\pi}^{kl} - qR}; \qquad (4.10)$$

as in Eq. (2.13) earlier.

4.3.1 *Simple relativistic particle analogy*

A simple toy model for this process is given by the relativistic particle, which satisfies a constraint analagous to Eq. (4.8),

$$- (p^0)^2 + \vec{p} \cdot \vec{p} + m^2 = 0. \qquad (4.11)$$

p^0 is the energy, and therefore the physical Hamiltonian is

$$E = p^0 = -p_0$$
$$= H = \sqrt{\vec{p} \cdot \vec{p} + m^2}. \qquad (4.12)$$

Hamilton's equations give the equations of motion for $\vec{p}$. $p^0 = H$ is determined in terms of the rest of the variables, just as $\tilde{\pi}$ is in Eq. (4.10). Therein the Hamiltonian density ($\propto \bar{H}$) even takes the relativistic square root form!

4.3.2   Reduced Hamiltonian and the lapse function

The Hamiltonian $\bar{H}/\beta$ of Eq. (4.10) generates $\ln q^{1/3}$ translations, and gives equations of motion for $\bar{q}_{ij}$ and $\bar{\pi}^{ij}$ with respect to this intrinsic time variable. We need to write qR in terms of $\bar{q}_{ij}$ and $\ln q^{1/3}$. This is quite straightforward. We do not have an equation for $\ln q$, since it is the time, but we do need a dynamical equation for $\tilde{\pi}$ which is given by

$$\tilde{\pi} + \frac{1}{\beta}\bar{H}(\bar{q}_{ij}, \bar{\pi}^{ij}, \ln q^{1/3}) = 0. \tag{4.13}$$

This is a rewriting of the Hamiltonian constraint, but it is no longer to be viewed as a constraint; rather, it is the definition for $\tilde{\pi}$. It can be interpreted respectively as the Hamilton-Jacobi equation and Schrödinger equation with many fingered time in the semi-classical and quantum regimes [16]. But it is physically more meaningful to discuss dynamics with respect to a global time parameter ($\propto \ln(V/V_0)$) rather than multi-fingered time.

   We choose the positive root for $\bar{H}$ when we take the square-root. Negative values of $\tilde{\pi}$ correspond, in line with current observations, to an expanding universe. Since the reduced Hamiltonian of the system is three-covariant, the evolution equations will propagate the momentum constraint. This is the only constraint left since the Hamiltonian constraint is solved exactly. Thus, we are left with a system that has the expected $2+2$ degrees of freedom.

   The symplectic potential of the pair $(\ln q^{1/3}, \tilde{\pi})$ contributes

$$\int\int\left(\tilde{\pi}\frac{\partial \ln q^{1/3}}{\partial t}\right)d^3x\,dt = -\int\left[\int\frac{\bar{H}}{\beta}\frac{\partial \ln q^{1/3}}{\partial t}d^3x\right]dt, \tag{4.14}$$

to the action. Besides this, the total Hamiltonian has an additional $\tilde{\pi}$ term from the generator of spatial diffeomorphisms, $H_i = -2q_{ik}\nabla_j\tilde{\pi}^{jk} = -2(q^{-\frac{1}{3}}q_{ik}\nabla_j\bar{\pi}^{jk} + \frac{1}{3}\nabla_i\tilde{\pi})$. If $\tilde{\pi}$ is eliminated as in Eq. (4.13), the effective Hamiltonian becomes

$$\begin{aligned}
H_f &= \int\left[-\left(\tilde{\pi}\frac{\partial \ln q^{1/3}}{\partial t}\right) + N^iH_i\right]d^3x \\
&= \int\left[-\tilde{\pi}\left(\frac{\partial \ln q^{1/3}}{\partial t} - \frac{2}{3}\nabla_iN^i\right) - 2N^iq^{-\frac{1}{3}}q_{ik}\nabla_j\bar{\pi}^{jk}\right]d^3x \\
&= \int\left[\frac{f}{\beta}\sqrt{\bar{q}_{ik}\bar{q}_{jl}\bar{\pi}^{ij}\bar{\pi}^{kl}} - qR - 2N_i(q^{-\frac{1}{3}}\nabla_j\bar{\pi}^{ij})\right]d^3x; \tag{4.15}
\end{aligned}$$

wherein integration by parts, $\int N^i(\nabla_i\tilde{\pi})\,d^3x = -\int\tilde{\pi}(\nabla_iN^i)\,d^3x$, has been carried out in the intermediate step, $\tilde{\pi}$ has been substituted as in Eq. (4.10);

and

$$f := \frac{\partial \ln q^{1/3}}{\partial t} - \frac{2}{3}\nabla_i N^i = \frac{\partial \ln q^{1/3}}{\partial t} - \mathcal{L}_{\vec{N}} \ln q^{1/3}. \tag{4.16}$$

The classical evolution of $(\bar{q}_{ij}, \bar{\pi}^{ij})$ w.r.t. the ADM time variable t can equivalently be then obtained from this effective Hamiltonian H_f.

On the other hand, by contracting Eq. (4.3) with q^{ij}, we find H_{ADM} yields

$$\frac{\partial \ln q^{1/3}}{\partial t} = -\frac{N\tilde{\pi}}{3\sqrt{q}} + \frac{2}{3}\nabla_i N^i. \tag{4.17}$$

It is clear that Eq. (4.16) agrees with Eq. (4.17) when

$$N = \frac{-3f\sqrt{q}}{\tilde{\pi}}. \tag{4.18}$$

If we solve for $\tilde{\pi}$ in Eq. (4.10), and obtain the lapse function N given by evaluating Eq. (4.17), Einstein's theory as formulated in ADM form yields the same emergent *a posteriori* value of the lapse N as in Eq. (3.5). It is then a straightforward exercise to show that the evolution equations arising from H_f agree completely with those from the ADM evolution equations derived from H_{ADM} in Eq. (4.2) with precisely this value of the lapse function.

Given any t-foliation of spacetime, we can thus find a corresponding reduced Hamiltonian that generates it.[3] This means that "many-fingered time" lives on in the intrinsic time picture but in a very different form by the way the emergent N, as given by Eq. (4.18), depends on the freely chosen f. Remarkably, a physically meaningful global time emerges out of infinitude many-fingered possibilities when spatial diffeomorphism invariance is factored in. To be more precise, by decomposing $N^i = N_T^i + q^{ij}\nabla_j W$ into transverse and longitudinal parts, we are free to set $f = \frac{\partial \ln q^{1/3}}{\partial t} - \frac{2}{3}\nabla^2 W = C$ to be a constant independent of time (this has the solution $W = -\frac{3}{2}\frac{1}{\nabla^2}(C - \frac{\partial \ln q^{1/3}}{\partial t})$ which fixes the longitudinal part of the shift vector).

Thus, taking the simplest and gauge-invariant solution, $f = C$, we get

$$N = \frac{-3C\sqrt{q}}{\tilde{\pi}} = \frac{3C}{2K}, \tag{4.19}$$

---

[3]Strictly speaking, the procedures in obtaining the effective reduced theory have yet to be completed. Elimination of $\tilde{\pi}$ in terms of other variables via the Hamiltonian constraint must entail a reduction of the conjugate $\ln q^{1/3}$. This will be realized and made concrete by the identification of time interval with the gauge invariant part of $\delta \ln q^{1/3} = dT = \frac{2}{3}d\ln V$.

where K is the trace of the extrinsic curvature. We can see the slicing we obtain is that generated by what the mathematicians call the "inverse mean curvature flow". In particular we pick the physical or gauge invariant part of intrinsic time interval $\delta \ln q^{1/3} = dT$ by setting $C = dT/dt$, and the Hamiltonian generating T-evolution reduces to

$$H_T = \int \left[\frac{1}{\beta} \sqrt{\bar{q}_{ik} \bar{q}_{jl} \bar{\pi}^{ij} \bar{\pi}^{kl}} - qR - 2N^i (\bar{q}_{ik} \nabla_j \bar{\pi}^{jk}) \right] d^3 x, \qquad (4.20)$$

which is of the same form advocated in earlier chapters. It is worth noting that in the above the generator of spatial diffeomorphisms for the barred variables $(\bar{q}_{ij}, \bar{\pi}^{ij})$ which emerges naturally from the reduction is $\mathcal{H}_i = -2\bar{q}_{ik} \nabla_j \bar{\pi}^{jk}$.

4.4   Moving away from General Relativity

Moving away from vacuum GR, how can we generalize this structure, while maintaining the 2 + 2 degrees of freedom and the 3-covariance?

There are changes we can make. We can multiply each of the three terms in the Hamiltonian constraint, as given by Eq. (4.8), with arbitrary constants. Since there is already an implicit coupling in the R for which we have set $2\kappa = 1$, this means deforming β in the π^2 term away from $\frac{1}{6}$, which is equivalent to deforming the parameter in the DeWitt supermetric in Eq. (2.10). More radically, we can replace the "potential", $-qR$, by any weight two scalar function $\mathcal{V}$ of the metric, and everything still works. The set of equations now reads,

$$H_T = \int \left[\frac{1}{\beta} \sqrt{\bar{q}_{ik} \bar{q}_{jl} \bar{\pi}^{ij} \bar{\pi}^{kl} + \mathcal{V}} + N^i \mathcal{H}_i \right] d^3 x, \qquad (4.21)$$

$$\tilde{\pi} := -\frac{1}{\beta} \sqrt{\bar{q}_{ik} \bar{q}_{jl} \bar{\pi}^{ij} \bar{\pi}^{kl} + \mathcal{V}}. \qquad (4.22)$$

With an explicit reduced Hamiltonian one can dynamically evolve the initial data set. However, if one insists on conventional wisdom of GR, with four constraints, prior to ITG the only known general way of solving and evolving the initial data $(q_{ij}, \tilde{\pi}^{ij})$ is the conformal method, which was initiated by Lichnerowicz [73]. There is a very comprehensive account in [74,76], especially in Chapter VII. Yet, these methods actually implicitly respect only three-dimensional or spatial diffeomorphisms for realistic evolutions. In what follows we recap the scheme and its extensions.

4.4.1 Constant mean curvature slice

The technique is to choose free data that consist of a base metric $\hat{q}_{ij}$, a tensor $\hat{\pi}_j^{iTT}$ that is both tracefree and divergence-free (or transverse) with respect to $\hat{q}_{ij}$, and a scalar $\hat{p}$. It is particularly simple if $\hat{p}$ is a constant, which is also a spatial diffeomorphism-invariant statement. This guarantees that the extrinsic curvature has constant trace, and thus we construct a "constant mean curvature"(CMC) slice. We make a conformal transformation

$$q_{ij} = \phi^4 \hat{q}_{ij}, \quad \tilde{\pi}^{ij} = \phi^{-4}\hat{\pi}^{ij}. \tag{4.23}$$

It follows that under this conformal scaling, $\hat{\pi}_j^{iTT}$ is invariant; it also remains TT with respect to q_{ij}. Thus we can assert, by contracting w.r.t. their respective metrics,[4]

$$\tilde{\pi}_j^{iTT} = \hat{\pi}_j^{iTT}. \tag{4.24}$$

By setting

$$\tilde{\pi} = \sqrt{q}\hat{p}, \tag{4.25}$$

we are guaranteed that

$$\tilde{\pi}_j^{iT} := (\tilde{\pi}_j^{iTT} + \sqrt{q}\delta_j^i \hat{p}/3) \tag{4.26}$$

is transverse i.e. it satisfies the momentum constraint for any conformal factor ϕ. In particular, for ITG we can specialise to

$$\phi^4 = q^{1/3}, \tag{4.27}$$

to get

$$(\hat{q}_{ij}, \hat{\pi}_j^{iTT}) = (\bar{q}_{ij}, \bar{\pi}_j^{iTT}). \tag{4.28}$$

This choice of ϕ in ITG actually eliminates the arbitrary conformal factor in the decomposition of the metric to zero in on unimodular $\bar{q}_{ij}$ and the physical TT momentric as in Eq. (4.28). The Hamiltonian constraint is solved by eliminating $\tilde{\pi}$ in terms of the other variables; whereas, as addressed below, the York scheme fixes ϕ with the Lichnerowicz-York equation.

---

[4]Intriguingly the resultant mixed index entity is precisely the momentric variable which will be further discussed in the next chapter.

4.4.2   *Solving the Lichnerowicz-York equation*

We now seek an appropriate ϕ in order to solve the Hamiltonian constraint. In GR, this reduces to solving the Lichnerowicz-York equation

$$8\hat{\nabla}^2\phi - \hat{R}\phi + \hat{q}^{-1}\hat{\pi}_j^{iTT}\hat{\pi}_i^{jTT}\phi^{-7} - \frac{1}{6}\hat{p}^2\phi^5 = 0, \qquad (4.29)$$

where $\hat{q}$ is the determinant of $\hat{q}_{ij}$. This is an extremely nice equation because it always has a unique, positive solution [75, 77].

As deformation of Einstein's theory, multiplication of the R and the $\tilde{\pi}^2$ terms in the Hamiltonian constraint by arbitrary positive constants yields the modified Lichnerowicz-York equation,

$$8\alpha^2\hat{\nabla}^2\phi - \alpha^2\hat{R}\phi + \hat{q}^{-1}\hat{\pi}_j^{iTT}\hat{\pi}_i^{jTT}\phi^{-7} - \beta^2\hat{p}^2\phi^5 = 0, \qquad (4.30)$$

which is just as nice as the original Lichnerowicz-York equation, Eq. (4.29), because it too always has a unique positive solution. Such a modification has been discussed in a different context in [78].

A higher curvature addition, such as an R^2 term in the potential leads to the Hamiltonian constraint,

$$-\alpha^2qR - \rho qR^2 + (\tilde{\pi}_j^i\tilde{\pi}_i^j) - \beta^2\tilde{\pi}^2 = 0, \qquad (4.31)$$

where ρ is another parameter. The new Lichnerowicz-York equation will be

$$8\alpha^2\hat{\nabla}^2\phi - \alpha^2\hat{R}\phi + \hat{q}^{-1}(\hat{\pi}_j^{iTT}\hat{\pi}_i^{jTT})\phi^{-7} - \beta^2\hat{p}^2\phi^5$$
$$= \rho\phi^{-5}(8\hat{\nabla}^2\phi - \hat{R}\phi)^2. \qquad (4.32)$$

This is not as pleasant an equation. If we linearize this equation about $\rho = 0$, a combination of the Fredholm alternative and the implicit function theorem [79] shows that the non-linear equation, Eq. (4.32), has a solution for a range of ρ's in a neighborhood of zero. Unfortunately, this technique gives us no estimate as to the size of this neighborhood, although it should work for any choice of metric potential.

The slices generated by a generic Hamiltonian will not stay CMC. We can use the same technique, i.e. the Fredholm alternative plus the implicit function theorem, to relax the condition that $\hat{p}$ is a constant. We can replace it by assuming

$$\hat{p} = \hat{p}_0 + \theta\hat{p}_1, \qquad (4.33)$$

where $\hat{p}_0$ is a nonzero constant, $\hat{p}_1$ is a function of position, and θ is a parameter. Now the conformal method gives us a system of 4 coupled nonlinear equations [77]. If we linearize about $\theta = 0$, we find that the equations decouple and we get a system with the necessary existence and uniqueness properties. This means we can construct non-CMC initial data, but it is not clear how large the deviation from CMC we can allow. What this illustrates is that departure from CMC initial data is possible, and the possibility that it may not play any fundamental role in the scheme.

4.5   An exceptional extension of Lichnerowicz-York equation: addition of Cotton-York term

If we add any other function of the metric to Eq. (4.29) we destroy the nice properties of the Lichnerowicz-York equation. For example, if we were to add an R^2 term, the Lichnerowicz-York equation would pick up the term $(8\hat{\nabla}^2\phi - \hat{R}\phi)^2$, which changes the nature of the Lichnerowicz-York equation completely and the existence and uniqueness results no longer hold. There is one exception. York, in [61], rediscovered a conformally invariant tensor which is a function of the metric. This is now known as the Cotton-York tensor (density) $\tilde{C}^i_j$. It is this mixed $(1,1)$ tensor density that is precisely conformal invariant i.e. $\hat{\tilde{C}}^i_j = \tilde{C}^i_j$ (this will be discussed further in Chapter 10). It is moreover transverse and traceless. So it possesses all the virtues of $\hat{\pi}^{iTT}_j$. Thus in analogy to $\hat{\pi}^{iTT}_j\hat{\pi}^{jTT}_i$, we can consider in Eq. (4.21) a generalized Hamiltonian with the inclusion of a Cotton-York term, and also a cosmological constant. The corresponding Hamiltonian constraint is

$$- \alpha^2 q(R - 2\Lambda) + \bar{q}_{ik}\bar{q}_{jl}\bar{\pi}^{ij}\bar{\pi}^{kl} + g'^2\tilde{C}^i_j\tilde{C}^j_i - \beta^2\pi^2 = 0, \qquad (4.34)$$

where g' is another coupling constant; and the generalized Lichnerowicz-York equation becomes

$$8\alpha^2\hat{\nabla}^2\phi - \alpha^2(\hat{R}\phi - 2\Lambda\phi^5) + \hat{q}^{-1}(\hat{\pi}^{iTT}_j\hat{\pi}^{jTT}_i + g'^2\hat{\tilde{C}}^i_j\hat{\tilde{C}}^j_i)\phi^{-7} = \beta^2\hat{p}^2\phi^5. \quad (4.35)$$

This is remarkably just as well-behaved as the original Lichnerowicz-York equation: it always possesses a positive unique solution. This conformal factor maps the free data onto a solution both of the generalized Hamiltonian constraint, Eq. (4.34), and the momentum constraint. Further motivations for the addition of the Cotton-York term to Einstein's theory with cosmological constant, and the physical ramifications, are fleshed out in other chapters.

It is worth emphasizing the addition of Cotton-York and other higher three-curvature terms to the potential explicitly breaks the 4-covariance of Einstein's theory. Despite this, the initial data formulation (with TT momentum and solving the Hamiltonian constraint with the Lichnerowicz-York equation) remains valid, and proves that we can consistently depart from Einstein's theory, keeping three-covariance and maintaining the same two d.o.f. as in GR.

4.6 Intrinsic time versus extrinsic time

A key advantage the intrinsic time formalism has over the extrinsic time formalism is that the Hamiltonian constraint can be easily put in the form of an algebraic equation for $\tilde{\pi}$. If we wanted to use extrinsic time, we would need to solve the Hamiltonian constraint for ϕ (hence $\sqrt{q}$) from the Lichnerowicz-York equation. Therefore the reduced Hamiltonian in the extrinsic time gauge is a non-local object. Another sign of non-locality is that the lapse function for a CMC foliation in pure GR is determined by [61]

$$(-\nabla^2 + \sigma^{\mu\nu}\sigma_{\mu\nu} + T^2/3)N = \text{const.}, \qquad (4.36)$$

where $\sigma^{\mu\nu}$ is the shear of the unit time-like normal field of the spatial hypersurface, and $T = 2\hat{p}/3$. This is obviously a nice elliptic equation for N, but clearly non-local as distinct from Eq. (4.18), the equation for the intrinsic time lapse. It is clear that using an intrinsic time gives a very clean Hamiltonian structure in classical gravity. These good properties carry over when we implement the canonical quantization program [16, 71]. This scheme may indeed capture all possible initial data in an ever-expanding closed Universe.

4.7 Penrose Weyl Curvature Hypothesis and Lichnerowicz-York equation with Cotton-York term

Penrose asserts that our classical Robertson-Walker Big Bang singularity corresponds to vanishing Weyl curvature which is "extraordinarily special", because initial data for black holes (which come with Weyl singularities) overwhelmingly dominate the phase space for gravitational degrees of freedom [25, 53]. In what follows, we study the initial data for the beginning of the universe within the context of ITG and the solution of the Lichnerowicz-York equation.

4.7.1 Asymptotic behavior of the Hamiltonian at early and late intrinsic times

Hodge decomposition for $\delta \ln q^{\frac{1}{3}}$ and its Heisenberg equation of motion lead to

$$\frac{d}{dT} \ln q^{\frac{1}{3}}(x,T) = \frac{\partial}{\partial T} \ln q^{\frac{1}{3}} + \frac{1}{i\hbar}[\ln q^{\frac{1}{3}}, H_T] = 1; \qquad (4.37)$$

with solution[5]

$$\ln\left[\frac{q(x,T)}{q(x,T_{\text{now}})}\right] = 3(T - T_{\text{now}}), \qquad (4.38)$$

and $-\infty < T < \infty$. Moreover,

$$\delta T = \frac{2}{3}\delta \ln V \qquad (4.39)$$

i.e.

$$T - T_{\text{now}} = \frac{2}{3}\ln(V/V_{\text{now}}). \qquad (4.40)$$

The potential in Eq. (4.21) has the special case of $\mathcal{V} = -\frac{q}{(2\kappa)^2}[R - 2\Lambda_{eff}]$ for Einstein's theory with cosmological constant. If we consider a tensor density of weight one, to third spatial derivatives it can be written as

$$\begin{aligned}
\tilde{\mathcal{W}}^i_j &= \sqrt{q}(\Lambda' + a'R)\delta^i_j + b'\sqrt{q}R^i_j + g'\tilde{C}^i_j \\
&= \sqrt{q}(\Lambda' + a'q^{-\frac{1}{3}}\bar{R})\delta^i_j + b'\sqrt{q}q^{-\frac{1}{3}}\bar{R}^i_j + g\hbar\tilde{C}^i_j \\
&\quad + \sqrt{q}(\partial_k \ln q \text{ terms}),
\end{aligned} \qquad (4.41)$$

wherein we have explicitly separated out the T-dependence (or q-dependence) from entities (labeled with overline) which depend only on $\bar{q}_{ij}$. The Cotton-York tensor density $\tilde{C}^i_j$ is conformally invariant, ergo q-independent. This allows us to compare how each term scales with q.

The Hamiltonian corresponding to Einstein's theory is explicitly (intrinsic)time-dependent, and not (intrinsic)time-reversal invariant, due to its Ricci scalar and effective cosmological constant in $\mathcal{V}$. Furthermore, the exponential scaling behaviour in Eq. (4.38) of q with intrinsic time implies in the limit

$$T - T_{\text{now}} \to -\infty, V/V_{\text{now}} \to 0$$

---

[5]The commutator $[\ln q^{\frac{1}{3}}, H_T]$ vanishes because H_T is not a function of the conjugate variable $\tilde{\pi}$.

(i.e. early times when the universe was very small in volume[6]), $\tilde{\mathcal{W}}^i_j$ was dominated by the Cotton-York term; whereas the limit

$$T - T_{\text{now}} \to \infty, V/V_{\text{now}} \to \infty$$

(i.e. late times when the universe becomes large) will be dominated by the Ricci scalar and especially the cosmological constant term, which is compatible with current observations and understanding of our ever expanding universe. In the middle period, curvature and cosmological terms will be comparable in importance. The upshot is that, in an ever expanding universe, a GR Hamiltonian with extra Cotton-York term, i.e. with $\mathcal{V} = -\frac{q}{(2\kappa)^2}[R - 2\Lambda] + g^2\hbar^2\tilde{C}^i_j\tilde{C}^j_i$, will be Cotton-York dominated near the Big Bang and at late times the theory becomes more and more like Einstein's GR.

It is interesting to analyse the behavior in the context of the modified Lichnerowicz-York equation of Eq. (4.35). At late intrinsic times or for large values of ϕ, the Cotton-York contribution vanishes rapidly as ϕ^{-7}; so, compared to other terms in $\mathcal{V}$, the Cotton-York addition and its attendant physical effects will be suppressed. However near the Big Bang, with $T \to -\infty$, or $\phi \to 0$, the divergence in Eq. (4.35) can be tamed only if $\hat{\pi}^{iTT}_j\hat{\pi}^{jTT}_i$ and the Cotton-York term, $\hat{\tilde{C}}^i_j\hat{\tilde{C}}^j_i$, *both* vanish. Whereas vanishing of the traceless part of the momentum is a statement on the extrinsic curvature, it says nothing of the intrinsic geometry. This poses less restriction on the initial singularity of the universe (for instance, as Penrose have pointed out, black holes singularities have divergent Weyl tensor whereas Robertson-Walker spacetimes have four-dimensional Ricci scalar singularity). In contradistinction, the new ingredient of vanishing Cotton-York tensor is the precise requirement of spatial conformal flatness on the intrinsic geometry! The only simply-connected closed conformally flat spatial manifold is the standard three-sphere[7] [80]. Together with vanishing $\hat{\pi}^{iTT}_j$ the non-triviality only resides in ϕ and $\hat{p}$. To wit, in the $\phi \to 0$ limit, Eq. (4.35) tends to

$$- \phi\hat{R} + 8\hat{\nabla}^2\phi + 2\Lambda\phi^5 = \frac{\beta^2}{\alpha^2}\hat{p}^2\phi^5. \tag{4.42}$$

---

[6]An example of the determination of global intrinsic time interval is its measurement through dimensionless redshift in homogeneous Friedmann-Lemaitre-Robertson-Walker (FLRW) cosmology.

[7]In ITG, $q = q_o e^{3(T-T_0)}$; validity of the Poincaré Conjecture [81] implies we may adopt q_0 to be proportional to the metric of the standard sphere, $ds^2_{S^3} = d\chi^2 + \sin^2\chi d\Omega^2$, with volume $2\pi^2$.

Corresponding to $\tilde{C}^i_j = \hat{C}^i_j = 0$, a standard sphere of radius a yields $R = 6/a^2 = \phi^{-5}(\phi\hat{R} - 8\hat{\nabla}^2\phi)$. This gives[8]

$$\hat{p}^2 = \frac{6a^2}{\beta^2}\left(\frac{\Lambda}{3} - \frac{1}{a^2}\right), \tag{4.43}$$

which is CMC; and the remaining equation,

$$-\hat{\nabla}^2\phi + \frac{\phi\hat{R}}{8} = \frac{3}{4a^2}\phi^5, \tag{4.44}$$

is still non-linear in flat coordinates with $\hat{R} = 0$, but it has $\phi = \sqrt{\frac{2a}{1+r^2}}$ as exact solution.[9] It is clear with the solution of ϕ and Eq. (4.43) that *the initial singularity is smooth* and metric singularities arise only from the scale factor $a \to 0$ (in the context of ITG, this singular limit merely corresponds to the beginning of intrinsic time at $T \to -\infty$). The new Cotton-York ingredient gives credence to Penrose's Hypothesis of initial vanishing Weyl curvature tensor and smooth Robertson-Walker Big Bang, whereas Einstein's theory, *sans* Cotton-York addition, does not single out our "extraordinarily special" universe.

It is also noteworthy that an initial data set of conformal flatness and vanishing $\hat{\pi}^{iTT}_j$ together with CMC (as per Eq. (4.43) and solution of Eq. (4.44)) is compatible with the initial data for Lorentzian de Sitter solution (with S^3 slicings) at the de Sitter throat (of size determined by $\hat{p} = 0$ i.e. $a = \sqrt{\frac{3}{\Lambda}}$). The complete vanishing of all momentum or extrinsic curvature at the throat is moreover the correct junction condition [56] for Euclidean-Lorentzian continuation. At even smaller values of a, the value of $\hat{p}$ becomes pure imaginary. Remarkably, an exact Hartle-Hawking quantum wave function can be found, with conformal flatness at its critical points and vanishing expectation value of the momentric. Its properties and ramifications will be taken up in other chapters.

---

[8]In Einstein's GR, $\alpha^2 = \frac{1}{(2\kappa)^2}$.

[9]ϕ is the solution of $\frac{1}{r^2}\partial_r r^2 \partial_r \phi = -\frac{3}{4a^2}\phi^5$. See also the discussion on conformal flatness in Chapter 10 and the conformal factor in Eq. (10.31).

Chapter 5

New commutation relations for Quantum Gravity

"Even in the dark times between experimental breakthroughs, there always continues a steady evolution of theoretical ideas, leading almost imperceptibly to changes in previous beliefs."
—Steven Weinberg

5.1 Motivation for momentric variables

A new set of fundamental commutation relations for quantum gravity will be presented. The basic variables are the unimodular spatial metric (or its corresponding dreibein) and eight generators which correspond to Klauder's momentric variables [39]. The commutation relations are not canonical, but they have well defined group theoretical meanings. The momentric can be realized as self-adjoint operators in the metric representation, and the kinetic operator in the Hamiltonian of ITG is a Casimir invariant of the group generated by the momentric variables.

In quantum field theories, 'equal time' fundamental commutation relations — part and parcel of microcausality — are predicated on the existence of a space-like hypersurface. Geometrodynamics bequeathed with positive-definite spatial metric is the simplest consistent framework to implement them.

In usual quantum theories, the fundamental *canonical* commutation relations,

$$[Q(x), P(y)] = i\hbar\delta^3(x, y),$$

implies $\frac{P}{\hbar}$ is the generator of translations of Q which are also symmetries of the 'free theories' in the limit of vanishing interaction potentials. For geometrodynamics, the corresponding canonical commutation relations are

$$[q_{ij}(x), \tilde{\pi}^{kl}(y)] = i\hbar\frac{1}{2}(\delta_i^k\delta_j^l + \delta_i^l\delta_j^k)\delta^3(x, y).$$

117

However, neither positivity of the spatial metric is preserved under arbitrary translations generated by conjugate momentum; nor is the 'free theory' invariant under translations when interactions are suppressed, because $G_{klmn}\tilde{\pi}^{kl}\tilde{\pi}^{mn}$ in the kinetic part of the Hamiltonian also contains the DeWitt supermetric [12], G_{ijkl}, which is dependent upon q_{ij}. In quantum gravity, states which are infinitely peaked at the flat metric, or for that matter any particular metric with its corresponding isometries, cannot be postulated ad hoc; consequently, the underlying symmetry of even the "free theory" is somewhat obscure.

5.2   New commutation relations

5.2.1   *The momentric variable and commutation relations*

Klauder's "momentric" variable [39], $\tilde{\pi}^i_j$, with one contravariant and one covariant index, is an intriguing entity which, in lieu of utilizing the momenta, confers many advantages. Classically

$$\tilde{\pi}^i_j := \bar{q}_{jk}\tilde{\pi}^{ik}; \tag{5.1}$$

and as a combination of momentum and metric, the "momentric" is aptly named.[1] Quantum mechanically it is advantageous to treat the momentric as fundamental rather than a composite of the metric and momentum.

The commutation relations come from restriction of Klauder's affine algebra to traceless momentric and unimodular part of the spatial metric. Consistent with the basic Poisson brackets of the spatial metric q_{ij} and its conjugate momentum $\tilde{\pi}^{ij}$, the following commutation relations can be postulated:

$$[\bar{q}_{ij}(x), \bar{q}_{kl}(y)] = 0, \tag{5.2}$$

$$[\bar{q}_{ij}(x), \tilde{\pi}^k_l(y)] = i\hbar \bar{E}^k_{l(ij)}\delta^3(x, y), \tag{5.3}$$

$$[\tilde{\pi}^i_j(x), \tilde{\pi}^k_l(y)] = \frac{i\hbar}{2}(\delta^k_j\tilde{\pi}^i_l - \delta^i_l\tilde{\pi}^k_j)\delta^3(x, y); \tag{5.4}$$

wherein,

$$\bar{E}^i_{j(mn)} = \frac{1}{2}(\delta^i_m\bar{q}_{jn} + \delta^i_n\bar{q}_{jm}) - \frac{1}{3}\delta^i_j\bar{q}_{mn} \tag{5.5}$$

satisfies

$$\delta^j_i\bar{E}^i_{j(mn)} = \bar{E}^i_{j(mn)}\bar{q}^{mn} = 0; \tag{5.6}$$

$$\bar{E}^i_{jil} = \bar{E}^i_{jli} = \frac{5}{3}\bar{q}_{jl}. \tag{5.7}$$

---

[1]Originally Klauder defined the momentric as $\tilde{\pi}^i_j := q_{jk}\tilde{\pi}^{ki}$, its traceless part is equivalent to the expression in (5.1) which is still a tensor density of weight one.

It is also true that $\bar{E}^i_{j(mn)}$ is the "vierbein for the supermetric", since

$$\bar{G}_{ijkl} = \bar{E}^m_{n(ij)}\bar{E}^n_{m(kl)}. \tag{5.8}$$

5.3 Self-adjoint momentric operators and pointwise $sl(3, R)$ algebra

Difficulties in implementing $\bar{\pi}^{ij}$ as a self-adjoint traceless operator in the metric representation can be surmounted by using the momentric.

Remarkably, the momentric Lie algebra in (5.4) is equivalent to saying that the eight momentric variables realize, at each spatial point, an $sl(3, R)$ algebra. Apart from redefinition by multiplicative constants, the momentric variables and algebra (5.4) correspond to the same ones displayed in Appendix A of Ref. [82]. The algebra implies the momentric has the transformation character of a mixed tensor with respect to real unimodular transformations in three spatial dimensions.

Among the advantages, the quantum momentric operator and commutation relations can be explicitly realised in the metric representation on wave-functionals of the metric $\Psi[\bar{q}] = \langle \bar{q}_{ij} | \Psi \rangle$ by functional differentiation

$$\bar{\pi}^i_j(x)\Psi[\bar{q}] = \frac{\hbar}{i}\bar{E}^i_{j(mn)}(x)\frac{\delta}{\delta\bar{q}_{mn}(x)}\Psi[\bar{q}]$$

$$= \left[\frac{\hbar}{i}\frac{\delta}{\delta\bar{q}_{mn}(x)}\bar{E}^i_{j(mn)}(x)\right]\Psi[\bar{q}] \tag{5.9}$$

with $\bar{\pi}^i_j$ being Hermitian on account of

$$\left[\frac{\delta}{\delta\bar{q}_{mn}(x)}, \bar{E}^i_{j(mn)}(x)\right] = 0. \tag{5.10}$$

Self-adjointness of $\bar{\pi}^i_j$ follows from two properties:
(i) that $\bar{\pi}^i_j$ is Hermitian;
(ii) given that $\Psi[\bar{q}]$ is in the domain of $\bar{\pi}^i_j$ i.e. $\Psi[\bar{q}] \in Dom(\bar{\pi}^i_j)$, then $\Psi[\bar{q}] \in Dom(\bar{\pi}^{\dagger i}_j)$ also.
Thus $\bar{\pi}^{\dagger i}_j = \bar{\pi}^i_j$ in (5.9) generates a *infinite-dimensional unitary representation* of pointwise $SL(3, R)$ which is non-compact.

It can be checked that the above definition of the momentric operator also leads to $\epsilon_{ijk}q^{jl}\bar{\pi}^k_l = 0$, the counterpart of the symmetry of the classical momentum $\bar{\pi}^{ij}$ in its indices.

The commutation relations in (5.3) imply that $\bar{\pi}^i_j$ generates infinitesimal $SL(3, R)$ transformations of the metric which exponentiates (with traceless parameter $\alpha_j^i(x)$) to

$$U^\dagger(\alpha)\bar{q}_{kl}(x)U(\alpha) = (e^{\frac{\alpha(x)}{2}})_k^m \bar{q}_{mn}(x)(e^{\frac{\alpha(x)}{2}})_l^n, \qquad (5.11)$$

wherein $e^{\frac{\alpha(x)}{2}} \in SL(3, R)$, and

$$U(\alpha) = e^{-\frac{i}{\hbar}\int \alpha_j^i \bar{\pi}_i^j d^3 y}. \qquad (5.12)$$

Positivity of the metric, and also the unimodular nature, of $\bar{q}_{ij}$ are preserved under $SL(3, R)$ transformations generated by the momentric; whereas the momentum variable generates arbitrary *translations* of the spatial metric which will not in general uphold its positivity [39].

5.3.1   *Relation to su(3) algebra*

The (unique) compact complex extension of $SL(3, R)$ is $SU(3)$ (of order eight, rank two, and semi-simple), obtained formally by multiplying the appropriate generators with i and letting some of the nonzero structure constants absorb a minus sign [82]. In fact, with 3×3 Gell-Mann matrices $\lambda^{A=1,\dots,8}$ [83], it can be checked that the corresponding

$$T^A(\mathbf{x}) := \frac{1}{\hbar}(\lambda^A)_i^j \hat{\bar{\pi}}_j^i(\mathbf{x}) \qquad (5.13)$$

generate, at each point, an algebra with $su(3)$ structure constants $f^{AB}{}_C$ i.e.

$$[T^A(x), T^B(y)] = -f^{AB}{}_C T^C(x)\delta^3(x, y). \qquad (5.14)$$

However, while the rest of the generators are self-adjoint, $T^{A=2,5,7}$ are anti-Hermitian (whereas the momentric is self-adjoint), so the algebra is really non-compact pointwise $sl(3, R)$ with the momentric as defined in (5.9) being generators of an irreducible infinite-dimensional unitary representation.

5.3.2   *Fundamental commutation relations of momentric and dreibien*

In (5.2) there is an apparent asymmetry in that there are only five independent components in the symmetric unimodular $\bar{q}_{ij}$ (likewise for the symmetric traceless $\bar{\pi}^{ij}$ in Eq. (2.6)), whereas the mixed-index traceless momentric variable, $\bar{\pi}_j^i$, has eight independent components. To incorporate fermions, it is necessary to introduce the dreibein which is more fundamental than the metric. Expressed in terms of dreibein, e_{ai}, the spatial metric,

$$q_{ij} := \delta^{ab} e_{ai} e_{bj} \qquad (5.15)$$

is automatically positive-definite if the dreibein is real and non-vanishing, modulo $SO(3,C)$ gauge rotations which leave q_{ij} invariant under these local Lorentz transformations. More precisely, e_{ai} is the inverse of the triad E^{ia} which transforms as a self-dual $SO(3,C)$ three-vector under the Lorentz group; the densitized triad $\tilde{E}^{ia} = \sqrt{q}E^{ia}$ is the conjugate momentum to the Ashtekar self-dual gauge variable [46]. With the introduction of this dreibein, both the unimodular dreibein $\bar{e}_{ai} := e^{-\frac{1}{3}}e_{ai}$ and traceless momentric variables, $\bar{\pi}^i_j$ possess eight components. They obey the fundamental commutation relations [70]

$$[\bar{e}_{ai}(x), \bar{e}_{bj}(y)] = 0, \tag{5.16}$$

$$\left[\bar{e}_{ai}(x), \bar{\pi}^j_k(y)\right] = \frac{i\hbar}{2}\left(\delta^j_i \bar{e}_{ak} - \frac{1}{3}\delta^j_k \bar{e}_{ai}\right)\delta^3(x,y), \tag{5.17}$$

$$\left[\bar{\pi}^i_j(x), \bar{\pi}^k_l(y)\right] = \frac{i\hbar}{2}\left(\delta^k_j \bar{\pi}^i_l - \delta^i_l \bar{\pi}^k_j\right)\delta^3(x,y). \tag{5.18}$$

The second commutation relation implies (c.f. (5.11) for the unimodular metric) the $\bar{e}_{ai}$ transforms as a vector,

$$U(\alpha)^\dagger \bar{e}_{ai}(x)U(\alpha) = (e^{\frac{\alpha}{2}})^j_i \bar{e}_{aj}(x). \tag{5.19}$$

5.4   Generator of local Lorentz symmetry

The local Lorentz group is gauged not as $SO(3,1)$, but more precisely (in a manner not predicated upon the existence of classical four-dimensional spacetime) as self-dual $SO(3,C)$ transformations of $\bar{e}_{ai}$, while two-component Weyl fermions transform as $SL(2,C)$ the double cover of $SO(3,C)$. This topic will be further addressed in Chapter 9.

Local $SO(3,C)$ Lorentz transformations are generated by the Gauss Law generator, and the corresponding constraint for local Lorentz symmetry is

$$G_a(x) := \epsilon_{abc}e^b_i \tilde{\pi}^{ci} - i\tilde{\pi}_\psi \frac{\tau_a}{2}\psi \tag{5.20}$$
$$= 0,$$

wherein $\tilde{\pi}^{ai}$ is the canonical conjugate momentum of the dreibein (classically it is equivalent to $2\tilde{\pi}^{ij}e^a_j$), $(\psi, \tilde{\pi}_\psi)$ denotes the conjugate pair of Weyl fermion and its momentum,[2] and τ_a are the Pauli matrices.

The determinent $e := \det(e_{ai})$ and $(\ln q^{\frac{1}{3}}, \tilde{\pi})$ are Lorentz singlets which commute with the Gauss Law constraint, and so do the momentric $\bar{\pi}^i_j$

---

[2]If there are more than one species, the Gauss Law will need to sum over the species.

variables. The constraint can be reexpressed with traceless momentric, rather than momentum, variable as

$$G_a := \epsilon_{abc}\bar{e}_i^b \bar{e}^{cj}\bar{\pi}_j^i - i\tilde{\pi}_\psi \frac{\tau_a}{2}\psi = 0. \tag{5.21}$$

That it generates $SO(3,C)$ local Lorentz rotations of the dreibein can be verified from

$$\left[\bar{e}_{ai}(x), \frac{i}{\hbar}\int \eta^b G_b d^3 y\right] = \epsilon_{abc}\eta^b(x)\bar{e}_i^c(x), \tag{5.22}$$

with η^b as the gauge parameter. In addition, it generates the associated local $SL(2,C)$ transformations of the fermionic degrees of freedom.

5.5 The kinetic operator, the free theory, and CSCO for momentric variables

The momentric has the transformation character of a mixed $(1,1)$ tensor with respect to $SL(3,R)$ real unimodular transformations in three dimensions. It follows that any invariant operator of the momentric formed by contraction with no free tensor indices commutes with all the generators. An upshot is the kinetic term in the Hamiltonian is thus a Casimir invariant of the pointwise $SL(3,R)$ group generated by the eight momentric variables. Classically $\bar{\pi}^{ij}\bar{G}_{ijkl}\bar{\pi}^{kl} = \bar{\pi}_j^i\bar{\pi}_i^j$. Furthermore, the symmetry of the free theory (obtained by setting $\mathcal{V}$ to zero in (3.15)) now becomes transparent. It is characterized by $SL(3,R)$ invariance generated by the momentric which commutes with the kinetic operator, because the Casimir invariant $T^A T^A$ is related to the kinetic operator through

$$\bar{\pi}_j^{i\dagger}\bar{\pi}_i^j = \bar{\pi}_j^i\bar{\pi}_i^j \tag{5.23}$$

$$= \frac{\hbar^2}{2}T^A T^A. \tag{5.24}$$

The spectrum of the free Hamiltonian can be thus labeled by eigenvalues of the complete set of commuting operators (CSCO) at each spatial point comprising the two Casimirs [82]

$$L^2(x) = \bar{\pi}_j^i\bar{\pi}_i^j, \tag{5.25}$$

$$C(x) = \det(\bar{\pi}_j^i) = \frac{1}{3!}\epsilon_{ijk}\tilde{\epsilon}^{lmn}\bar{\pi}_l^i\bar{\pi}_m^j\bar{\pi}_n^k \propto d_{ABC}T^A T^B T^C, \tag{5.26}$$

with $d_{ABC} = \frac{1}{4}Tr(\{\lambda_A, \lambda_B\}\lambda_C)$; the Cartan subalgebra $T^3(x) \propto \bar{\pi}_1^1 - \bar{\pi}_2^2$, $T^8(x) \propto \bar{\pi}_1^1 + \bar{\pi}_2^2 - 2\bar{\pi}_3^3$, and "isospin"

$$I(x) = \sum_{B=1}^{3} T^B T^B. \tag{5.27}$$

Basis states for the CSCO can be labeled by $|L^2, C, I, m_3, m_8\rangle_x$ corresponding to the five respective eigenvalues at each spatial point.

An underlying group structure has the advantage the action of the kinetic and other momentric operators on wave functions by functional differentiation can be traded for the well defined group action of generators of $SL(3, R)$ on states expanded in this CSCO basis. For instance, in the metric representation the momentric has the action[3]

$$\hat{\bar{\pi}}^i_j(x) \left\langle \bar{q}_{kl} \left| \prod_y \right| L^2, C, I, m_3, m_8 \right\rangle_y$$

$$= \frac{\hbar}{i} \bar{E}^i_{j(mn)}(x) \frac{\delta}{\delta \bar{q}_{mn}(x)} \left\langle \bar{q}_{kl} \left| \prod_y \right| L^2, C, I, m_3, m_8 \right\rangle_y$$

$$= \left\langle \bar{q}_{kl} \left| \bar{\pi}^i_j(x) \prod_y \right| L^2, C, I, m_3, m_8 \right\rangle_y , \qquad (5.28)$$

in which the intermediate step may be skipped in favor of the final group theoretic action of the generator $\bar{\pi}^i_j$ on $|L^2, C, I, m_3, m_8\rangle$.

5.5.1 $SL(3, R)$ symmetry generated by momentric versus diffeomorphism symmetry generated by momentum constraint

$SL(3, R)$ transformations generated at each point by $\bar{\pi}^i_j$ mimic the transformations of the unimodular metric under general coordinate changes. However there are only five independent parameters in $\bar{q}_{ij}$ and a field theory of the metric with $SL(3, R)$ realized as a local Yang-Mills gauge group would be subjected to too many constraints rendering it trivial (unless it is embellished with other d.o.f. such as torsion and/or non-metricity). The criterion of the symmetry generators annihilating physical states would mean that all these states are singlets of $SL(3, R)$, and all the eigenvalues in $|L^2, C, I, m_3, m_8\rangle$ would have to be null. Geometrodynamics as a dynamical field theory of unembellished pure metric and its conjugate momentum cannot be a naive Yang-Mills theory of local $GL(3, R)$ gauge group. In contradistinction, the momentum, $H_i := -2q_{ik}\nabla_j \tilde{\pi}^{jk}$, which generates spatial diffeomorphisms (infinitesimal changes which are Lie derivatives) of

---

[3]It can be checked that in the dreibein representation $\bar{\pi}^i_j := \frac{\hbar}{2i}(\bar{e}_{aj}\frac{\delta}{\delta \bar{e}_{ai}} - \frac{1}{3}\delta^i_j \bar{e}_{ak}\frac{\delta}{\delta \bar{e}_{ak}})$ is self-adjoint and realizes (5.17) and (5.18). This is an infinite-dimensional unitary representation of non-compact $SL(3, R)$.

the metric and its conjugate variable, has only three components. It can be decomposed as

$$H_i = \mathcal{H}_i - \frac{2}{3}\nabla_i\tilde{\pi}, \qquad (5.29)$$

with[4]

$$\mathcal{H}_i = -2\nabla_j\bar{\pi}_i^j = -2(\partial_j\bar{\pi}_i^j + \bar{\Gamma}_{jk}^j\bar{\pi}_i^k - \bar{\Gamma}_{ij}^k\bar{\pi}_k^j) \qquad (5.30)$$

separately generating spatial diffeomorphisms of $(\bar{q}_{ij}, \bar{\pi}^{ij})$; and $\nabla_i\tilde{\pi} = \sqrt{q}\partial_i\frac{\tilde{\pi}}{\sqrt{q}}$ responsible for diffeomorphisms of $(\ln q, \tilde{\pi})$. In the above, $\bar{\Gamma}_{jk}^j$ is the Christoffel symbol of $\bar{q}_{ij}$.

In ITG the Hamiltonian constraint is solved by expressing $\tilde{\pi}$ in terms of the rest of the variables and the corresponding auxiliary condition sets $\delta\ln q$ to be its diffeomorphism-invariant change $2\delta\ln V$. The remaining constraints $\mathcal{H}_i = 0$ and its auxiliary conditions of transverse unimodular metric excitations eliminate precisely three degrees of freedom at each spatial point,[5] leaving pure Geometrodynamics with two local d.o.f. Furthermore, as expressed in the momentum constraint and its auxiliary conditions, transversality which enforces the absence of longitudinal modes, is a prerequisite and salient feature of long range interactions. For the unimodular metric and momentric, $\mathcal{H}_i = 0$ is the analog of the transversality encoded in the Gauss Law constraint, $(D_i^A\tilde{\pi}_A^i)^a = 0$, in Yang-Mills theory. The field theoretic physical correspondence of Geometrodynamics and Yang-Mills is not grounded in the naive substitution of the fiber[6] or Yang-Mills gauge group by $GL(4, R)$ (or some other conjectured geometrical group), but in the commonality between momentum constraint and Yang-Mills Gauss Law which enforce transversality and make of long range interactions viable[7] even when the fundamental field excitation which is pure spin-2 in metric GR is a completely different entity from the pure spin-1 vector boson in unbroken Yang-Mills theory.

5.5.2   *Why is the universe intuitively "metric", and not "conjugately realized"?*

The momentric variables obey the algebra of Eq. (5.18) and thus do not totally commute among themselves. A consequence is that quantum wave

---

[4]No extraneous non-metricity entities are introduced in ITG.

[5]This will be elaborated in Chapter 8.

[6]As discussed in Sec. 1.7.7, unlike Yang-Mills Gauss Law constraints, diffeomorphism constraints do not generate a pointwise product Lie algebra, so an exact Yang-Mills fiber bundle structure analog for geometrodynamics is incongruous.

[7]This does not mean that non-Abelian self interactions cannot lead to confinement (as in quantum chromodynamics) or gravitational collapse.

functions cannot be chosen as functionals of all the momentric variables, but this restriction does not apply to the metric or dreibein representation. This differs starkly from the usual *canonical* commutation relations which allow wave functionals to be realized in either of the conjugate representations. As discussed, there are in terms of components, 8 for momentric and 5 for the unimodular metric. However, there are the precisely 5 commuting entities at each point which can be obtained from the momentric. In this sense the unimodular metric-momentric asymmetry is "rectified". For the unimodular dreibein with 8 components, local Lorentz invariance and the fact that $\bar{q}_{ij}$ defines the $\bar{e}_{ai}$ up to Lorentz transformations, imply that in pure Geometrodynamics without fermions, gauge invariant observables of $\bar{e}_{ai}$ will be manifested in only terms of $\bar{q}_{ij}$. Thus the selection rule of commuting metric over non-commuting momentric variables does not seem to be the cause of why our universe seems fundamentally and intuitively 'metric' in nature, and not 'conjugately realized'. Rather, the requirement of spacelike separation in microcausality may be a more compelling reason. In the metric representation, hypersurfaces on which the fundamental commutation relations are defined will remain spacelike even when superpositions and weighted averages are taken in any $\Psi[\bar{q}_{ij}(x)]$, whereas arbitrary $\Psi[\bar{\pi}^{ij}(x)]$ does not seem make definite predictions as to whether the separation between any two points is spacelike or not. Theories of other variables besides the metric simply presuppose quantum wave functionals are defined for fields on spacelike hypersurfaces.

From the technical point of view, the presence of the Cotton-York term and the scalar Ricci potential, both of which depends on the metric and its inverse, make the momentric representation intractable. The problem of ordering an operator such as the Cotton-York tensor in the momentric representation, with the additional complication the momentric operators (unlike momentum) do not commute among themselves, practically rule out the momentric representation of the Hamiltonian.

5.5.3   *Two gravitational degrees of freedom*

Group theoretical considerations also lead to a succinct description of graviton d.o.f. and their associated quantum excitations. In geometrodynamics, local $SL(3,R)$ transformations of $\bar{q}_{kl}$ are generated through $U(\alpha)$ as in Eq. (5.11) and Eq. (5.12). Spatial diffeomorphisms for the momentric and unimodular spatial d.o.f. on the other hand are generated by

$$\mathcal{H}_i = -2\nabla_j \bar{\pi}_i^j. \tag{5.31}$$

Integrated with smearing function ξ^i, this works out to be

$$\int \xi^i \mathcal{H}_i d^3x = \int (2\nabla_j \xi^i) \bar{\pi}_i^j d^3x \qquad (5.32)$$

after integration by parts. The action of spatial diffeomorphisms can thus be subsumed into $SL(3, R)$ by specialization of α_j^i in (5.12) to

$$\alpha_j^i = 2\nabla_j \xi^i, \qquad (5.33)$$

with the upshot that $SL(3, R)$ transformations which are *not* spatial diffeomorphisms are parametrized by α_j^i complement to $2\nabla_j \xi^i$. Given a background metric

$$q_{ij}^{\mathcal{B}} = q^{\frac{1}{3}} \bar{q}_{ij}^{\mathcal{B}}, \qquad (5.34)$$

this complement is precisely characterized by the choice of transverse traceless (TT) parameter

$$(\alpha_{TT})_j^i := (q^{\mathcal{B}})^{jk} \alpha_{(kj)}^{Phys}, \qquad (5.35)$$

because the condition

$$\nabla_{\mathcal{B}}^j (\alpha_{TT})_j^i = 0 \qquad (5.36)$$

excludes non-trivial ξ^i through

$$\nabla_{\mathcal{B}}^2 \xi^i = 0 \qquad (5.37)$$

if $(\alpha_{TT})_j^i$ were of the form $2\nabla_j^{\mathcal{B}} \xi^i$. This imposes four restrictions on the symmetric α_{ij}^{Phys} (tracelessness $(q^{\mathcal{B}})^{ij} \alpha_{(ij)}^{Phys} = 0$, and transversality $\nabla_{\mathcal{B}}^j \alpha_{(ji)}^{Phys} = 0$), leaving exactly two free parameters.

The above analysis is in completely agreement with the unique decomposition of an infinitesimal change of the metric into mutually orthogonal physical and gauge changes. With respect to any q_{ij},

$$\delta \bar{q}_{ij} = \alpha_{(ij)}^{Phys} + \mathcal{L}_\xi \bar{q}_{ij} \qquad (5.38)$$

wherein

$$\mathcal{L}_\xi \bar{q}_{ij} = \nabla_i \xi_j + \nabla_j \xi_i - \frac{2}{3} \bar{q}_{ij} \nabla_k \xi^k$$

$$= \left\{ \bar{q}_{ij}, \int \xi^i \mathcal{H}_i d^3x \right\}_{P.B.}, \qquad (5.39)$$

$$\nabla^i \alpha_{(ij)}^{Phys} = 0, \quad \text{and} \quad q^{ij} \alpha_{(ij)}^{Phys} = 0. \qquad (5.40)$$

The orthogonality $\int \bar{G}^{ijkl} \alpha_{(ij)}^{Phys} (\mathcal{L}_\xi \bar{q}_{kl}) \sqrt{q} d^3x = 0$ holds for all $\xi_l := q_{kl} \xi^k$ after integration by parts, by virtue of the TT nature of $\alpha_{(ij)}^{Phys}$. The study of these gravitational excitations will be taken up in the next chapter.

Chapter 6

Gravitational waves in Intrinsic Time Geometrodynamics

"...'experts' were confused.
That is what comes from looking for conserved energy tensors,
etc. instead of asking 'can the waves do work?'"
—Richard Feynman

"If you want to find the secrets of the universe, think in terms
of energy, frequency and vibration."
—Nikola Tesla

6.1 Introductory remarks

The Hulse-Taylor binary pulsar system yielded the first observational, albeit indirect, evidence of the existence of gravitational waves [84]. Measurements of orbital changes since the system's discovery are in precise agreement with the loss of energy due to emitted gravitational waves described by Einstein's theory [85]. Predicated upon instantaneous action-at-a-distance and the associated Poisson equation, Newton's Law of Gravitation is unable to produce these waves. Einstein and Rosen initially came to the conclusion gravitational waves could not exist in General Relativity because any such solution would not be singularity-free. The work concluded that the solutions were not genuine radiation capable of transmitting energy or having measurable physical effects. Upset by rejection, Einstein withdrew the manuscript of 1936, *Do gravitational waves exists?* He was never to submit again to *The Physical Review*. Convinced later by his assistant, Leopold Infeld, that the anonymous referee (who was actually H. P. Robertson) was indeed correct to assert the caustics in question were merely due to coordinate choices, a thorough reworking by Einstein resulted in a revision supporting the physical nature of what would later came to be known as Einstein-Rosen gravitational waves. Re-titled as *On gravitational waves* the new version was published in *The Journal of the Franklin Institute* in 1937 [86]. Rosen however remained skeptical, and through pseudotensor constructions argued the energy of the gravitational waves can be made to vanish at every point. Feynman, with his impeccable physical intuition and characteristic flair, posed the "sticky bead argument" at the Chapel Hill *Conference on the Role of Gravitation in Physics* in 1957. The thought experiment imagined an antenna with two beads which can slide on a rod, albeit not completely free to do so due to some "stickiness". A passing gravitational wave would cause the proper distance between the beads to oscillate, the beads would then rub against the rod, thus generating friction and heat; the energy must surely come from the gravitational radiation. In his letter to Victor Weisskopf, Feynman recalls the Chapel Hill meeting, with discerning comments quoted at the beginning of this chapter.

In recent years LIGO and Virgo detectors have reported, in agreement with Einstein's theory, the spectacular detection and confirmation of the gravitational wave forms emitted from black hole mergers, as well as from black hole-neutron star and binary neutron star sources [45]. It is fair to assert by now that any respectable theory of gravity must be able to produce gravitational waves, and in the domain in which GR holds sway, these perturbations must agree with Einstein's theory. ITG has a non-vanishing and spatial diffeomorphism-invariant reduced Hamiltonian generating intrinsic time development in our expanding closed universe. Equipped with a physical Hamiltonian, it resolves 'the problem of time' and bridges the deep divide between usual classical and quantum mechanics and conventional canonical formulations of gravity. Gravitational waves will thus carry energy at each point. The form and nature of gravitational waves in ITG warrant further study and discussion.

6.2   Gravitational waves in our spatially closed expanding universe

In ITG, the central tenet is that gravitational interactions are governed by a physical Hamiltonian [16, 26, 72, 87] which dictates time evolution with respect to the expansion of the universe. ITG advocates a spatially closed expanding universe. With strong evidence for positive cosmological constant and accelerated expansion in our universe [7], the usual exposition of gravitational waves in Minkowski background needs to be supplanted. Motivated by cosmological observations, the preeminent substitute is to consider perturbations in spatially closed $k = +1$ expanding Lorentzian de Sitter universe (it is conformally flat both spatially and also as a four-dimensional manifold of vanishing Weyl curvature). Gravitational waves as perturbations in de Sitter spacetime for various topologies have been discussed before.[1] This study of gravitational waves in compact $k = +1$ cosmological de Sitter spacetime is in contradistinction to, and complements, previous $k = -1$ investigations and other more familiar $k = 0$ works. In addition, possible non-four-covariant Horava gravity contribution, specifically, a Cotton-York term, is taken into consideration; hence the use of canonical Hamiltonian, rather than Lagrangian, methods. The perturbations are expanded in terms of eigenfunctions of the Laplacian operator which has a discrete spectrum in a spatially closed universe; and explicit S^3 transverse-traceless modes and their spectrum [51] are deployed to complete the discussion.

A main result of the Hamiltonian provided by the approach is that three dimensional spatial diffeomorphism invariance, in conjunction with its clean separation of physical degrees of freedom, admits a natural decomposition in terms of discrete modes of transverse-traceless tensor spherical harmonics [51] neatly capturing the two degrees of freedom of gravitational fluctuations [26]. It will be demonstrated that linearization of Hamilton's equations about the spatially compact de Sitter solution produces transverse traceless excitations, with the physics of gravitational waves in Einstein's GR recovered in the low curvature low frequency limit. A noteworthy feature is that, unlike descriptions which rely on boundary Hamiltonians, even for closed spatial slicings, ITG is equipped with a physical Hamiltonian, and gravitational waves carry *local positive energy density*.

6.3   Einstein's General Relativity as a special case of a wider class of "non-projectable" Horava gravity theories

The physical Hamiltonian H_{Phys} which generates evolution with respect to intrinsic time T has been elucidated earlier in Chapter 2. It takes the

---

[1]The case of $k = -1$ for GR has been considered previously in Ref. [88]; compact $k = +1$ case without non-covariant Cotton-York contribution and without explicit $Y^{Klm}_{(4,5)ij}$ transverse traceless modes has also been discussed in Ref. [89] (this was submitted to arXiv later than the work which was subsequently published as Ref. [52]); Refs. [90,91] also considered gravitational waves in modified gravity theories with higher curvature contributions.

simple form,

$$H_{\mathrm{Phys}} = \frac{1}{\beta} \int \bar{H}(x) d^3 x, \tag{6.1}$$

with a local Hamiltonian of density weight one[2]

$$\bar{H} = \sqrt{\bar{\pi}^i_j \bar{\pi}^j_i + \alpha q(R - 2\Lambda) + g^2 \tilde{C}^{ij} \tilde{C}_{ij}}, \tag{6.2}$$

and

$$\alpha = -\frac{1}{(2\kappa)^2}. \tag{6.3}$$

The Cotton-York tensor density (of weight one) is denoted by $\tilde{C}^{ij}$ and g (dimensionless) is a coupling constant of the theory. It follows that H_{Phys} is invariant under spatial diffeomorphisms generated by the momentum constraint, $H_i = -2\nabla_j \bar{\pi}^j_i = 0$. H_{Phys} is not a local constraint, but a true non-vanishing Hamiltonian which generates physical evolution of the variables $(\bar{q}_{ij}, \bar{\pi}^i_j)$ with respect to the change

$$\delta T = \frac{2}{3} \delta \ln V, \tag{6.4}$$

wherein V is the spatial volume of our universe [16, 72, 87]. This is a very physical description of dynamics which resolves the 'problem of time' in GR and its extensions, and renders them amenable to the usual rules of classical and quantum dynamics. However, unlike many Horava gravity theories [31] with an extra ambient time parameter, here T is constructed from the intrinsic geometry of the three-metric. A degree of freedom $(\ln q/q_0, \tilde{\pi})$ has been used to solve the Hamiltonian constraint and to provide a clock: the determinant of the metric, q obeys the Heisenberg equation of motion,

$$\frac{d \ln q^{1/3}}{dT} = 1; \tag{6.5}$$

and the trace of the momentum, $\tilde{\pi}^i_i$. The latter is totally absent in H_{Phys}. Thus, despite of not having a local Hamiltonian constraint, only two degrees of freedom,

$$(\bar{q}_{ij}, \bar{\pi}^i_j) \quad \text{with} \quad H_i = 0 \tag{6.6}$$

are subjected to fluctuations, whereas in the "projectable" Horava gravity theories there exists an extra, and possibly pathological mode because the

---

[2]As will be elaborated on in the next chapters, the Hamiltonian density is the square root of a positive semidefinite, self-adjoint operator that governs the evolution of the theory; and Einstein's Ricci scalar potential and cosmological constant term emerge after regularization and renormalization of a coincident commutator term [92].

Hamiltonian constraint is not solved. H_{Phys} being an integral, rather than a local constraint, also ensures that addition of higher spatial curvature terms to this Hamiltonian does not lead to intractable further constraints or inconsistencies in the constraint algebra. Einstein's GR is the limit $\beta^2 = 1/6$ and $g = 0$, i.e. when the potential term in $\bar{H}$ reduces to just the spatial Ricci scalar and cosmological constant terms. Without four-covariance and arbitrary *a priori* lapse function, N, Einstein's theory is recaptured in the sense that H_{Phys} produces an effective or emergent lapse function. The EOM and constraints of GR lead to precisely this same *a posteriori* value of N;[3] and the square-root form of the Hamiltonian in (6.2) is needed for this agreement [16, 72, 87]. The presence of higher *spatial* curvature terms needed for improved ultra-violet convergence and completion of the theory signal the explicit loss of 4-covariance. A corresponding Lagrangian of the Baierlein-Sharp-Wheeler type was written down in Eq. (2.95), but it is rather cumbersome to work with, and not really needed. The Hamiltonian description of mechanics is both complete and consistent. The use of canonical Hamiltonian, rather than Lagrangian methods, is more suitable for taking into account possible non-four-covariant contribution such as the Cotton-York tensor density term. This also provides a sound canonical prescription for the study of gravitational waves in Horava gravity theories maintaining two gravitational degrees of freedom. A comparison of intrinsic time geometrodynamics with other approaches such as York extrinsic time and scalar field time can be found in the remarks section of Ref. [26]. Fermionic matter and Yang-Mills fields can also be incorporated into the ITG formalism (see, for instance, comments relating to the gauging of local Lorentz invariance in Ref. [70]); and modifications to the Hamiltonian and momentum constraints and their solutions are explicitly addressed in Ref. [26]. In ITG, adoption of intrinsic time in an ever-expanding closed universe is fundamental to the framework of our physical universe and the resolution of the problem of time in classical and quantum GR.

6.4   Hamilton's equations for ITG

The classical gravitational wave equation will be derived from the Hamiltonian of ITG. To wit, the Heisenberg equations of motion for the unimodular

---

[3]For any ADM decomposition of the metric, ITG leads to $N = \frac{\sqrt{q}(\partial_t \ln q^{1/3} - \frac{2}{3}\nabla_i N^i)}{4\beta\kappa\bar{H}}$. This in fact ensures the Hamiltonian constraint is satisfied classically in the form $\frac{(TrK)^2}{9} = \frac{4\kappa^2\beta^2}{q}\bar{H}^2$. Further details on the equivalence between GR and H_{Phys}^{ADM} had already been fully discussed in Chapter 2.

metric variable and momentric,

$$\frac{\partial \bar{q}_{ij}(x)}{\partial T} = \frac{1}{i\hbar}[\bar{q}_{ij}(x), H_{Phys}]; \tag{6.7}$$

$$\frac{\partial \bar{\pi}_l^k(x)}{\partial T} = \frac{1}{i\hbar}[\bar{\pi}_l^k(x), H_{Phys}], \tag{6.8}$$

lead,[4] in the classical limit of setting $\hbar \to 0$, to

$$\frac{\partial \bar{q}_{ij}(x)}{\partial T} = \frac{1}{\beta \bar{H}(x)} \bar{E}^l_{kij}(x) \bar{\pi}_l^k(x), \tag{6.9}$$

and

$$\frac{\partial \bar{\pi}_l^k(x)}{\partial T} = -\frac{1}{\beta} \bar{E}^k_{lij}(x) \cdot$$
$$\int \frac{1}{\bar{H}(x')} \left[\frac{\alpha q(x')}{2} \frac{\delta R(x')}{\delta \bar{q}_{ij}(x)} + \tilde{C}_{mn} \frac{\delta \tilde{C}^{mn}}{\delta \bar{q}_{ij}(x)} \right] d^3x'. \tag{6.10}$$

In the above, we have used the fact that traceless part of the momentric commutes with the kinetic operator, $\bar{\pi}_n^m \bar{\pi}_m^n$ of H_{Phys}, which is a Casimir invariant of the $sl(3, R)$ algebra generated by $\bar{\pi}_j^i$. Following the derivations in [52],

$$\frac{\partial \bar{\pi}_l^k(x)}{\partial T} = -\frac{\alpha}{2\beta} q^{1/3}(x) \int \frac{q(x')}{\bar{H}(x')} \bar{E}^k_{lij}(x) \cdot$$
$$\left[-R^{ij}(x') + \nabla_{x'}^i \nabla_{x'}^j - q^{ij}(x') \nabla^2 \right] \delta^3(x, x') d^3x'$$
$$- \frac{1}{\beta} \bar{E}^k_{lij}(x) \int \frac{1}{\bar{H}(x')} \tilde{C}_{mn} \frac{\delta \tilde{C}^{mn}}{\delta \bar{q}_{ij}(x)} d^3x'. \tag{6.11}$$

Besides projecting out the traceless part of the Ricci tensor,

$$\bar{R}_l^k = R_l^k - \frac{1}{3} \delta_l^k R, \tag{6.12}$$

the traceless projector, $\bar{E}^k_{lij}$, also commutes with ∇^2. Upon integration, the result is

$$\frac{\partial \bar{\pi}_l^k(x)}{\partial T} = \frac{\alpha q}{2\beta \bar{H}} \bar{R}_l^k - \frac{\alpha}{2\beta} q^{4/3} \bar{E}^k_{lij} \nabla^i \nabla^j \frac{1}{\bar{H}}$$
$$- \frac{1}{\beta} \bar{E}^k_{lij}(x) \int \frac{1}{\bar{H}(x')} \tilde{C}_{mn} \frac{\delta \tilde{C}^{mn}}{\delta \bar{q}_{ij}(x)} d^3x'. \tag{6.13}$$

---

[4]An explicit metric representation of the momentric operator is $\bar{\pi}_j^i = \frac{\hbar}{i} \bar{E}^i_{jmn} \frac{\delta}{\delta \bar{q}_{mn}}$.

6.5   Gravitational waves in de Sitter background spacetime

6.5.1   *Background solution and linearization*

We shall consider background solutions of the Hamilton equations with constant spatial 3-curvature geometries compatible with the Cosmological Principle. Explicitly, these Friedmann-Lemaitre-Robertson-Walker three-metrics are

$$d\ell^2 = a^2(T) \left[\frac{dr^2}{1 - kr^2} + r^2(d\theta^2 + \sin^2\theta d\phi^2) \right] \tag{6.14}$$

with $k = +1, 0, -1$ respectively for the compact S^3, and non-compact $\mathbb{R}^3$ and H^3 intrinsic 3-geometries. The metric is Einstein,

$$R_{ij} = \frac{1}{3}q_{ij}R \quad \text{with} \quad R = \frac{6k}{a^2}; \tag{6.15}$$

hence $\bar{R}_{ij}$ and the Cotton-York tensor $\tilde{C}_{ij}$ both vanish. In addition,

$$\bar{q}_{ij} = q^{-1/3}q_{ij} \tag{6.16}$$

is independent of a (and thus of T). For the background extrinsic geometry, we set the traceless momentric variable $\bar{\pi}^i_j$ to zero. These considerations lead to spatially covariantly constant Hamiltonian density

$$\bar{H} = \sqrt{\alpha q(R - 2\Lambda)}, \tag{6.17}$$

and each term in Eq. (6.13) vanishes, resulting in

$$\dot{\bar{\pi}}^i_j = 0. \tag{6.18}$$

Thus, the pair of Hamilton equations Eq. (6.9) and Eq. (6.10) are identically satisfied, the initial data is preserved; and the Friedmann-Lemaitre-Robertson-Walker three-geometry with vanishing momentric indeed constitutes a background solution of the theory, even when H_{Phys} contains higher curvature terms $\tilde{C}^{mn}\tilde{C}_{mn}$ in addition to the scalar potential of Einstein's GR. The constant spatial curvature solution with vanishing momentric is also a saddle point of the exact vacuum solution in the Cotton-York era [71].

It is noteworthy that $\bar{H}$ involves only the square of $\tilde{C}^{mn}$ (which is identically zero for any constant 3-curvature metric). Thus we may assert a simple fact.

Theorem[5]: *Any spatially conformally flat classical solution of GR is also a solution of the EOM of the Hamiltonian of Intrinsic Time Geometrodynamics.*

In the ADM decomposition of any four-dimensional classical solution with coordinate time variable t, the lapse function takes the form

$$N = \frac{\sqrt{q}\,\partial_t \ln q^{1/3}}{4\beta\kappa\bar{H}} \tag{6.19}$$

modulo spatial diffeomorphisms [16, 87]. We can therefore recast the background solution into the usual four-dimensional Friedmann-Lemaitre-Robertson-Walker form by reparametrizing the cosmic time interval as

$$dt' := N dt = \frac{(\partial_t \ln a)dt}{\sqrt{6}\beta\sqrt{\frac{\Lambda}{3} - \frac{k}{a^2}}}. \tag{6.20}$$

To wit,

$$ds^2 = -dt'^2 + a^2(t')\left[\frac{dr^2}{1 - kr^2} + r^2(d\theta^2 + \sin^2\theta d\phi^2)\right], \tag{6.21}$$

and the above relation between a and t' reduces to

$$\frac{da}{dt'} = \sqrt{6}\beta\sqrt{\frac{\Lambda}{3}a^2 - k}. \tag{6.22}$$

This yields, for the GR value of $\beta = \sqrt{\frac{1}{6}}$, the de Sitter solution with

$$a(t') = \begin{cases} \sqrt{\frac{3}{\Lambda}}\cosh\left[\sqrt{\frac{\Lambda}{3}}(t' - t_0')\right], & \text{for } k = +1, \\ Ae^{\sqrt{\frac{\Lambda}{3}}(t'-t_0')}, & \text{for } k = 0, \\ \sqrt{\frac{3}{\Lambda}}\sinh\left[\sqrt{\frac{\Lambda}{3}}(t' - t_0')\right] & \text{for } k = -1, \end{cases} \tag{6.23}$$

for the respective spatial slicings.

We shall linearize the Hamilton equations about the de Sitter background $({}^*\bar{q}_{ij}, {}^*\bar{\pi}^i_j)$ with $k = +1, S^3$ slicings, and expand the variables as

$$\bar{q}_{ij} = {}^*\bar{q}_{ij} + \bar{h}_{ij} \tag{6.24}$$

and

$$\bar{\pi}^i_j = {}^*\bar{\pi}^i_j + \Delta\bar{\pi}^i_j. \tag{6.25}$$

In addition, on account of spatial diffeomorphism symmetry, we require the physical fluctuations to be transverse i.e.

$${}^*\nabla^i\bar{h}_{ij} = 0; \tag{6.26}$$

---

[5]See also Ref. [67].

and $\bar{h}_{ij}$, being perturbations of the unimodular $\bar{q}_{ij}$, are traceless

$$^*q^{ij}\bar{h}_{ij} = 0 \tag{6.27}$$

as well. The * prefix denotes the background value.

Differentiating with respect to T, the Euler-Lagrange equation for the metric fluctuation,

$$\frac{\partial^2 \bar{q}_{ij}}{\partial T^2} = \frac{\partial}{\partial T}\left(\frac{1}{\beta\bar{H}}\bar{E}^l_{kij}\right)\bar{\pi}^k_l + \frac{1}{\beta\bar{H}}\bar{E}^l_{kij}\frac{\partial\bar{\pi}^k_l}{\partial T}, \tag{6.28}$$

yields the linearized identity

$$\frac{\partial^2 \bar{h}_{ij}}{\partial T^2} =^* \left[\frac{\partial}{\partial T}\left(\frac{1}{\beta\bar{H}}\bar{E}^l_{k(ij)}\right)\right]\Delta\bar{\pi}^k_l + \Delta\left(\frac{\partial}{\partial T}\left(\frac{1}{\beta\bar{H}}\bar{E}^l_{kij}\right)\right)^*\bar{\pi}^k_l$$
$$+^*\Delta\left(\frac{1}{\beta\bar{H}}\bar{E}^l_{kij}\right)^*\left(\frac{\partial\bar{\pi}^k_l}{\partial T}\right) +^*\left(\frac{1}{\beta\bar{H}}\bar{E}^l_{kij}\right)\frac{\partial(\Delta\bar{\pi}^k_l)}{\partial T}. \tag{6.29}$$

On account of vanishing background value of $\bar{\pi}^i_j$ and its time derivative, only the first and last terms remain. Linearization of Eq. (6.9) yields

$$(\beta^*\bar{H})\frac{\partial\bar{h}_{ij}(x)}{\partial T} =^* \bar{E}^l_{kij}\Delta\bar{\pi}^k_l, \tag{6.30}$$

and substituting for $^*\bar{E}^l_{kij}\Delta\bar{\pi}^k_l$ in the first term of Eq. (6.29) leads to the EOM for $\bar{h}_{ij}$ which is

$$\frac{\partial^2 \bar{h}_{ij}}{\partial T^2} = -\frac{\partial\ln{}^*\bar{H}}{\partial T}\left(\frac{\partial\bar{h}_{ij}}{\partial T}\right) +^*\left(\frac{1}{\beta\bar{H}}\bar{E}^l_{kij}\right)\frac{\partial(\Delta\bar{\pi}^k_l)}{\partial T}. \tag{6.31}$$

6.5.2 *Gravitational wave equation*

Explicit calculations lead to

$$\frac{\partial\ln{}^*\bar{H}}{\partial T} = \frac{(R - 3\Lambda)}{(R - 2\Lambda)}, \tag{6.32}$$

and

$$\frac{\partial(\Delta\bar{\pi}^k_l)}{\partial T} = -\frac{\alpha q}{4\beta^*\bar{H}}q^{km}\left(\nabla^2 - \frac{1}{3}R\right)h_{ml} + \dots. \tag{6.33}$$

wherein by ... we mean the higher curvature contribution arising from $\tilde{C}^{mn}\tilde{C}_{mn}$. This shall be addressed later. The background Hamiltonian density with zero momentric, $^*\bar{H}$, is covariantly constant, and

$$\frac{\alpha q}{^*\bar{H}^2} = \frac{1}{R - 2\Lambda}, \tag{6.34}$$

with

$$R = \frac{6k}{a^2}. \tag{6.35}$$

Henceforth we drop the $*$ label when there is no confusion, and it is understood that, apart from the perturbations $(\bar{h}_{ij}, \tilde{\pi}^i_j)$, all other metric entities refer to the de Sitter background.

With the use of the above identities, Eq. (6.31) gives the resultant gravitational wave equation on de Sitter background, and expressed with respect to intrinsic time T, as

$$\frac{\partial^2 \bar{h}_{ij}}{\partial T^2} + \frac{(R - 3\Lambda)}{(R - 2\Lambda)} \frac{\partial \bar{h}_{ij}}{\partial T} + \frac{1}{4\beta^2 (R - 2\Lambda)} \left(\nabla^2 - \frac{R}{3} \right) \bar{h}_{ij} + \cdots = 0 \tag{6.36}$$

for physical, transverse traceless perturbations $\bar{h}_{ij}$. Without the higher order curvature terms, this reproduces the gravitational wave equation for GR on a de Sitter background for $\beta^2 = 1/6$. The factor,

$$(R - 2\Lambda) = 6 \left(\frac{k}{a^2} - \frac{\Lambda}{3} \right) \tag{6.37}$$

vanishes only at the de Sitter "throat" at $t' = t'_0$ for $k = +1$, but this factor is otherwise always negative regardless of whether $k = +1, 0, -1$. Likewise, the coefficient $\frac{(R-3\Lambda)}{(R-2\Lambda)}$ is positive definite, and the $\frac{\partial \bar{h}_{ij}}{\partial T}$ term plays the role of frictional force, tempered by the expansion of the universe. Bearing in mind

$$\frac{da}{dt'} = \sqrt{6}\beta \sqrt{\frac{\Lambda}{3} a^2 - k}, \tag{6.38}$$

with $R = \frac{6k}{a^2}$ and

$$dT = 2d \ln a(t'), \tag{6.39}$$

conversion of Eq. (6.36) into variation w.r.t. t' yields

$$\left(\frac{1}{24\beta^2 (\frac{\Lambda}{3} - \frac{k}{a^2})} \frac{\partial^2}{\partial t'^2} - \frac{1}{24\beta^2 (\frac{\Lambda}{3} - \frac{k}{a^2})} \nabla^2 \right) \bar{h}_{ij} + \ldots = 0 \tag{6.40}$$

which implies (as the coefficients of the time and spatial derivative terms are the same) the speed of the GR wave is 1 (in units c $= 1$ since we have previously used the notation $-dt'^2$ instead of $-c^2 dt'^2$ in the de Sitter metric[6]).

---

[6]Note that intrinsic time T is truly dimensionless while cosmic time coordinate t or t' carries the dimension of length.

6.5.3 Higher curvature Cotton-York contribution

The higher curvature contribution to the wave equation which arises from Eq. (6.13) is

$$-\frac{1}{\beta^2}\frac{1}{*\bar{H}(x)}*(\bar{E}^l_{kij}\bar{E}^k_{luv})(x)\int\frac{1}{*\bar{H}(x')}\left(\Delta\tilde{C}_{mn}(x')*\frac{\delta\tilde{C}^{mn}(x')}{\delta\bar{q}_{uv}(x)}\right)d^3x',$$

(6.41)

and it can be explicitly computed as in [52]. This may in turn be expressed as

$$-\int\frac{q^{-\frac{1}{3}}}{\beta\bar{H}(x')}\Delta\tilde{C}_{mn}(x')\mathcal{O}^{mn}{}_{ij}\delta^3(x,x')d^3x',$$

(6.42)

with

$$\mathcal{O}^{ijkl}=-\frac{1}{\beta\bar{H}}\left[\frac{g}{8}(q^{ik}\epsilon^{jlm}+q^{jk}\epsilon^{ilm}+q^{il}\epsilon^{jkm}+q^{jl}\epsilon^{ikm})\nabla_m\right]\left(\nabla^2-\frac{R}{3}\right),$$

(6.43)

for the background. Apart from the extra $\frac{1}{q^{\frac{1}{3}}\beta\bar{H}}$ factor, $\mathcal{O}^{ijkl}\delta^3(x,x')=\frac{1}{q^{\frac{1}{3}}\beta\bar{H}}\frac{\delta\tilde{C}^{ij}(x')}{\delta\bar{q}_{kl}(x)}$ is essentially the Hessian in the second variation of the Chern-Simons functional with respect to the spatial metric. The significance of the relation between the Chern-Simons functional, the Cotton-York tensor, and the Hessian operator will be elucidated in Chapter 10.

The resultant wave equation with higher curvature contribution is therefore

$$\frac{\partial^2\bar{h}_{ij}}{\partial T^2}+\frac{(R-3\Lambda)}{(R-2\Lambda)}\frac{\partial\bar{h}_{ij}}{\partial T}+\frac{1}{4\beta^2(R-2\Lambda)}\left(\nabla^2-\frac{R}{3}\right)\bar{h}_{ij}+\mathcal{O}^{\dagger mn}{}_{ij}\mathcal{O}_{mn}{}^{kl}\bar{h}_{kl}=0,$$

(6.44)

wherein explicit the higher curvature term is

$$\mathcal{O}^{\dagger mn}{}_{ij}\mathcal{O}_{mn}{}^{kl}\bar{h}_{kl}=\frac{q}{\beta^2\bar{H}^2}\left(-\frac{g^2}{4}\left(\nabla^2-\frac{R}{2}\right)\right)\left(\nabla^2-\frac{R}{3}\right)^2\bar{h}_{ij}.$$

(6.45)

At this level of approximation, we may assume $dT=\frac{2}{3}d\ln V=2d\ln a$, or

$$e^{T-T_{now}}=\left(\frac{a}{a_{now}}\right)^2=(1+z)^{-2}.$$

(6.46)

Thus the equation may be expressed equivalently in terms of T, a, or z-development. While all $k=0,\pm1$ are valid descriptions in this work wherein $dT=2d\ln a$, in general intrinsic time geometrodynamics advocates $dT=\frac{2}{3}d\ln V$ and favors a spatially closed universe.

6.6 S^3 tensor harmonics and physical excitations

It is known that there are six independent mutually orthogonal symmetric second rank tensor harmonics on S^3. For the Friedmann-Lemaitre-Robertson-Walker metric with closed S^3 spatial slicing,

$$ds^2 = -dt^2 + a^2(t)[d\chi^2 + \sin^2(\chi)(d\theta^2 + \sin^2\theta d\phi^2)], \qquad (6.47)$$

these pertinent modes are denoted, following Ref. [51], by $Y^{Klm}_{(A)ij}$ with $A = 0, 1, 2, 3, 4, 5$. They are associated with a set of discrete eigenvalues $\{Klm\}$, and they can be explicitly expressed as

$$Y^{Klm}_{(0)ij} = \frac{1}{\sqrt{3}} Y^{Klm} q_{ij}, \qquad (6.48)$$

with scalar Laplacian eigenfunctions $\nabla^2 Y^{Klm} = -\frac{K(K+2)}{a^2(t)} Y^{Klm}$;

$$Y^{Klm}_{(1)ij} = \frac{a(t)}{\sqrt{2(K-1)(K+3)}} (\nabla_i Y^{Klm}_{(1)j} + \nabla_j Y^{Klm}_{(1)i}), \qquad (6.49)$$

$$Y^{Klm}_{(2)ij} = \frac{a(t)}{\sqrt{2(K-1)(K+3)}} (\nabla_i Y^{Klm}_{(2)j} + \nabla_j Y^{Klm}_{(2)i}), \qquad (6.50)$$

with $\quad Y^{Klm}_{(1)i} \quad := \quad \dfrac{a^2(t)}{\sqrt{l(l+1)}} \epsilon_i{}^{pq} \nabla_p Y^{Klm} \nabla_q \cos\chi, \quad Y^{Klm}_{(2)i} \quad :=$

$\dfrac{a(t)}{(K+1)} \epsilon_i{}^{pq} \nabla_p Y^{Klm}_{(1)q}$, and $\nabla^i Y^{Klm}_{(1,2)i} = 0$;

$$Y^{Klm}_{(3)ij} = \frac{\sqrt{3}a(t)}{\sqrt{2(K-1)(K+3)}} \left(\nabla_i Y^{Klm}_{(0)j} + \frac{\sqrt{K(K+2)}}{\sqrt{3}r} Y^{Klm}_{(0)ij} \right); \quad (6.51)$$

$$Y^{Klm}_{(4)ij} = a(t) \sqrt{\frac{(l-1)(l+2)}{2K(K+2)}} \left\{ \frac{1}{2} E^{Kl} (\nabla_i F^{lm}_j + \nabla_j F^{lm}_i) \right. \qquad (6.52)$$

$$\left. + \csc^2\chi \left[\frac{1}{2}(l-1)\cos\chi E^{Kl} + C^{Kl} \right] (F^{lm}_i \nabla_j \cos\chi + F^{lm}_j \nabla_i \cos\chi) \right\};$$

$$Y^{Klm}_{(5)ij} = \frac{a(t)}{2(K+1)} (\epsilon_i{}^{pq} \nabla_p Y^{Klm}_{(4)qj} + \epsilon_j{}^{pq} \nabla_p Y^{Klm}_{(4)qi}); \qquad (6.53)$$

with $C^{Kl} := \dfrac{Y^{Klm}}{Y^{lm}(\theta, \phi) \sin^l \chi}$ wherein $Y^{lm}(\theta, \phi)$ are spherical harmonics of S^2,

$$E^{Kl} := -\frac{2\csc^{l+1}\chi}{(l-1)(l+2)} \frac{d}{d\chi} (\sin^{l+2}\chi C^{Kl}), \text{ and}$$

$$F^{lm}_i := \frac{a^2(t)}{\sqrt{l(l+1)}} \epsilon_i{}^{pq} \nabla_p (\sin^l \chi Y^{lm}) \nabla_q \cos\chi.$$

By construction, the modes $Y_{(4,5)ij}^{Klm}$ are transverse and traceless i.e. $\nabla^i Y_{(4,5)ij}^{Klm} = 0$ and $q^{ij} Y_{(4,5)ij}^{Klm} = 0$. In addition, all the modes satisfy orthogonality conditions

$$\frac{1}{a^3} \int q^{ip} q^{jq} Y_{(A)ij}^{Klm} Y_{(B)pq}^{*K'l'm'} \sqrt{q} \, d^3x = \delta_{(A)(B)} \delta^{KK'} \delta^{ll'} \delta^{mm'}. \tag{6.54}$$

The Lie derivative of the unimodular metric is of the form

$$\mathcal{L}_\xi \bar{q}_{ij} = (\mathcal{L}_\xi q^{\frac{1}{3}}) q_{ij} + q^{\frac{1}{3}} (\mathcal{L}_\xi q_{ij})$$

$$= q^{-\frac{1}{3}} \left(\nabla_i \xi_j + \nabla_j \xi_i - \frac{2}{3} (\nabla^a \xi_a) q_{ij} \right) \tag{6.55}$$

$$= q^{-\frac{1}{3}} (\mathcal{L}_\xi q_{ij} - \frac{1}{3} tr[\mathcal{L}_\xi q_{pq}] q_{ij}).$$

This constitutes gauge variation under spatial diffeomorphisms. $Y_{(0)ij}^{Klm}$ is just a trace term of no relevance to traceless changes in $\bar{q}_{ij}$; while the other three modes (as in Eq. (6.49), Eq. (6.50) and Eq. (6.51)) correspond essentially to non-physical Lie derivative changes since

$$Y_{(1)ij}^{Klm} \propto \mathcal{L}_{\vec{Y}_{(1)}^{Klm}} q_{ij}, \text{ and } \nabla^i Y_{(1)i}^{Klm} = 0;$$

$$Y_{(2)ij}^{Klm} \propto \mathcal{L}_{\vec{Y}_{(2)}^{Klm}} q_{ij}, \text{ and } \nabla^i Y_{(1)i}^{Klm} = 0; \tag{6.56}$$

$$Y_{(3)ij}^{Klm} \propto \left(\mathcal{L}_{\vec{Y}_{(0)}^{Klm}} q_{ij} - \frac{1}{3} q_{ij} tr \left[\mathcal{L}_{\vec{Y}_{(0)}^{Klm}} q_{pq} \right] \right).$$

Thus $q^{-\frac{1}{3}} Y_{(1,2,3)ij}^{Klm}$ are all of the form $\mathcal{L}_{\vec{\xi}_{(1,2,3)}} \bar{q}_{ij}$ or mere diffeomorphism gauge transformations of $\bar{q}_{ij}$, whereas

$$\bar{Y}_{(I=4,5)ij}^{Klm} := q^{-\frac{1}{3}} Y_{(I=4,5)ij}^{Klm} \text{ with } K \geq 2, \tag{6.57}$$

span the transverse traceless space of physical changes orthogonal to diffeomorphisms. So the physical excitations $\bar{h}_{ij}$ can be expanded in terms of these explicit S^3 harmonics.[7] They are also eigenfunctions of Laplacian operator, ∇^2, of S^3 with (negative) eigenvalues

$$-(\lambda_K)^2 = -\frac{K(K+2) - 2}{a^2}. \tag{6.58}$$

A similar expansion can be carried out for the transverse traceless $\Delta \bar{\pi}_b^a$. Through the expansion

$$\bar{h}_{ij} = \sum_{I=4,5} C_{Klm}^{(I)} \bar{Y}_{(I)ij}^{Klm}, \tag{6.59}$$

---

[7]Note that $\bar{Y}_{(A=4,5)ij}^{Klm}$ are independent of the background scale factor $a(t)$ or intrinsic time T. The eigenvalues of the Laplacian operator are not affected by multiplication with any power of q which is covariantly constant.

and orthogonality of the eigenfunctions, Eq. (6.44) reduces to just the equation for intrinsic time-dependence for the mode coefficients which are associated with the discrete eigenvalues $\{Klm\}$. As they are independent, $C_{Klm}^{(4)}$ and $C_{Klm}^{(5)}$ must each satisfy the same equation. The resultant equation encodes full-fledged information of all the time development of the physical modes which can arise from gravitational perturbations during different epochs of the expanding de Sitter universe. It reduces to

$$\ddot{C}_{\{K\}} + \frac{(R - 3\Lambda)}{(R - 2\Lambda)}\dot{C}_{\{K\}} + \frac{E_K}{4\beta^2(R - 2\Lambda)}C_{\{K\}} - \frac{g^2(E_K - \frac{R}{6})}{4\beta^2\alpha(R - 2\Lambda)}E_K^2 C_{\{K\}} = 0;$$

$$(6.60)$$

wherein, for simplicity, we have denoted $C_{\{K\}} := C_{Klm}^{(I=4,5)}$, and have defined

$$E_K := -(\lambda_K)^2 - \frac{R}{3} = -\frac{K(K+2)R}{6}, \qquad (6.61)$$

which is the eigenvalue of $\nabla^2 - \frac{R}{3}$; and $R = \frac{6}{a^2}$ is the relation between the Ricci scalar and Robertson-Walker scale factor.

6.7 Time development of transverse traceless modes

The solution of (6.60) has already been tackled in Ref. [90]. The analysis will be discussed in this section. To wit, by introducing the variable x so that

$$x^2 := \frac{a(t)^2}{\gamma^2} - 1, \quad \gamma := \sqrt{\frac{3}{\Lambda}} \Rightarrow dT = d\ln(1 + x^2), \qquad (6.62)$$

and in terms of the intrinsic time variable T,

$$\frac{d}{dT} = \frac{1 + x^2}{2x}\frac{d}{dx} = \frac{1}{2}\left(x + \frac{1}{x}\right)\frac{d}{dx}$$

$$\frac{d^2}{dT^2} = \frac{1}{4}\left[\left(\frac{1 + x^2}{x}\right)^2\frac{d^2}{dx^2} + \left(x - \frac{1}{x^3}\right)\frac{d}{dx}\right]. \qquad (6.63)$$

The upshot is (6.60) leads to the master equation for the time development of the mode functions to assume the form

$$\frac{d}{dx}(1+x^2)^2\frac{d}{dx}C_{\{K\}} + \frac{K(K+2)}{6\beta^2}C_{\{K\}} + \frac{1}{6\beta^2}\left(\frac{\eta K(K+1)(K+2)}{1+x^2}\right)^2 C_{\{K\}} = 0.$$

$$(6.64)$$

where, $\eta = 16\pi l_p^2/\gamma^2 g$ (which is $\approx 5 \times 10^{-121}g$, with the present observational value of the cosmological constant, $\frac{l_p}{\gamma} \approx 10^{-61}$. In the early universe, the effective cosmological constant could have been much higher, say, of Planck scale, in value), and g is the dimensionless coupling constant of the Cotton-York contribution. Thus except near the Big Bang or in the very early universe, we can safely neglect contributions from the Cotton-York term.

6.7.1 Einstein's theory and gravitational waves in expanding closed de Sitter universe

Setting the Cotton-York term to zero, and with $\beta^2 = \frac{1}{6}$, the mode equation of gravitational waves for Einstein's GR in the de Sitter Robertson-Walker background is given by,

$$\frac{d}{dx}(1+x^2)^2\frac{d}{dx}C_{\{K\}} + K(K+2)C_{\{K\}} = 0, \qquad K \geq 2. \qquad (6.65)$$

Defining $\dfrac{c_K}{\sqrt{1+x^2}} := C_K := C_{\{K\}}$ and using just K to represent $\{K\}$ for simplicity, the time development of the mode equation for c_K can be simplified to,

$$\frac{d}{dx}(1+x^2)\frac{d}{dx}c_K - \left[2 - \frac{(K+1)^2}{1+x^2}\right]c_K = 0 \qquad (6.66)$$

which can be compared with the *associated Legendre equation*,

$$\frac{d}{dx}(1-x^2)\frac{d}{dx}c_K + \left[l(l+1) - \frac{m^2}{1-x^2}\right]c_K = 0. \qquad (6.67)$$

Our equation of interest is therefore an associated Legendre equation, but with $l = 1$, $m = K + 1$, and with the variable $x \to ix$. For real variables, Rodrigues' formula,

$$P_l^m = \frac{(-1)^m}{2^l l!}(1-x^2)^{m/2}\frac{d^{l+m}}{dx^{l+m}}(x^2-1)^l,$$

implies vanishing values for $m = K + 1 \geq 3 > l = 1$. However, our variable of relevance is imaginary; this implies that analytic continuation of the hypergeometric functions should be adopted, and moreover, associated Legendre polynomials of the second kind (which blow up at real $x = \pm 1$) should not *a priori* be dropped as physical solutions.

Expressed in the hypergeometric functions, analytic continuation to complex values yields generalised Legendre polynomial of the first kind as

$$P_\lambda^\mu = \frac{1}{\Gamma(1-\mu)}\left[\frac{1+z}{1-z}\right]^{\mu/2}{}_2F_1\left(-\lambda, \lambda+1, 1-\mu, \frac{1-z}{2}\right);$$

$$_2F_1(\alpha, \beta, \gamma, z) = \frac{\Gamma(\gamma)}{\Gamma(\alpha)\Gamma(\beta)}\sum_{n=0}^{\infty}\frac{\Gamma(n+\alpha)\Gamma(n+\beta)}{\Gamma(n+\gamma)n!}z^n.$$

Substituting $\lambda = 1$, $\mu = K + 1$, $z = ix$, gives

$$P_1^{K+1} = \frac{1}{\Gamma(-1)\Gamma(2)} \left[\frac{1+ix}{1-ix}\right]^{(K+1)/2} \sum_{n=0}^{\infty} \frac{\Gamma(n-1)\Gamma(n+2)}{\Gamma(n+K)n!} \left(\frac{1-ix}{2}\right)^n$$

$$x =: tan\theta \Rightarrow$$

$$\left[\frac{1+ix}{1-ix}\right]^{(K+1)/2} = \left[\frac{cos\theta + isin\theta}{cos\theta - isin\theta}\right]^{(K+1)/2} = e^{i(K+1)\theta}$$

$$\left(\frac{1-ix}{2}\right)^n = \left(\frac{cos\theta - isin\theta}{2cos\theta}\right)^n = \left(\frac{e^{-i\theta}}{2cos\theta}\right)^n$$

$$\left| cos\theta = \frac{1}{\sqrt{1+x^2}} \right.$$

$$= \left(\frac{\sqrt{1+x^2}e^{-i\theta}}{2}\right)^n$$

$$= \frac{e^{i(K+1)\theta}}{\Gamma(-1)\Gamma(2)} \sum_{n=0}^{\infty} \frac{\Gamma(n-1)\Gamma(n+2)}{\Gamma(n+K)n!} \left(\frac{\sqrt{1+x^2}e^{-i\theta}}{2}\right)^n.$$

The above expression has positive powers of n in the sum, and blows up at $x \to \infty$. Although c_K may diverge at $x \to \infty$, we are interested in $C_K = c_K/\sqrt{1+x^2}$ with a nontrivial denominator. To obtain the complete solution to Eq. (6.66), we proceed to Legendre polynomials of the second kind. For real variable x, they are traditionally expressed as

$$Q_n^m(x) = (1-x^2)^{m/2}\frac{d^m}{dx^m}Q_n(x);$$

$$Q_n(x) = \begin{cases} Q_0(x) = \frac{1}{2}\ln\left[\frac{1+x}{1-x}\right], \\ Q_1(x) = xQ_0(x) - 1, \\ Q_{n\geq2}(x) = \frac{2n-1}{n}xQ_{n-1}(x) - \frac{n-1}{n}Q_{n-2}(x). \end{cases} \quad (6.68)$$

It is easy to observe that Q_0 diverges at $x = \pm1$, but not necessarily for our pure imaginary variable that is relevant. Ignoring normalization factors in Eq. (6.68), continuation of x to ix yields Legendre functions of second kind,

$$Q_1(ix) = \frac{ix}{2} \ln\left[\frac{1+ix}{1-ix}\right] - 1$$

$$x =: \tan\theta$$

$$\left| \Rightarrow \ln\left[\frac{1+ix}{1-ix}\right] = \ln\left[\frac{1+i\tan\theta}{1-i\tan\theta}\right] \right. \tag{6.69}$$

$$= i2\tan^{-1}x$$

$$= -(1 + x\tan^{-1}x).$$

Moreover, $i\tan^{-1}x = Q_0(ix)$, for $Q_1(ix)$ in $Q_1^m(ix)$, and we arrive at the new functions (denoted by $\tilde{\ })$

$$\tilde{Q}_1^{K+1}(x) = (1+x^2)^{(k+1)/2}\frac{d^{K+1}}{dx^{K+1}}(x\tan^{-1}x)$$

$$\left| \frac{d^{K+1}}{dx^{K+1}}(x\tan^{-1}x) \sim \frac{x^{K-1}}{(1+x^2)^{K+1}} + ... \right. \tag{6.70}$$

$$= \frac{q_K}{(1+x^2)^{(K+1)/2}};$$

wherein q_K is a polynomial whose highest power in x is $K-1$. And the relevant solution for C_K becomes

$$\frac{\tilde{Q}_1^{K+1}(x)}{\sqrt{1+x^2}} = \frac{q_{K-1}}{(1+x^2)^{1+K/2}}. \tag{6.71}$$

Noting that there is a factor of $\frac{1}{\sqrt{1+x^2}}$ difference between C_K and c_K; we shall express $C_k := \frac{y}{(1+x^2)^{1+k/2}}$ and perform a power series expansion in $y(x) = \sum\limits_{n=0}^{\infty} a_n x^n$. A truncation condition of the power series solution can then be obtained and through the properties of the above Legendre polynomials; and the two independent solutions of Eq. (6.65) can then be arrived at [90]. To wit, the general solution is

$$C_K(x) = c_1\frac{\tilde{P}_1^{K+1}(x)}{\sqrt{1+x^2}} + c_2\frac{\tilde{Q}_1^{K+1}(x)}{\sqrt{1+x^2}}$$

$$= c_1\frac{p_K(x)}{(1+x^2)^{1+K/2}} + c_2\frac{q_K(x)}{(1+x^2)^{1+K/2}} \tag{6.72}$$

$$=: c_1 P_K(x) + c_2 Q_K(x);$$

wherein p_K and q_K are polynomials are essentially determined by the recurrence relations of their coefficients.

p_K is a polynomial of order $(K + 2)$, with recurrence relation

$$a_{n+2} = -\frac{(K - n + 2)(K - n - 1)}{(n + 2)(n + 1)}a_n,$$

q_K is a polynomial of order $(K - 1)$, with the same recurrence relation. Also, p_K has the same odd and even behaviour as the value of K, while q_K has odd and even behaviour opposite to the value of K. From the recurrence relations, and for p_K and q_K to correspond to associated Legendre polynomials of the first and second kind respectively we may set a_0 for p_K and a_1 for q_K to equal $(-1)^{K/2+1}$ when K is even (and a_0 for q_K and $a_1 = (-1)^{(K+1)/2+1}$ for p_K when K is odd). The first few polynomials are then

$$p_2(x) = 1 - 2x^2 - \frac{1}{3}x^4,$$

$$p_3(x) = -x + \frac{2}{3}x^3 + \frac{1}{15}x^5,$$

$$p_4(x) = -1 + 9x^2 - 3x^4 - \frac{1}{5}x^6,$$

$$q_2(x) = x,$$

$$q_3(x) = -1 + 5x^2,$$

$$q_4(x) = -x + \frac{5}{3}x^3.$$

With $p_K(x)$ and $q_K(x)$ as above, the first few $P_K(x)$ and $Q_K(x)$ are

$$P_2(x) = \frac{1 - 2x^2 - \frac{1}{3}x^4}{(1 + x^2)^2},$$

$$P_3(x) = \frac{-x + \frac{2}{3}x^3 + \frac{1}{15}x^5}{(1 + x^2)^{5/2}},$$

$$P_4(x) = \frac{-1 + 9x^2 - 3x^4 - \frac{1}{5}x^6}{(1 + x^2)^3},$$

$$Q_2(x) = \frac{x}{(1 + x^2)^2},$$

$$Q_3(x) = \frac{-1 + 5x^2}{(1 + x^2)^{5/2}},$$

$$Q_4(x) = \frac{-x + \frac{5}{3}x^3}{(1 + x^2)^3}.$$

The large x or large T intrinsic time limits are,

$$P_K(x) = \frac{p_K(x)}{(1+x^2)^{1+K/2}}$$

$$\xrightarrow{x \ large} \sim \frac{x^{K+2}}{x^{K+2}} \longrightarrow const,$$

$$Q_K(x) = \frac{q_K(x)}{(1+x^2)^{1+K/2}} \tag{6.73}$$

$$\xrightarrow{x \ large} \sim \frac{x^{K-1}}{x^{K+2}} \sim \frac{1}{x^3} \longrightarrow 0.$$

It is interesting to observe that the $P_K(x)$ solution does not decay completely and remains non-vanishing even when $x \to \infty$ (note that we are dealing with boundary effect in the time development domain, not the spatial boundary effect which is not applicable to a spatially closed universe. All spatial dependence are already accounted for by the expansion in of $\bar{h}_{ij}$ terms of $\bar{Y}^{Klm}_{(4,5)ij}$). This is an explicit indication of memory effects in gravitational waves which may lend itself to be tested.

6.8 Gravitational waves in the Cotton-York era

In the very early universe, we expect the Cotton-York term to dominate over the Einstein contribution. Keeping, for concreteness, $\beta^2 = \frac{1}{6}$ in Eq. (6.64), leads to

$$\frac{d}{dx}(1+x^2)^2 \frac{d}{dx}C_K + K(K+2)C_K + \left(\frac{\eta K(K+1)(K+2)}{1+x^2}\right)^2 C_K = 0. \tag{6.74}$$

To compare the effects of Cotton-York term versus Einstein's GR, we may define the ratio,

$$\zeta(k,\eta,x) := \frac{\eta^2 K(K+1)^2(K+2)}{(1+x^2)^2} \tag{6.75}$$

which parametrizes the relative importance of the 3 variables. Cotton-York contributions dominate over pure GR whenever $\zeta \gg 1$; and Einstein's theory holds whenever $\zeta \to 0$.

We itemize below the behavior of the mode functions in different domains of $\zeta(k,\eta,x)$:

- $\zeta \to 0$

  The domain $\zeta \to 0$ can be obtained by taking large x and/or small η and non-extreme values of K. The Cotton-York term has a $\dfrac{1}{(1+x^2)^2}$ dependence; so for small η and/or large x the Cotton-York contributions can be ignored (unless extremely energetic k modes will be involved). This situation reduces to the wave equation for standard Einstein GR era. In the limit of large x, the equation becomes,

$$\frac{d}{dx}x^4\frac{d}{dx}C_K + K(K+2)C_K \sim 0$$

$$C_K =: \frac{y}{x} \Rightarrow \frac{d}{dx}x^2\frac{d}{dx}y - 2y = 0 \tag{6.76}$$

$$\Rightarrow C_K \sim c_1 + \frac{c_2}{x^3}.$$

This is just the asymptotic behaviour of the general linear combination of P_k and Q_k solutions discussed in the earlier in Sec. 6.7.1.

- Large ζ

  There will be three different domains in the large ζ limit.

  (i) K is large enough but x is not very large

  If $K^2 \gg \dfrac{1}{\eta}$ $(\eta = 16\pi(\dfrac{l_p}{\gamma})^2 g)$, we can neglect the K^2 term, but keeping x not exceedingly large; the solutions are oscillatory and do not decay,

$$C_K = c_1 \cos\left[\frac{K^3\eta(x + \tan^{-1}x + x^2\tan^{-1}x)}{2(1+x^2)}\right]$$
$$+c_2 \sin\left[\frac{K^3\eta(x + \tan^{-1}x + x^2\tan^{-1}x)}{2(1+x^2)}\right].$$

  The K^3 dependence is the salient characteristic of the Cotton-York dominated era.

  (ii) Both K and x are large

  This is for highly energetic modes at late times. If K and x are both large, and $\dfrac{\eta^2 K^6}{x^4} \gg K^2 \Rightarrow \eta^2 K^4 \gg x^4$, we obtain,

$$\frac{d}{dx}x^4\frac{d}{dx}C_K + \frac{\eta^2 K^6}{x^4}C_K = 0.$$

$$\Rightarrow C_K = c_1 \cos\left[\frac{K^3\eta}{3x^3}\right] + c_2 \sin\left[\frac{K^3\eta}{3x^3}\right].$$

  (iii) ζ is not extremely small but x is small

  This will be the early universe if $x \ll 1$. By redefining $C_K =: \dfrac{y}{1+x^2}$, we arrive at,

$$y''(x) + \left[-\frac{2}{1+x^2} + \frac{K(K+2)}{(1+x^2)^2} + \frac{(\eta K(K+1)(K+2))^2}{(1+x^2)^4}\right]y(x) = 0,$$

$$\stackrel{x\ll 1}{\Longrightarrow} y''(x) + [K(K+2) - 2 + (\eta K(K+1)(K+2))^2]y(x) = 0.$$

Because $K \geq 2$, $K(K+2) - 2 + (\eta K(K+1)(K+2))^2 \geq 0$, the solution becomes

$$\Rightarrow C_K \simeq c_1 \cos\left(\sqrt{K(K+2) - 2 + (\eta K(K+1)(K+2))^2}\, x\right)$$
$$+ c_2 \sin\left(\sqrt{K(K+2) - 2 + (\eta K(K+1)(K+2))^2}\, x\right).$$

The leading K dependence in the sin and cos functions is $\sim K^3$, which again is the characteristic effect of Cotton-York dominance. If K is large the solution will become

$$C_K \simeq c_1 \cos(K^3 x) + c_2 \sin(K^3 x).$$

6.8.1 *Summary on time dependence of gravitational waves in expanding $k = +1$ de Sitter universe*

The gravitation wave mode equations and solutions may be summarized with the following observations:

- Without the Cotton-York term
  For each value of K, the mode function $C_K(x)$ behaves exactly as the associated Legendre polynomial (with variable ix) divided by $\sqrt{1 + x^2}$. Solutions of the usual second kind which diverge at $x = \pm 1$ for real x are in fact non-singular and allowed one to perform an analytic continuation to ix.
- At late times, the Cotton-York term will drop out ($\zeta \to 0$ as $x \to \infty$), and the two independent solutions of the second-order-in-time wave equation are such that $Q_K \sim \dfrac{1}{x^3} \to 0$, while $P_k \to constant$. So only P_K solutions remain for large times with asymptotic constant values. This is the time domain where Einstein's GR holds sway. This implies either
  a) a memory effect captured in the P_K modes which do not decay away completely and can be observationally verified, or
  b) observations at late times may rule out the P_K modes simply because they are barred by, say, Dirichlet boundary condition such that $P_K \to 0$ as $x \to \infty$. Further studies on how much energy will be involved and how these modes are produced may reveal whether they are indeed physically viable.
- In the oscillatory era, which happens at early times, ($\zeta \gg 1$ as $x \to 0$), the Cotton-York contribution is significant and it leads to a K^3

dependence which is the characteristic salient feature of the Cotton-York era. The Cotton-York contribution also makes the oscillatory era lasts longer than if there were no such an addition.

- Regardless of the late or early times, contributions of the Cotton-York modification are always revealed by the very high K modes. At extremely high K values (leading to ζ (very large K, η, x) $\gg 1$), the Cotton-York addition will overcome the Einstein term. These extremely high K modes features characteristic K^3 dependence, unlike the Einstein modes which are precisely governed by the associated Legendre polynomial dependence on x.

Matter can be included and added to the square root Hamiltonian of Eq. (6.2). Adaptation of the preceding de Sitter background to a generic FLRW background can be performed. To wit,

- The usual $\rho_{matter}(t) \propto a(t)^{-3(1+\omega_I)}$ wherein $\omega_I = 0$ for non-relativistic matter and $\omega_I = 1/3$ for relativistic matter (more general equations of state for exotic matter can also be adopted if desired) can be appended inside the square-root of Eq. (6.2).
- $\rho_{matter}(t)$ will change the T-dependent coefficients in Eq. (6.60), but the result will still be a second order ordinary differential equation in x.
- The solution will no longer be the associated Legendre polynomials, but series solutions can still be found by working out the corresponding recurrence relations.

6.9   Energy of gravitational perturbations, and further remarks

Intrinsic time geometrodynamics, as in Horava-type gravity theories, introduces only higher order spatial, but not time, derivatives into the wave equation through higher order spatial curvature terms which improve the ultra-violet convergence of the theory without compromising unitarity; yet the theory captures the physics of Einstein's GR in long wavelength low curvature circumstances. From the wave equation, we can also see that the propagator for flat background will contain additional terms (up to the highest order of $1/p^6$ in momentum dependence from the square of Cotton-York tensor in H_{Phys}), but there will be no additional poles for p_0 in the absence of higher time derivatives. In the modified wave equation, the ratio of the higher curvature contribution to that of Einstein's GR is explicitly

$$-\frac{[g^2(E_K - \frac{R}{6})]E_K}{\alpha} = (16\pi)^2[g^2(K+1)^2]K(K+2)\left(\frac{l_{\text{Planck}}}{a}\right)^4, \quad (6.77)$$

which can be computed for any given set of K, a and g. In the current epoch, this ratio too small to be of significance at LIGO's characteristic detection wavelengths. However, departures from Einstein's theory can become significant in the regime of large curvatures in the early universe and/or for large values of K. In the era of $a \to 0$, all physics is dominated by the Cotton-York term, the de Sitter solution is a saddle point of the exact vacuum state [71] (this topic will be further discussed in Chapter 10); and, instead of Einstein's GR, Eq. (6.60) will be dominated by the last term associated with linearized excitations in the Cotton-York era.

There is no contradiction in having both *physical local energy density* and *spatial* diffeomorphism invariance. Without the paradigm of four-covariance, the total Hamiltonian density is not required to vanish; at each point, two physical d.o.f. remain even after spatial diffeomorphisms are taken into account. In perturbative excitations, the remaining physical d.o.f. are precisely the transverse traceless $(\bar{h}_{ij}, \Delta\bar{\pi}^i_j)$ modes. A noteworthy feature of intrinsic time geometrodynamics in which time change is identified with variation in the logarithm of (finite) spatial volume is that gravitational waves always carry physical positive energy *density*, even for compact spatial slicing without any energy contribution from boundary Hamiltonian. The energy for the gravitational wave excitation is

$$H_{\text{Phys}}[\bar{\pi}^i_j, q_{ij}] - H_{\text{Phys}}[^*\bar{\pi}^i_j, {}^*q_{ij}]$$
$$\approx \int \frac{1}{2\beta^*\bar{H}}\left[\Delta\bar{\pi}^m_n\Delta\bar{\pi}^n_m + \frac{\alpha q}{4}\bar{q}^{ik}\bar{q}^{jl}\bar{h}_{ij}\left(\nabla^2 - \frac{R}{3}\right)\bar{h}_{kl}\right.$$
$$\left. + (\beta\bar{H})^2(q^{\frac{1}{3}}\mathcal{O}_{mn}{}^{kl}\bar{h}_{kl})(q^{\frac{1}{3}}\mathcal{O}^{mnij}\bar{h}_{ij})\right]d^3x. \quad (6.78)$$

The expression is positive-definite for flat and compact spatial topologies since $\alpha < 0$, but the $R = -6/a^2$ term is negative for open spatial topologies. Again by expanding in eigenmodes of the Laplacian operator, the expression of the Hamiltonian density can be computed explicitly in terms of the mode coefficients. In the classical theory, the momentric is related to the time change of the metric via Eq. (6.30). The energy density of the excitations obeys the dispersion relation dictated by (6.78).

Chapter 7

Intrinsic time gravity, heat kernel regularization, and emergence of Einstein's theory

"Space and time are the framework within which the mind is constrained to construct its experience of reality."
—Immanuel Kant

"God does not care about our mathematical difficulties. He integrates empirically."
—Albert Einstein

7.1 Overview and motivations

The physical Hamiltonian of ITG dictates dynamical evolution with respect to the fractional change in the spatial volume of the expanding Universe. The energy density is in the relativistic form of a square-root; this chapter elucidates its properties and addresses the extensions and interactions that can be incorporated within the framework of ITG. Gravitational interactions are introduced by extending Klauder's momentric variable with similarity transformations, and explicit spatial diffeomorphism invariance is enforced via similarity transformation with exponentials of spatial integrals. It is well-known that the potential term in the Yang-Mills Hamiltonian is the trace of the square of magnetic field $\tilde{B}^{ia}$ which is the functional derivative of the Yang-Mills Chern-Simons functional w.r.t. the gauge potential. In analogy with Yang-Mills theory, a Cotton-York term is obtained from the Chern-Simons functional of the affine connection; this belies the naive paradigm of spatial curvature in GR as the counterpart to Yang-Mills magnetic field when it is the Cotton-York tensor density $\tilde{C}^i_j$ that plays that role. The essential difference is the fundamental variable for geometrodynamics is the metric rather than a gauge connection; in the case of Yang-Mills, there is also no analog of the integral of the spatial Ricci scalar curvature. It is also noteworthy that this Ricci scalar potential of Einstein's theory emerges in ITG as the regularized coincident limit of a commutator, analagous to the zero point energy in a simple quantum harmonic oscillator. Heat kernel regularization is employed to isolate the divergences of coincidence limits; and a prescription in which Einstein's Ricci scalar potential emerges naturally from the positive-definite self-adjoint Hamiltonian of the theory is demonstrated.

7.2   Hamiltonian of ITG

Quantum geometrodynamics with first-order Schrödinger evolution in intrinsic time, $i\hbar\frac{d\Psi}{dT} = H_{\text{Phys}}.\Psi$, has been advocated in earlier chapters and also in a series of articles [26, 52, 72, 87]. In an ever-expanding spatially closed universe, $dT = \frac{2}{3}d\ln V$ (wherein V is the spatial volume of the universe which is assumed to be spatially compact [8]) serves as the preeminent and concrete physical time interval to discuss dynamics and evolution. A crucial difference between ITG and other projectable Horava gravity theories [31] (which lack a local Hamiltonian constraint and thus are plagued by a possible extra degree of freedom) is that in the latter, time is an "ambient" parameter; whereas in ITG a d.o.f. in geometrodynamics, namely dT, the gauge-invariant part of $\delta\ln q^{\frac{1}{3}}$, is used as an internal clock interval to deparametrize the theory. The Hamiltonian constraint, is exactly solved, as in Chapters 3 and 4, by eliminating the trace of the momentum in terms of the rest of the variables. Thus ITG makes use of the Hamiltonian constraint and its auxiliary condition.[1] to eliminate the one local degree of freedom contained in the pair $(\ln q^{\frac{1}{3}}, \tilde{\pi})$ to yield a reduced physical Hamiltonian,

$$H_{\text{Phys}} = \int \tilde{\pi}(x)d^3x = \frac{1}{\beta}\int \bar{H}(x)d^3x, \qquad (7.1)$$

which generates evolution in global cosmic intrinsic time. A full canonical analysis of the resultant two physical degrees of freedom and the reduced phase space, together with a discussion of quantising the system of second class Dirac constraints in terms of Dirac Brackets instead of first class constraints in terms of Poisson Brackets, and the associated Faddeev-Popov determinants and functional integral measures will be carried out in Chapter 8 [26].

---

[1] This will be addressed explicitly in Chapter 8 which will also compare and contrast the reduced Hamiltonians of scalar field time, York extrinsic time, and of intrinsic time.

Spatial diffeomorphism invariance is maintained provided the Hamiltonian density,

$$\bar{H}(x) = \sqrt{\bar{G}_{ijkl}\bar{\pi}^{ij}\bar{\pi}^{kl} + \mathcal{V}(q_{ij})}, \tag{7.2}$$

is a scalar density of weight one.[2] Horava's bold proposal [31] to eschew 4-covariance in favour of just spatial diffeomorphism symmetry and its advantages is adopted in intrinsic time geometrodynamics in a consistent modification and extension of Einstein's theory within this framework which yields a hitherto unexplored vista to overcome the many technical and conceptual obstacles in quantum gravity [9, 12]. It has been shown elsewhere [16, 72] and in earlier chapters that this Hamiltonian produces the equations of motion of Einstein's theory, with an emergent lapse function (as in Chapters 2, 3 and 4) which agrees with the a posteriori lapse function of GR when $\mathcal{V}$ is of the form of a Ricci scalar plus consmological constant term. The physical Hamiltonian yields a ordered time-development operator $U(T, T_0)$; and in an ever expanding universe, this time-ordering is well-defined and also spatially diffeomorphism invariant.

As extensions of the Hamiltonian constraint, matter and Yang-Mills interactions can be added under the square root to $\mathcal{V}$ as

$$\frac{1}{2\kappa}(H_{matter} + H_{YM}) \tag{7.3}$$

Ref. [26]; for instance, in the case of scalar field, we have the usual

$$H_\phi(x) = \frac{1}{2}(\tilde{\pi}_\phi\tilde{\pi}_\phi + qq^{ij}\nabla_i\phi\nabla_j\phi) + qV(\phi), \tag{7.4}$$

written as a tensor density of weight two, and H_{YM} will be brought up in the next section. A main objective of this chapter, which shall be elucidated below, is to demonstrate the emergence of the Einstein Ricci potential from a Hamiltonian density of the form

$$\bar{H} = \sqrt{Q_j^{\dagger i}Q_i^j + q\mathcal{K}}. \tag{7.5}$$

7.3 Momentric, and self-adjoint positive-definite Hamiltonian density

It is desirable to express the square-root form of the Hamiltonian density as a the square-root of a positive-definite and self-adjoint entity.[3] The traceless

---

[2]The square-root form of the Hamiltonian density is needed to reproduce the correct equations of motion [16]. Some discussions on the existence and uniqueness of the square root operators, for both bounded and unbounded cases, can be found in Ref. [93].

[3]Reference [94] provides an elementary proof that infinite dimensional positive-definite self-adjoint operators have complete sets of eigenfunctions.

momentum variable, $\bar{\pi}^{ij}$, cannot be easily implemented upon quantization as a self-adjoint traceless operator in the metric representation. Klauder's "momentric" variable [39], $\bar{\pi}^i_j$, with one contravariant and one covariant index, is an intriguing variable which, in lieu of utilizing the momenta, confers many advantages, as has been discussed in Chapter 5. One of which is the kinetic operator can be interpreted to be a Casimir invariant of $SL(3,R)$. Classically,

$$\bar{\pi}^i_j = q_{jk}\bar{\pi}^{ik};\tag{7.6}$$

consequently the kinetic term is expressible as

$$\bar{G}_{ijkl}\bar{\pi}^{ij}\bar{\pi}^{kl} = \bar{\pi}^i_j\bar{\pi}^j_i.\tag{7.7}$$

The fundamental commutation relations for the spatial metric and momentric have been addressed in detail in Chapter 5. Among other advantages, the quantum momentric operator and commutation relations can be explicitly realised in the metric representation on wave-functionals of the unimodular metric $\psi[\bar{q}] = \langle \bar{q}_{ij}|\psi\rangle$ by functional differentiation

$$\bar{\pi}^i_j(x)\psi[\bar{q}] = \frac{\hbar}{i}\bar{E}^i_{j(mn)}(x)\frac{\delta\psi[\bar{q}]}{\delta\bar{q}_{mn}(x)}$$

$$= \left[\frac{\hbar}{i}\frac{\delta}{\delta\bar{q}_{mn}(x)}\bar{E}^i_{j(mn)}(x)\right]\psi[\bar{q}] = \bar{\pi}^{\dagger i}_j(x)\psi[\bar{q}],\tag{7.8}$$

with $\bar{\pi}^i_j$ being Hermitian on account of

$$\left[\frac{\delta}{\delta\bar{q}_{mn}(x)}, \bar{E}^i_{j(mn)}(x)\right] = 0,\tag{7.9}$$

as well as traceless ($\bar{\pi}^i_i = 0$). Self-adjointness of $\bar{\pi}^i_j$ follows from two properties demonstrated above; namely that

- $\bar{\pi}^i_j$ is Hermitian; and secondly,
- given that $\psi[\bar{q}]$ is in the domain of $\bar{\pi}^i_j$ i.e. $\psi[\bar{q}] \in Dom(\bar{\pi}^i_j)$ implies that $\psi[\bar{q}] \in Dom(\bar{\pi}^{\dagger i}_j)$ also.

The commutation relations between the unimodular metric and momentric variables discussed in Chapter 5 imply that $\bar{\pi}^i_j$ generates infinitesimal $SL(3,R)$ transformations which exponentiates (with traceless α^i_j parameter) to

$$U^\dagger(\alpha)\bar{q}_{kl}(x)U(\alpha) = (e^{\frac{\alpha(x)}{2}})^m_k\bar{q}_{mn}(x)(e^{\frac{\alpha(x)}{2}})^n_l; \quad e^{\frac{\alpha(x)}{2}} \in SL(3,R);\tag{7.10}$$

with $U(\alpha) = e^{-\frac{i}{\hbar}\int \alpha^i_j\bar{\pi}^j_i d^3y}$. Thus the momentric generates transformations which preserve the positivity of the eigenvalues of the metric, whereas the momentum variable generates arbitrary translations of the spatial metric which will not in general uphold this positivity [39]. The commutation relations between the momentric operators generate an $sl(3, R)$ algebra at each spatial point. The kinetic operator is self-adjoint and the free theory with $\mathcal{V} = 0$, is defined by this Casimir invariant.

7.4 Introducing interactions with similarity transformations

The immediate advantage of starting with the momentric kinetic operator for the free theory and introducing interactions via similarity transformations of the momentric variable is an explicit positive definite Hamiltonian density. This method of bringing in the interactions is not only preserves the positivity and self-adjointness of $\bar{H}^2(x)$ but is also explicitly spatially diffeomorphism invariant. From a conventional point of view this is not an obvious route to adopt, especially when the Einstein Ricci scalar in the potential should be $-R$ to agree with GR; but the case of pure Yang-Mills theory demonstrates that it is not only without precedence, but also motivates the introduction of the Cotton-York term via the corresponding Chern-Simons functional. To this end, positivity of the Hamiltonian and explicit spatial diffeomorphism invariance can be achieved by defining the operator Q^i_j and its adjoint $Q^{\dagger i}_j$ as

$$Q^i_j := e^{W_T} \bar{\pi}^i_j e^{-W_T} = \frac{\hbar}{i} \bar{E}^i_{j(mn)} \left[\frac{\delta}{\delta \bar{q}_{mn}} - \frac{\delta W_T}{\delta \bar{q}_{mn}} \right], \qquad (7.11)$$

$$Q^{\dagger i}_j := e^{-W_T} \bar{\pi}^i_j e^{W_T} = \frac{\hbar}{i} \bar{E}^i_{j(mn)} \left[\frac{\delta}{\delta \bar{q}_{mn}} + \frac{\delta W_T}{\delta \bar{q}_{mn}} \right]; \qquad (7.12)$$

and constructing the Hamiltonian density as

$$\bar{H}(x) = \sqrt{Q^{\dagger i}_j(x) Q^j_i(x) + q(x)\mathcal{K}}, \qquad (7.13)$$

with positive coupling $\mathcal{K}$ for the bare cosmological constant allowed by the symmetries of the theory. Thus, interactions are introduced by extending the momentric with similarity transformations, and explicit spatial diffeomorphism invariance is enforced via similarity transformation with exponentials of spatial integrals, wherein

$$W_T = \alpha W_{EH} - g W_{CS} \qquad (7.14)$$

$$= \alpha \int \sqrt{q} R d^3 x - \frac{g}{2} \int \tilde{\epsilon}^{ikj} \left(\Gamma^l_{im} \partial_j \Gamma^m_{kl} + \frac{2}{3} \Gamma^l_{im} \Gamma^m_{jn} \Gamma^n_{kl} \right) d^3 x. \qquad (7.15)$$

Here W_{EH} is the spatial Einstein-Hilbert action with coupling constant of inverse length dimension, and W_{CS} the Chern-Simons functional of the affine connection, Γ^i_{jk}, of the spatial metric [95], with dimensionless coupling[4] constant g. With W_T, this yields[5]

$$Q^i_j = \tfrac{\hbar}{i}\bar{E}^i_{j(mn)}\tfrac{\delta}{\delta\bar{q}_{mn}} + i\alpha\hbar\sqrt{q}\overline{R}^i_j - ig\hbar\tilde{C}^i_j,$$

$$Q^{\dagger i}_{\;\;j} = \tfrac{\hbar}{i}\bar{E}^i_{j(mn)}\tfrac{\delta}{\delta\bar{q}_{mn}} - i\alpha\hbar\sqrt{q}\overline{R}^i_j + ig\hbar\tilde{C}^i_j; \tag{7.16}$$

wherein $\overline{R}^j_i = R^i_j - \tfrac{1}{3}\delta^i_j R$ is the traceless part of the spatial Ricci tensor; and the Cotton-York tensor density is expressible, remarkably, as $\tilde{C}^i_j = \bar{E}^i_{j(mn)}\tfrac{\delta W_{CS}}{\delta\bar{q}_{mn}}$. That it is traceless ($\tilde{C}^i_i = 0$), and transverse ($\nabla_i\tilde{C}^i_j = 0$) can be established from

$$\tilde{C}^{ij} = \frac{\delta W_{CS}}{\delta q_{ij}}$$

$$= \tilde{\epsilon}^{imn}\nabla_m\left(R^j_n - \frac{1}{4}R\delta^j_n\right)$$

$$= \frac{1}{2}(\tilde{\epsilon}^{imn}\nabla_m R^j_n + \tilde{\epsilon}^{jmn}\nabla_m R^i_n); \tag{7.17}$$

and the last equality holds because the antisymmetric part of the intermediate entity vanishes on account of the Bianchi identity,

$$\nabla_m\left(R^m_n - \frac{1}{2}R\delta^m_n\right) = 0. \tag{7.18}$$

In three dimensions, the manifold is conformally flat iff the Cotton-York tensor vanishes identically. More discussions on $\tilde{C}^i_j$ will be taken up in Chapter 10.

7.5 Analogy between Yang-Mills magnetic field and the Cotton-York tensor

It is also of relevance to note that the appearance of the Cotton-York term is entirely natural. It has the analogy of the magnetic field term in Yang-Mills Hamiltonian which can in similar fashion be introduced by extending the momentum conjugate to the gauge connection, $\hat{\tilde{\pi}}^{ia}_A$ by similarity transformation with the exponential of the Chern-Simons functional, $W_{CS}[A]$,

---

[4]Higher curvature invariants in W_T come with (undesirable) coupling constants of negative mass dimensions; they have also been discussed in Ref. [92].

[5]Due to the $\bar{E}^i_{j(mn)}$ projector only the traceless part of the Ricci tensor $\bar{R}^i_j$ survives in Eq. (7.15); and any volume term $\lambda\int\sqrt{q}d^3x$ in W_T commutes with $\bar{\pi}^i_j$.

of the Yang-Mills connection $A_a = A_{ia}dx^i$, so that

$$W_{CS}[A] := \frac{1}{2} \int \left(A^a dA_a + \frac{1}{3} f_{abc} A^a \wedge A^b \wedge A^c \right), \qquad (7.19)$$

$$\hat{Q}^{ia} = e^{W_{CS}[A]} \hat{\tilde{\pi}}_A^{ia} e^{-W_{CS}[A]} \qquad (7.20)$$

$$= \hat{\tilde{\pi}}_A^{ia} + i\hbar \tilde{B}^{ia}, \qquad (7.21)$$

and similarly for its adjoint $\hat{Q}^{ia\dagger}$. The second equality above follows from the conjugate momentum to the gauge potential and the magnetic field being, respectively,

$$\hat{\tilde{\pi}}_A^{ia} = \frac{\hbar}{i} \frac{\delta}{\delta A_{ia}}, \qquad \tilde{B}^{ia} = \frac{\delta W_{CS}[A]}{\delta A_{ia}}. \qquad (7.22)$$

In Yang-Mills theory, the Hamiltonian density is thus expressible as

$$H_{YM} = \frac{1}{2} q_{ij} (\hat{\tilde{\pi}}_A^{ia} \hat{\tilde{\pi}}_A^{ja} + \hbar^2 \tilde{B}^{ia} \tilde{B}^{ja}) \qquad (7.23)$$

$$= \frac{1}{2} q_{ij} \hat{Q}^{\dagger\,ia} \hat{Q}^{ja}, \qquad (7.24)$$

but

$$q_{ij}[\hat{\tilde{\pi}}_A^{ia}, \tilde{B}^{ja}] = 0, \qquad (7.25)$$

with the Yang-Mills squared electric field playing the role of the kinetic and the squared magnetic field playing the role of the potential term. As explained, the Cotton-York tensor density in Eq. (7.17) is the functional derivative with respect to the spatial metric of the Chern-Simons functional W_{CS} of the affine connection. The essential difference with Yang-Mills theory is that the fundamental variable for geometrodynamics is the metric rather than a gauge connection.

In addition, in the case of Yang-Mills theory, there is no analog of the spatial Einstein-Hilbert action term, $\int \sqrt{q} R d^3 x$, which can be added to W_T. The Einstein–Hilbert term W_{EH}, is the term which will eventually lead, through regularisation of the quantum commutator, to the Einstein Ricci scalar potential in the Hamiltonian density. Without the quantum correction from non-commutativity, within the context of the current formulation, the Einstein Ricci scalar will simply be absent from the theory. The appearance of the Einstein Ricci scalar potential hints at the importance of quantum zero point energy in shaping the ultimate dynamics of the cosmos.

To summarize, *instead of postulating "detailed balance" terms* as in Ref. [31], the interactions can be thought of as being introduced through

Q_j^i (via similarity transformation of the momentric) with exponential of the Chern-Simons functional associated with dimensionless coupling constant (as in Yang-Mills theory) and the additional spatial Einstein-Hilbert action available to geometrodynamics. It should also be remarked that introduction of interactions via similarity transformations, or conjugation, has been carried out before [96].

7.6 Regulating the quantum commutator

It follows from the above discussion and Eq. (7.16) that the Hamiltonian density can be put into an obvious positive definite form,

$$\bar{H} = \sqrt{Q_j^{\dagger i} Q_i^j + q\mathcal{K}} \tag{7.26}$$

$$= \sqrt{\bar{\pi}_i^{\dagger j} \bar{\pi}_j^i + \hbar^2 (\alpha \sqrt{q} \overline{R}_j^i - g\tilde{C}_j^i)(\alpha \sqrt{q} \overline{R}_i^j - g\tilde{C}_i^j) + i\alpha\hbar\sqrt{q}[\bar{\pi}_j^i, \overline{R}_i^j] + q\mathcal{K}}.$$

Moreover, it can be verified explicitly that

$$[\bar{\pi}_j^i, \tilde{C}_i^j] = 0 \tag{7.27}$$

(analogously $q_{ij}[\hat{\bar{\pi}}_A^{ia}, \tilde{B}^{ja}] = 0$), so there is no Cotton-York contribution in the commutator term in Eq. (7.26).

While $Q_j^{\dagger i}$ and Q_j^i are related to $\bar{\pi}_j^i$ by $e^{\mp W_T}$ similarity transformations, they are non-Hermitian and generate two unitarily inequivalent representations of the non-compact group $SL(3, R)$ at each spatial point; whereas the momentric

$$\bar{\pi}_j^i = \frac{1}{2}(Q_j^{\dagger i} + Q_j^i) \tag{7.28}$$

is self-adjoint and thus generates an infinity dimensional unitary representation of $\prod_x SL(3, R)_x$. In fact, the commutator term in $\bar{H}$ can be identified as

$$i\alpha\hbar\sqrt{q}[\bar{\pi}_j^i, \overline{R}_i^j] = \frac{1}{2}[Q^{\dagger i}_j, Q_i^j]. \tag{7.29}$$

In the functional Schrödinger representation, the commutator at coincident spatial position can be expressed as

$$\frac{1}{2}[Q^{\dagger i}_j(x), Q_i^j(x)] = \alpha\hbar^2 \lim_{x \to y} \sqrt{q(x)} \bar{E}^i_{j(kl)}(x) \frac{\delta \overline{R}_i^j(y)}{\delta \bar{q}_{kl}(x)}. \tag{7.30}$$

Using the following formula for the infinitesimal variation of the Ricci tensor and the following useful relation involving the projector

$$\delta R_{ik} = \frac{1}{2} q^{jl}(\nabla_j \nabla_i \delta q_{kl} + \nabla_j \nabla_k \delta q_{il} - \nabla_j \nabla_l \delta q_{ik} - \nabla_i \nabla_k \delta q_{jl}); \tag{7.31}$$

$$\frac{\partial \bar{q}_{ij}}{\partial \bar{q}_{kl}} = \frac{1}{2}(\delta_i^k \delta_j^l + \delta_j^k \delta_i^l) - \frac{1}{3} \bar{q}_{ij} \bar{q}^{kl} \tag{7.32}$$

wherein $\bar{q}^{kl}$ is the inverse unimodular metric, the point-split result is

$$\bar{E}^i_{j(kl)}(x)\frac{\delta \overline{R}^j_i(y)}{\delta \bar{q}_{kl}(x)} = -\left(\frac{5}{6}\nabla^2 + \frac{5}{3}R\right)\delta^3(x,y). \tag{7.33}$$

Thus

$$\frac{1}{2}[Q^{\dagger i}_j(x), Q^j_i(y)] = -c_1\alpha\hbar^2\sqrt{q(x)}(c_2\nabla^2_x + R)\delta^3(x,y); \tag{7.34}$$

with $c_1 = \frac{5}{3}$, $c_2 = \frac{1}{2}$. The $x \to y$ coincident limit is divergent, and requires regularization which will be addressed below.

7.7 Heat kernel regularization, and emergence of Einstein's theory

The heat kernel, $K(\epsilon; x, y)$ with

$$\lim_{\epsilon \to 0} K(\epsilon; x, y) = \delta^3(x, y), \tag{7.35}$$

presents the means to regularize the coincident limit in Eq. (7.34) for generic metrics. It satisfies the heat equation,

$$\nabla^2 K(\epsilon; x, y) = \frac{\partial K(\epsilon; x, y)}{\partial \epsilon}; \tag{7.36}$$

and in terms of Seeley-DeWitt coefficients a_n, $2\sigma(x, y)$ the square of the geodesic length, and Δ_V the Van Vleck determinant,

$$K(\epsilon; x, y) = (4\pi\epsilon)^{-\frac{3}{2}}\Delta_V^{\frac{1}{2}}(x, y)e^{-\frac{\sigma(x,y)}{2\epsilon}}\sqrt{q}\sum_{n=0}^{\infty} a_n(x, y; \nabla^2)\epsilon^n, \tag{7.37}$$

wherein ϵ is of dimension L^2. In the coincidence $x = y$ limit, the coefficients for closed manifolds are [97]

$$a_0 = 1, \ a_1 = \frac{R}{6}; \tag{7.38}$$

$$a_2 = \frac{1}{180}(R^{ijkl}R_{ijkl} - R^{ij}R_{ij}) + \frac{1}{30}\nabla^2 R + \frac{R^2}{72}; \dots \tag{7.39}$$

The coincidence limit of $\lim_{\epsilon \to 0} K(\epsilon; x, y) = \delta^3(x, y)$ is a regularization of $\delta^3(x, x)$, with

$$K(\epsilon; x, x) = \sqrt{q}(4\pi)^{-3/2}\left(a_0\epsilon^{-3/2} + a_1\epsilon^{-1/2} + a_2\epsilon^{1/2} + \dots\right) \tag{7.40}$$

and as regularization of $\nabla^2\delta^3(x, x)$ we may take the coincidence limit of the heat equation,

$$\nabla^2\delta^3(x, x) = \lim_{\epsilon \to 0}\left[\nabla^2 K(\epsilon; x, x) = \frac{\partial K(\epsilon; x, x)}{\partial \epsilon}\right]. \tag{7.41}$$

To wit, at the level before removal of the regulator we have

$$\frac{\partial K(\epsilon; x, x)}{\partial \epsilon} = \left(-\frac{3}{2\epsilon} + \frac{a_1 + 2a_2\epsilon + \dots}{\sum_0^\infty a_n \epsilon^n} \right) K(\epsilon; x, x)$$

$$= \sqrt{q}(4\pi)^{-\frac{3}{2}} \left(-\frac{3}{2}\epsilon^{-\frac{5}{2}} - \frac{a_1}{2}\epsilon^{-\frac{3}{2}} + \frac{a_2}{2}\epsilon^{-\frac{1}{2}} + \dots \right) \quad (7.42)$$

as a Laurent series. Substituting, as regularizations, $\lim_{\epsilon \to 0} K(\epsilon; x, x)$ for $\delta^3(x, x)$ and $\lim_{\epsilon \to 0} \frac{\partial K(\epsilon; x, x)}{\partial \epsilon}$ for $\nabla^2 \delta^3(x, x)$, the regularized coincident limit of Eq. (7.34) takes the form

$$\frac{1}{2}[Q^{\dagger i}{}_j(x), Q^j_i(x)] = \lim_{\epsilon \to 0} - \alpha c_1 \hbar^2 \sqrt{q} \left[c_2 \frac{\sqrt{q}}{(4\pi)^{\frac{3}{2}}} \left(-\frac{3}{2}\epsilon^{-\frac{5}{2}} - \frac{a_1}{2}\epsilon^{-\frac{3}{2}} \right. \right.$$

$$\left. + \frac{a_2}{2}\epsilon^{-\frac{1}{2}} + \dots \right) + \frac{\sqrt{q}}{(4\pi)^{\frac{3}{2}}}(\epsilon^{-\frac{3}{2}}a_0 + a_1\epsilon^{-\frac{1}{2}} + \dots)R \Big]$$

$$= \lim_{\epsilon \to 0} - \frac{\alpha c_1 \hbar^2 q}{(4\pi\epsilon)^{\frac{3}{2}}} \left[-\frac{3c_2}{2\epsilon} + \frac{(12 - c_2)R}{12} \right.$$

$$\left. + \left(a_1 R - c_2 \frac{a_2}{2} \right) \epsilon + \text{terms with } \epsilon^{n \geq 2} \right]; \quad (7.43)$$

wherein $\frac{12 - c_2}{12} R = a_0 R - \frac{c_2}{2} a_1$. This yields the Hamiltonian density

$$\bar{H} = \sqrt{Q^{\dagger i}_j Q^j_i + q\mathcal{K}}$$

$$= \lim_{\epsilon \to 0} \left\{ \bar{\pi}^{\dagger j}_i \bar{\pi}^i_j + \hbar^2 (\alpha \sqrt{q}\bar{R}^i_j - g\tilde{C}^i_j)(\alpha \sqrt{q}\bar{R}^j_i - g\tilde{C}^j_i) + q \left(\mathcal{K} \right. \right.$$

$$\left. \left. + \frac{3\alpha c_1 c_2 \hbar^2}{16\pi^{\frac{3}{2}}\epsilon^{\frac{5}{2}}} \right) - \frac{\alpha c_1 \hbar^2 q}{(4\pi\epsilon)^{\frac{3}{2}}} \left(\frac{12 - c_2}{12} R + (a_1 R - c_2 \frac{a_2}{2})\epsilon \right) \right\}^{\frac{1}{2}}. \quad (7.44)$$

7.8 Renormalising the coupling constants

7.8.1 *Newtonian limit fixes the renormalised coupling phenomenologically*

In the theory there are three fundamental coupling constants: $g, \mathcal{K}$ and α. In the limit of regulator removal with $\epsilon \to 0$, the divergent $\epsilon^{-\frac{3}{2}}$ term can be countered by α. To yield the correct Newtonian limit, the renormalized finite value is identified phenomenologically, with

$$\frac{1}{(2\kappa)^2} = \frac{\alpha c_1 \hbar^2}{(4\pi\epsilon)^{\frac{3}{2}}} \left(\frac{12 - c_2}{12} \right) \longrightarrow \alpha = \frac{12(4\pi\epsilon)^{\frac{3}{2}}}{c_1 \hbar^2 (12 - c_2)(2\kappa)^2} \quad (7.45)$$

and finiteness of κ implies $\alpha \to 0$ as $\epsilon \to 0$.

Furthermore, it is noteworthy that the theory actually produces all the Seeley-DeWitt coefficients a_n, but these higher curvature terms associated with higher powers of ϵ in Eq. (7.44) all disappear upon regulator removal, leaving behind just the first Einstein Ricci scalar term with the $-\frac{1}{(2\kappa)^2}$ coupling constant.

7.8.2 *Positive bare cosmological constant term to counter negative fermionic vacuum energy*

The naive effective cosmological constant Λ_{eff} can be read off from

$$\frac{2\Lambda_{\text{eff}}}{(2\kappa)^2} = \mathcal{K} + \frac{3\alpha c_1 c_2 \hbar^2}{16\pi^{\frac{3}{2}}\epsilon^{\frac{5}{2}}}. \tag{7.46}$$

For positive $\mathcal{K}$, this naive value of Λ_{eff} diverges upon regulator removal; however, zero point energies of other bosonic and fermionic fields (which have, respectively, positive and negative zero point energies [98]) have not been taken into account, and they all contribute to the net cosmological constant. In the $SU(3) \times SU(2) \times U(1)$ Standard Model with bosonic Higgs and gauge fields, and *three generations of fermions*, it is well-known that there are *"too many fermions"*, resulting in the overabundance of net *negative* zero point energies. With all zero point energies, together with the contribution from a positive $\mathcal{K}$ taken into account, the net renormalized Λ_{eff} may well be positive and finite by countering with $\mathcal{K}$.

It should be noticed that from a mathematical point of view it is possible to choose $\alpha \propto \epsilon^{\frac{5}{2}}$, so that no divergences would appear in the limit of vanishing ϵ and with finite $\mathcal{K}$. Consequently, there would also be no Einstein Ricci scalar and other curvature terms, and apart from the cosmological constant contribution, only the Cotton-York term remains. Following our earlier arguments on the introduction of interactions, it is noteworthy that this would be, for geometrodynamics, the counterpart of renormalizable Yang-Mills theory with dimensionless coupling constant. Such a theory however would be missing the Einstein scalar curvature term and its Newtonian limit and physical effects (and it would also fail to yield the correct equations for gravitational waves, as discussed in Ref. [52] and in Chapter 6). In the earlier and adopted prescription, starting from the proposed Hamiltonian to its regularization, an interesting relation and physical motivation emerges: that keeping the cosmological constant contribution divergent at this stage (by keeping $\alpha/\epsilon^{\frac{3}{2}}$ finite) retains just the Einstein term and also yields the opportunity to counter net *negative* Standard Model zero point energy with divergent positive ($\mathcal{K} + \frac{3\alpha c_1 c_2 \hbar^2}{16\pi^{\frac{3}{2}}\epsilon^{\frac{5}{2}}}$), even if this does not explain why the resultant Λ_{eff} from the cancellation would be naturally as small as the current empirical value.

The final Hamiltonian in the limit of regulator removal in Eq. (7.44) becomes

$$H_{\text{Phys}} = \frac{1}{\beta} \int \bar{H}(x) d^3x,$$
$$\bar{H} = \sqrt{Q_j^{\dagger i} Q_i^j + q\mathcal{K}}$$
$$= \sqrt{\bar{\pi}_i^{\dagger j} \bar{\pi}_j^i + g^2 \hbar^2 \tilde{C}_j^i \tilde{C}_i^j - \frac{q}{(2\kappa)^2}(R - 2\Lambda_{\text{eff}})}. \qquad (7.47)$$

Apart from the extra Cotton-York potential,[6] Einstein's theory emerges wonderfully, and in a whole new light.

7.9   Further remarks

Within the paradigm of ITG with spatial diffomorphism invariance, Einstein's GR is a particular realization of a wider class of theories. In the attempt to reconcile unitarity with renormalizability, the availability of the Cotton-York tensor was crucial to Horava's advocacy of forgoing four-covariance in favour of just spatial diffeomorphism symmetry and higher spatial derivatives in the Hamiltonian [31]. The upshot is a higher spatial derivative theory with dimensionless coupling constant, and improved ultraviolet convergence without higher time derivatives. Spatial compactness, which is part and parcel of the framework of ITG, curbs infrared divergence.

7.9.1   *Simple harmonic oscillator analogy*

While Sakharov's induced gravity from matter loops [99] can produce similar emergent terms, the origin of the terms in this work is very different indeed. The basis of the work here has a simple analogy. In the harmonic oscillator, the Hamiltonian is

$$H = \left(a^\dagger a + \frac{1}{2}\right) \hbar\omega, \qquad (7.48)$$

---

[6]Note that the Heat kernel can never produce the Cotton-York term which has an extra scale invariant symmetry.

wherein

$$a^\dagger a = \frac{m\omega}{2\hbar}q^2 + \frac{1}{2m\omega\hbar}p^2 - \frac{1}{2}; \tag{7.49}$$

$$a := \sqrt{\frac{m\omega}{2\hbar}}q + i\sqrt{\frac{1}{2m\omega\hbar}}p,$$

$$a^\dagger := \sqrt{\frac{m\omega}{2\hbar}}q - i\sqrt{\frac{1}{2m\omega\hbar}}p.$$

Without the reassuring L.H.S., it is not easy to surmise that the R.H.S. of Eq. (7.49) is actually positive-definite; moreover, the negative term which contributes to the ZPE of $+\frac{1}{2}\hbar\omega$ in the Hamiltonian actually arose from

$$-\frac{1}{2} = \frac{i}{2\hbar}[q,p] = \frac{1}{2}[a^\dagger, a]. \tag{7.50}$$

The analog in Eq. (7.34) has a similar origin, but unlike the case of simple harmonic oscillator in quantum mechanics, the commutator at coincident point in a quantum field theory needs regularization, with the consequent emergence of Einstein's scalar curvature from such a framework.

7.9.2 *Implication of additional Cotton-York term on the initial state*

In the limit of large volume and long wavelengths, the Cotton-York term in Eq. (7.47) will be negligible compared to qR which would dominate; this corresponds to the familiar Einstein-Hilbert theory and era at later times provided $\beta = \sqrt{\frac{1}{6}}$. At the opposite extreme of vanishing volume or at the earliest intrinsic times with $q \to 0$, the qR term will be small compared to the Cotton-York potential since $\tilde{C}^i_j$ is invariant under scaling and independent of q. A natural question is what governs physics in this era, known as the Big-Bang era. With the availability of the Cotton-York term, the upshot is an initial Cotton-York dominated era associated with a quantum vacuum state which satisfies

$$Q^i_j|0\rangle = 0. \tag{7.51}$$

Since $Q^i_j = e^{-gW_{CS}}\bar{\pi}^i_j e^{gW_{CS}}$, it follows that the wave function is of the form

$$\langle \bar{q}_{ij}|0\rangle \sim e^{-gW_{CS}}, \tag{7.52}$$

which has critical points precisely at the vanishing of the Cotton-York tensor (which is the statement of spatial conformal flatness). Classically, it is intriguing that the lowest energy state is both $\bar{\pi}^{ij}$ and $\tilde{C}^{ij}$ vanishing, which is exactly the phase space initial data compatible with Friedmann-Lemaitre-Robertson-Walker spacetimes and Penrose's Weyl Curvature Hypothesis [23, 24]. Chapter 4 analyzed the initial data in the context of a generalized Lichnerowicz-York equation with the addition of a Cotton-York term, and arrived at a solution of an initial CMC classical data with conformal flatness and vanishing momenta. Chapter 10 will elaborate on the nature and the explicit initial quantum state of the universe; it will address how the Penrose Weyl curvature condition can be achieved at the level of expectation values of the Chern-Simons state, and discuss in detail the quantum fluctuations and correlations that are manifested.

Chapter 8

Cosmic time and reduced phase space of General Relativity

"The universe needs no, and has no, external clock."[1].

—Lee Smolin

*"To some one who could grasp the universe from
a unified standpoint, the entire creation would
appear as a unique truth and necessity."*

—J. D'Alembert

8.1 An ever-expanding spatially closed universe needs no, and has no, external clock

General Relativity is a theory of space and time; yet in usual expositions of Einstein's theory, time is a riddle, mysterious in its origins and elusive to pin down. Naive canonical decomposition of the Einstein-Hilbert action reveals a set of constraints in place of a true Hamiltonian to dictate evolution in ADM coordinate time. "Time must be determined intrinsically" was De-Witt's suggestion to adopt a d.o.f. in the theory as the clock [12]. However, solutions of Einstein's field equations are known to exhibit all sorts of fascinating, and even bizarre, causal structures that would preclude any simple universal clock and preferred slicing of spacetime. The first exact solution to allow closed timelike curves dates back to Kurt Gödel's work of 1949 [2]; it revealed a universe in which a traveller can, without superluminal motion, choose to end up anywhere at any time, including in the past. In his short but profound festschrift devoted to Einstein [3], Gödel went as far as to say "... it seems that one obtains an unequivocal proof for the view of those philosophers who, like Parmenides, Kant, and the modern idealists,

---

[1] *Time in Cosmology* Conference, Perimeter Institute (June 27–30, 2016).

deny the objectivity of change and consider change as an illusion or an appearance due to our special mode of perceptions." Yet the objective reality of the relations "earlier than" and "later than" is essential to the notion and reality of time in the description of the physical world.[2] To date, all our successful quantum descriptions, be it the Schrödinger or Heisenberg picture, is founded upon the concept of time-ordered evolution. Indeed, for quantum field theories in Minkowski spacetime, fundamental (anti)commutation relations enforcing (anti)commutation of fields for spacelike separations[3] are so constructed to guarantee time-ordering in the development operator, $U(t, t_0) = \mathcal{T}(e^{-\frac{i}{\hbar} \int_{t_0}^{t} H(t')dt'})$, remains well-defined [98]. To the great logician Einstein had replied "Kurt Gödel's essay constitutes, in my opinion, an important contribution to the general theory of relativity, especially to the analysis of the concept of time. The problem here involved disturbed me already at the time of the building up of the general theory of relativity, without my having succeeded in clarifying it [...] It will be interesting to weigh whether these are not to be excluded on physical grounds." The great physicist had appealed to Nature for a resolution. Einstein's faith was not misplaced. The central thesis of our work is that it is not incidental that all observational data points to an ever-expanding, and even spatially closed, universe; and unless and until the theory of gravitation is synthesized with this precise input it cannot be logically and physically sound.

An ever-expanding spatially closed universe needs no, and has no, external clock. It is (pun intended) "the universal clock", all encompassing and infinitely robust;[4] the fractional change of the volume is the preeminent intrinsic time interval to describe dynamical evolutions in GR. It is all the more amazing that auxiliary condition on the metric to arrive at this global time variable is the natural complement to the Hamiltonian constraint. It is a subsidiary condition to transform Einstein's theory from a Dirac first class to a second class constrained system when the physical degrees of freedom are also identified with transverse traceless excitations.

---

[2]In Minkowski spacetime, time-ordering of timelike separated events is Lorentz invariant.

[3]This guarantees the commutation of the Hamiltonian densities at spacelike separated points even when time ordering of these events themselves fails to be Lorentz invariant.

[4]In ITG, the global intrinsic time variable T is not subject even to quantum fluctuations; because the conjugate $\tilde{\pi}$ is eliminated in terms of other variables, and also absent in the reduced Hamiltonian. Just as in all quantum mechanics, T in the reduced theory is not subject to commutation relations, in accord with Pauli's concerns and conclusions [1].

8.2   Framing the physical contents of GR: Dirac first class versus second class constrained systems

For GR, the Hamiltonian and spatial diffeomorphism (or momentum) constraints (respectively $H = 0$, $H_i = 0$) form the first class Dirac algebra of Eqs. (1.18)–(1.20). Conversely, it has been shown the Dirac algebra is the hallmark of space-time 4-covariance and embeddability of hypersurface deformations, from which the *explicit forms* of H_i and H of Einstein's theory of geometrodynamics can be regained [33]. Einstein's theory is the only (time-reversible) canonical representation of the algebra if the intrinsic metric of the hypersurface and a conjugate momentum are the sole canonical variables. This can be interpreted as a no-go theorem against deformations of GR.

In the quantum context, following Dirac's own prescription, first class constraints can consistently be imposed to annihilate quantum states; but closure of the quantum algebra and divergences in Einstein's theory also need to be resolved meaningfully. While H_i can rigorously be interpreted as generators of spatial diffeomorphisms, the role of H which is quadratic in the momentum is not as transparent, and the constraints do *not* generate four-diffeomorphisms *off-shell*. What is clear is that *these constraints are needed to capture the physical contents of Einstein's theory*, at least at the classical level.

In what follows two complementary perspectives will be adopted to frame the physical contents of Einstein's theory in an ever-expanding spatially closed universe:

(1) casting Einstein's theory in an ever expanding universe as a *theory with second class constraints* [100],

(2) studying the *physical* canonical phase space functional integral of GR.

In the process, we shall isolate the true degrees of freedom, solve all constraints, resolve the problem of time and construct the corresponding reduced physical Hamiltonian. Freed from the shackles of the Dirac algebra and first class constraints,[5] it will also be revealed that four-covariance is not really needed to capture the physical contents of Einstein's theory [16, 72, 87]; and modifications of GR are in fact allowed within the same framework. This also provides a rigorous canonical foundation for

---

[5]In Chapter 2, Master Constraint was introduced as a mathematical device that can consistently retain the first class nature of the constraints even when GR is deformed. However, second class constraints with Dirac brackets offer a better paradigm to frame and understand the physical contents and, especially, observables of GR.

the consistency of a class of Horava gravity theories [31] which are obtained by modifying GR.

8.2.1   *Reduced phase space of GR*

With the addition of auxiliary condition for an ever-expanding closed universe, the total restrictions no longer form a first class system, but constitute a perfect set of second class constraints in Dirac's terminology [100] when the two physical degrees of freedom are identified with transverse-traceless excitations. The Hamiltonian constraint is solved by eliminating $\bar{\pi}$ in terms of the other variables, and spatial diffeomorphism symmetry is tackled explicitly by imposing transversality. The theorems of Maskawa-Nishijima [101] appositely relate the reduced phase space to the physical variables in canonical functional integral and Dirac's criterion for second class constraints to nonvanishing Faddeev-Popov determinants [102] in the phase space measures. A reduced physical Hamiltonian for intrinsic time evolution of the two physical degrees of freedom emerges. Freed from the first class Dirac algebra, deformation of the Hamiltonian constraint is permitted, and natural extension of the reduced Hamiltonian while maintaining spatial diffeomorphism invariance leads to a theory with Cotton-York term as the ultraviolet completion of Einstein's theory; and spatial compactness cures infrared divergence.

In a Dirac second class system, it is the Dirac, rather than Poisson, brackets which rule; *all variables, by construction, Dirac commute with all the constraints.* Thus when all second class constraints are satisfied, *all variables of the reduced phase are observables.* The prescription for quantization is to promote fundamental Dirac brackets to quantum commutation relations. In contradistinction, observables in a first class system are required to Poisson commute with the constraints;[6] which makes the identification of all observables of the theory a formidable task. This is particularly true of the formulation, be it classical or quantum, of GR as a first class system with a Hamiltonian constraint quadratic in the momenta.

8.2.2   *Recap on second class systems and the Dirac bracket*

Dirac was a pioneer, when quantum mechanics was first formulated, in the realization of the Poisson bracket $\{\,,\,\}_{P.B.} \rightarrow \frac{1}{i\hbar}[\,,\,]$ prescription. This

---

[6]These are usually referred to as "Dirac observables", even though the terminology makes much more sense applied to second class systems when all phase space variables have identically zero Dirac brackets with all the constraints.

worked for simple unconstrained systems; a systematic study of the quantization of constrained systems came later, also with Dirac [30]. In the quantization of gravity one should be wary of the naive application of Poisson bracket to commutator prescription and the perplexities that arise in the identification of physical states and observables. In Dirac second class systems, it is Dirac, not Poisson, bracket that takes center stage in quantization and in the identification of physical observables.

Dirac starts with a very simple example in which one of the configuration variable is constrained to vanish i.e. $q^1 = 0$. This is first-class; but clearly the Poisson bracket-commutator rule would lead to nonsensical result. It is reasonable to eliminate the conjugate variable by setting $p_1 = 0$; however $\{q^1, p_1\}_{P.B.} = 1$ is obviously at odds with $q^1\Psi = 0$, $p_1\Psi = 0$ and the Poisson bracket-commutator rule. In this simple case the task is to invent a new rule to effectively discard the extraneous pair (q^1, p_1) from playing any role. The generalization of this leads to second class systems and the Dirac bracket.

A set of constraints $\{C_\alpha\}$ is deemed to be second class if $\det(\{C_\alpha, C_\beta\}_{P.B.}) \neq 0$. The Dirac bracket is defined as

$$\{C_\alpha, C_\beta\}_{D.B.} := \{C_\alpha, C_\beta\}_{P.B.} - \{C_\alpha, C_\gamma\}_{P.B.} I_{\gamma\epsilon} \{C_\epsilon, C_\beta\}_{P.B.}; \qquad (8.1)$$

wherein $I_{\alpha\beta}$ is the inverse of the matrix of Poisson brackets between the constraints; it obeys

$$I_{\alpha\beta}\{C_\beta, C_\gamma\}_{P.B.} = \delta_{\alpha\gamma}. \qquad (8.2)$$

The Dirac brackets satisfy the same properties as Poisson brackets in their antisymmetry, law of commutation with composites, and in obeying the Jacobi identity.

8.2.3   *Dirac bracket, reduced phase space; and observables*

What is important for GR with its first class Hamiltonian and diffeomorphism constraints is that the system can be turned into second-class. A $2N$-dimensional phase space, with first class constraints $\phi_i = 0$ (with $i = 1, ..., \mathcal{M} < N$) can be turned into a second class system by introducing an equal number of commuting supplementary or auxiliary conditions, $\chi_j = 0$, provided

$$\det(\{\chi_i, \phi_j\}_{P.B.}) \neq 0. \qquad (8.3)$$

Taken together $\{\phi_{i=1,...,\mathcal{M}}; \chi_{j=1,...,\mathcal{M}}\} = \{C_{\alpha=1,...,2\mathcal{M}}\}$ form a second-class system. We can set $\{\chi_i\}$ as the first $\mathcal{M}$ variables of a new set of general coordinates of the same phase space, and denote this by $*$ superscript s.t. $\chi_i =: q^{i*} = 0$. Then the nonvanishing determinant translates into

$$\{q^{i*}, \phi_j\}_{P.B.} = \frac{\partial \phi_j}{\partial p_i^*} \neq 0; \tag{8.4}$$

which allows the determination of the corresponding p_i^*. Maskawa and Nakajima [101] proved that for these second class systems, canonical variables exist for the resultant restricted submanifold. The total canonical phase space $\{(q^I, p_I); I = 1, ..., N\}$ can be decomposed into the sum of the reduced physical phase space $\{(q^r, p_r); r = \mathcal{M} + 1, ..., N\}$ and $\{(q^{i*}, p_i^*); i = 1, ..., \mathcal{M}\}$; and the constrained surface can be obtained by setting canonical variables $\{(q^{i*}, p_i^*); i = 1, ..., \mathcal{M}\}$ to zero (which is equivalent to enforcing $\phi_i = 0, \chi_i = 0, \forall i = 1, ..., \mathcal{M}$). Thus the resultant manifold is a canonical phase space, and it is of dimension $2(N - \mathcal{M})$.

Remarkably, the Poisson bracket in the reduced phase space is just the Dirac bracket, since

$$\{A, B\}_{\text{reduced}} = \sum_{I=1}^{N} \left(\frac{\partial A}{\partial q^I} \frac{\partial B}{\partial p_I} - \frac{\partial A}{\partial p_I} \frac{\partial B}{\partial q^I} \right) - \sum_{j=1}^{\mathcal{M}} \left(\frac{\partial A}{\partial q^{j*}} \frac{\partial B}{\partial p_j^*} - \frac{\partial A}{\partial p_j^*} \frac{\partial B}{\partial q^{j*}} \right)$$

$$= \{A, B\}_{P.B.} - \sum_{j=1}^{\mathcal{M}} \left(\{A, p_j^*\}_{P.B.} \{q^{j*}, B\}_{P.B.} - \{q^{j*}, A\}_{P.B.} \{B, p_j^*\}_{P.B.} \right)$$

$$= \{A, B\}_{P.B.} - \sum_{j=1}^{\mathcal{M}} \sum_{L=1}^{N} \left(\left(\frac{\partial A}{\partial q^L} \frac{\partial p_j^*}{\partial p_L} - \frac{\partial A}{\partial p_L} \frac{\partial p_j^*}{\partial q^L} \right) \{q^{j*}, B\}_{P.B.} \right.$$

$$\left. + \{A, q^{j*}\}_{P.B.} \left(\frac{\partial B}{\partial q^L} \frac{\partial p_j^*}{\partial p_L} - \frac{\partial B}{\partial p_L} \frac{\partial p_j^*}{\partial q^L} \right) \right)$$

$$= \{A, B\}_{P.B.} + \sum_{j=1}^{\mathcal{M}} \sum_{L=1}^{N} \left(\left(\frac{\partial A}{\partial q^L} \frac{\partial \phi_k}{\partial p_L} - \frac{\partial A}{\partial p_L} \frac{\partial \phi_k}{\partial q^L} \right) \left(-\frac{\partial p_j^*}{\partial \phi_k} \right) \{q^{k*}, B\}_{P.B.} \right.$$

$$\left. + \{A, q^{j*}\}_{P.B.} \left(\frac{\partial p_j^*}{\partial \phi_k} \right) \left(\frac{\partial B}{\partial q^L} \frac{\partial \phi_k}{\partial p_L} - \frac{\partial B}{\partial p_L} \frac{\partial \phi_k}{\partial q^L} \right) \right)$$

$$= \{A, B\}_{P.B.} + \left(\{A, \phi_j\}_{P.B.} \frac{1}{\{\phi_j, \chi_k\}_{P.B.}} \{\chi_k, B\}_{P.B.} \right.$$

$$\left. + \{A, \chi_j\}_{P.B.} \frac{1}{\{\chi_j, \phi_k\}_{P.B.}} \{\phi_k, B\}_{P.B.} \right)$$

$$= \{A, B\}_{D.B.}. \tag{8.5}$$

The correct evolution of an operator in a constrained system should be w.r.t. to the physical reduced phase space; this implies Dirac, rather than

Poisson, bracket should be used to arrive at the resultant EOM i.e. $\dot{A} = \{A, H_{phys}\}_{D.B.}$, wherein H_{phys} is the Hamiltonian of the system. In second-class systems, all the constraints are strong equalities, and one is no longer dealing with weakly vanishing entities of first class systems.

Moreover, all variables (subject to $\{C_\alpha = 0\}$) automatically Dirac *commute with all the constraints*,

$$
\begin{aligned}
&\{f(q^I, p_I), C_\alpha\}_{D.B.} \\
&= \{f(q^I, p_I),\ C_\alpha\}_{P.B.} - \{f(q^I, p_I), C_\beta\}_{P.B.} I_{\beta\gamma} \{C_\gamma,\ C_\alpha\}_{P.B.} \\
&= \{f(q^I, p_I), C_\alpha\}_{P.B.} - \{f(q^I, p_I),\ C_\beta\}_{P.B.} \delta_{\beta\alpha} = 0.
\end{aligned}
\tag{8.6}
$$

So, unlike Poisson brackets and first class systems, with Dirac brackets the addition of constraints to H_{phys} to make up a total Hamiltonian really adds nothing to the EOM.

In first class systems, "Dirac observables" are required to Poisson commute with all the constraints. As diffeomorphisms are gauge symmetries of GR, this leads to the suspicions, and even outright claims, that there can be no local (or spatially dependent) observables in the theory; and efforts have been directed to studying quasilocal entities and/or non-perturbative observables, despite the fact that by any reasonable method of counting there are two *local* d.o.f. in GR. By turning GR into a Dirac second class system, this quandary finds a resolution. As will be demonstrated, the needed auxiliary conditions are the transversality of the unimodular metric excitations. The resultant system is second class; the Maskawa-Nakajima theorem guarantees a reduced phase space which is canonical. All the constraints can be solved; and the true d.o.f. are *local* and the infinitesimal physical excitations are contained in transverse-traceless metric fluctuations. From a starting metric, finite unimodular metric excitations can be built up, from an infinite number of infinitesimal ones, each transverse and traceless w.r.t. to the preceding metric.

8.3 Supplementary condition in an ever-expanding universe

An ever-expanding spatially closed universe leads to a natural supplementary condition, which is to impose spatially independent[7]

$$
\delta \ln q^{\frac{1}{3}} = \delta T,
\tag{8.7}
$$

---

[7]In the Hodge decomposition $\delta \ln q^{\frac{1}{3}} = \delta T + \nabla_i \delta Y^i$, the $\nabla_i \delta Y^i = \mathcal{L}_{\frac{3}{2}\delta\vec{Y}} \ln q^{\frac{1}{3}}$ part is precisely a gauge variation while $\delta T = \frac{2}{3} d \ln V$ is gauge invariant.

or equivalenty,[8]

$$\chi = \frac{1}{3} \ln \left[\frac{q}{q_0} \right] - (T - T_0) = 0. \tag{8.8}$$

The Hamiltonian constraint is expressible in an interesting manner as [16]

$$\mathcal{H} := \beta^2 \tilde{\pi}^2 - \bar{H}^2 = (\beta \tilde{\pi} + \bar{H})(\beta \tilde{\pi} - \bar{H}) = 0, \tag{8.9}$$

$$\bar{H}^2 := \bar{G}_{ijkl} \bar{\pi}^{ij} \bar{\pi}^{kl} + \mathcal{V}(\bar{q}_{ij}, q)$$

$$= \frac{1}{2} [\bar{q}_{ik} \bar{q}_{jl} + \bar{q}_{il} \bar{q}_{jk}] \bar{\pi}^{ij} \bar{\pi}^{kl} + \mathcal{V}(q_{ij});$$

wherein Einstein's GR corresponds to

$$\mathcal{V}(\bar{q}_{ij}, q) = -\frac{q}{(2\kappa)^2} (R - 2\Lambda_{eff})$$

and $\beta^2 := \frac{1}{6}$. In addition, $\bar{H}(\ln q^{1/3}, \bar{q}_{ij}, \bar{\pi}^{ij})$ commutes with $\ln q^{\frac{1}{3}}$; so Eq. (8.8) and Eq. (8.9) lead to

$$\{\chi, \mathcal{H}\}_{P.B.} = 2\beta^2 \tilde{\pi} = \pm 2\beta \bar{H}, \tag{8.10}$$

on the constraint surface. In the last chapter we have detailed how the value of $\bar{H}$ can in fact be a positive-definite entity, thus making $\{\chi, \mathcal{H}\}$ nonvanishing (this is another motivation for casting GR as a theory with second class constraints) in an ever-expanding closed universe.[9] Some may argue an ever-expanding universe is the *consequence* of Einstein's GR given the distribution of matter and the effective value of the cosmological constant, rather than an input. However, current observations do not preclude the counterargument this work explores: an ever-expanding closed universe is fundamental to the framework of our physical universe and the resolution of the problem of time in classical and quantum GR.

8.4 Second class constraints and canonical phase space functional integral of GR

As explained in Ref. [102], to reduce the phase space of a system with $2n$ degrees of freedom subject to m constraints to a well-defined $2(n-m)$-dimensional canonical phase space with $(n-m)$ physical degrees of freedom, it is necessary to introduce m additional conditions. These m conditions are often looked upon as "gauge-fixing", but this terminology is somewhat

---

[8]As explained in Sec. 4.7.1 this is also compatible with the EOM.

[9]The on-shell relation $-\kappa\tilde{\pi} = \sqrt{q}tr(K) = \frac{3\sqrt{q}}{2N}(\frac{\partial \ln q^{\frac{1}{3}}}{\partial t} - \frac{2}{3}\nabla_i N^i)$ together with Eq. (8.12) show that $\tilde{\pi}$ is nonvanishing and negative in an ever-expanding universe.

obfuscating as these conditions (the transversality of physical gauge potentials as an example[10]) may stem from restrictions to gauge-invariant parts of natural decompositions, as well as from physical conditions compatible with empirical observations of the system. The crucial point is that the original constraints, supplemented by these extra restrictions can turn the system into one with $2m$ second class constraints and well-defined $(n - m)$ reduced canonical physical degrees of freedom. For GR, spatial diffeomorphism invariance in an ever expanding spatially compact universe delivers the upshot of imposing $\chi = 0$, which, together with transversality of the unimodular perturbations, yield not only the correct 2 degrees of freedom, but also a nonvanishing physical Hamiltonian generating true (intrinsic) time translations.

We recall that the momentum constraint can be decomposed as

$$H_i = \mathcal{H}_i - \frac{2}{3}\nabla_i\tilde{\pi} = 0.$$

Moreover, $-\frac{2}{3}\nabla_i\tilde{\pi}$ and $\mathcal{H}_i := -2\bar{q}_{ik}\nabla_j\bar{\pi}^{jk}$ separately, and respectively, generate spatial diffeomorphisms of $(\tilde{\pi}, \ln q^{\frac{1}{3}})$ and $(\bar{\pi}^{ij}, \bar{q}_{ij})$; specifically,

$$\int N^i\mathcal{H}_i d^3x = \int \bar{\pi}^{ij}\mathcal{L}_{\vec{N}}\bar{q}_{ij}d^3x$$

with corresponding algebra,

$$\{\mathcal{H}_i[M^i], \mathcal{H}_j[M'^j]\}_{P.B.} = \mathcal{H}_k[(\mathcal{L}_{\vec{M}}\vec{M}')^k].$$

Focusing on the $\tilde{\pi}$ terms in the action, the contribution in H_i combine with the symplectic term to yield[11]

$$
\begin{aligned}
S &= \int dt \int \left[\left(\tilde{\pi}\frac{\partial \ln q^{\frac{1}{3}}}{\partial t}\right) - N^iH_i + ...\right]d^3x \\
&= \int dt \int \left[\tilde{\pi}\left(\frac{\partial \ln q^{\frac{1}{3}}}{\partial t} - \frac{2}{3}\nabla_iN^i\right) + 2N^i\bar{q}_{ik}\nabla_j\bar{\pi}^{jk} + ...\right]d^3x \\
&= \int \left(\int \tilde{\pi}d^3x\right)dT - \int\int N^i\mathcal{H}_i d^3x dt + ... \; ;
\end{aligned}
\tag{8.11}
$$

---

[10]For instance, in Electrodynamics, an arbitrary gauge potential has the decomposition $A_i = A_i^T + \nabla_i\alpha$. Transversality of A_i^T leads to $\nabla^2\alpha = \nabla^iA_i$; consequently $A_i^T = A_i - \nabla_i\frac{1}{\nabla^2}\nabla^jA_j$ is *explicitly gauge invariant* under $A_i \mapsto A_i + \nabla_i\eta$ $\forall\eta$. So while the condition $\nabla^iA_i^{\text{phys}} = 0$ is often called "gauge-fixing", the solution $A_i^{\text{phys}} = A_i^T$ is, as shown, really *invariant* under gauge transformations. The Coulomb "gauge" is thus a condition to extract the *gauge invariant physical degrees of freedom* from the original configuration space.

[11]See also the discussions in Chapter 4 related to Eq. (4.15).

wherein integration by parts has been carried out in the intermediate step; and the choice of

$$\frac{\partial \ln q^{1/3}}{\partial t} - \frac{2}{3}\nabla_i N^i = \frac{dT}{dt} \tag{8.12}$$

has been invoked in the final step. This can always be achieved by using the gauge freedom in the longitudinal part of the shift vector, since $N^i = N^i_T + N^i_L = N^i_T + q^{ij}\nabla_j W$ leads to

$$\nabla^2 W = -\frac{3}{2}\left(\frac{dT}{dt} - \frac{\partial \ln q^{1/3}}{\partial t}\right), \tag{8.13}$$

with W as the solution (unique up to a constant) of the elliptic equation. Thus the theory reduces to one in which $-(\int \tilde{\pi} d^3 x)$ (which is constrained to be $\pm\frac{1}{\beta}\int \bar{H} d^3 x$ by the Hamiltonian constraint) is the global Hamiltonian generating T translations, with remaining spatial diffeomorphism invariance of the $(\bar{\pi}^{ij}, \bar{q}_{ij})$ variables enforced by $\mathcal{H}_i = 0$.

Together with Eq. (8.8) and Eq. (8.9) the explicit additional conditions, we need for $\mathcal{H}_i$ an additional set of auxiliary conditions

$$\chi_i = 0 \tag{8.14}$$

which will render $\{\chi, \mathcal{H}, \chi_i, \mathcal{H}_i\}$ into a set of second class constraints. This will be explicitly addressed; we first note the Poisson brackets of the set of constraints $\{\chi, \mathcal{H}, \chi_i, \mathcal{H}_i\}$ are as tabulated below:

$\{,\}_{P.B.}$	χ	$\mathcal{H}$	χ_j	$\mathcal{H}_l$
χ	0	$2\beta^2\tilde{\pi}$	0	0
$\mathcal{H}$	$-2\beta^2\tilde{\pi}$	X	Y_j	Z_l
χ_i	0	$-Y_i$	0	$\{\chi_i, \mathcal{H}_l\}_{P.B.}$
$\mathcal{H}_k$	0	$-Z_k$	$\{\mathcal{H}_k, \chi_j\}_{P.B.}$	$\{\mathcal{H}_k, \mathcal{H}_l\}_{P.B.}$

The matrix is invertible when

$$\{\chi, \mathcal{H}\}_{P.B.} = 2\beta^2\tilde{\pi} \neq 0, \quad \text{and} \quad \det[\{\chi_i, \mathcal{H}_j\}_{P.B.}] \neq 0. \tag{8.15}$$

8.4.1 *Maskawa-Nakajima theorems, Dirac's criterion for second class, Faddeev-Popov determinants, and reduced variables*

In Dirac's terminology, $\{\chi, \mathcal{H}, \chi_i, \mathcal{H}_i\}$ perfectly constitutes a set of "second class constraints" [100] when Eq. (8.15) holds. Maskawa and Nakajima [101] showed that any set of canonical variables governed by second class constraints is canonically equivalent to $\{Q^r, P_r; \mathcal{Q}^\alpha, \mathcal{P}_\alpha\}$ such that the constraints read

$$\mathcal{Q}^\alpha = \mathcal{P}_\alpha = 0. \tag{8.16}$$

Moreover, the Dirac bracket [30] is equal to the Poisson bracket calculated in terms of the reduced set of unconstrained variables i.e.

$$\{A, B\}_{\text{Dirac}} = \frac{\partial A}{\partial Q^r}\frac{\partial B}{\partial P_r} - \frac{\partial B}{\partial Q^r}\frac{\partial A}{\partial P_r}. \tag{8.17}$$

Thus all physics and computations become transparent and simplify greatly if the reduction to true degrees of freedom can be completed and the reduced Hamiltonian constructed explicitly. In the canonical functional integral one can view

$$\mathcal{Q}^\alpha = (\chi, \chi_i); \tag{8.18}$$

and with

$$\mathcal{H}_\alpha = (\mathcal{H}, \mathcal{H}_i), \tag{8.19}$$

it follows that

$$\det[\{\mathcal{Q}_\alpha, \mathcal{H}_\beta\}_{P.B.}] = \det\left[\frac{\partial \mathcal{H}_\beta}{\partial \mathcal{P}_\alpha}\right] \neq 0 \tag{8.20}$$

allows

$$\mathcal{H}_\beta = 0 \tag{8.21}$$

to be solved in terms of $\mathcal{P}_\alpha$. Since

$$\delta(\mathcal{Q}^\alpha)\det[\{\mathcal{Q}^\alpha, \mathcal{H}_\beta\}_{P.B.}]\delta(\mathcal{H}_\beta(\mathcal{P}_\alpha)) = \delta(\mathcal{Q}^\alpha)\delta(\mathcal{P}_\alpha), \tag{8.22}$$

our previous factors, $\det[\{\chi, \mathcal{H}\}_{P.B.}]$ and $\det[\{\chi_i, \mathcal{H}_j\}_{P.B.}]$, are just the required Faddeev-Popov ghost determinants [102] in $\det[\{\mathcal{Q}^\alpha, \mathcal{H}_\beta\}_{P.B.}]$ to enforce the restriction to the true physical phase space (Q^r, P_r).

The canonical phase space functional integral is

$$\int D\mu\Theta(-\tilde{\pi})\exp\left[\frac{i}{\hbar}\int dt\int_\Sigma \left(\tilde{\pi}^{ij}\frac{\partial \tilde{q}_{ij}}{\partial t} + \tilde{\pi}\frac{\partial \ln q^{1/3}}{\partial t}\right)d^3x\right]. \tag{8.23}$$

In the functional integral we have inserted a theta function to pick out the choice of positive-definite $\bar{H}$ (corresponding to negative value for $\tilde{\pi}$) compatible with an expanding universe; whereas $\mathcal{H} = 0$ is satisfied by

$$\tilde{\pi} = \pm\frac{\bar{H}}{\beta}. \tag{8.24}$$

As the total determinant of the Poisson brackets in the table factorizes, the total measure will be associated with a direct product $\det[\{\chi, \mathcal{H}\}] \det[\{\chi_i, \mathcal{H}_j\}]$. To wit, including the determinants and constraints, the complete measure is

$$D\mu = D\bar{\mu} \prod_x \frac{1}{2\pi\hbar} \delta \ln q^{\frac{1}{3}} \delta\tilde{\pi} \det\{\chi, \mathcal{H}\}_{P.B.} \delta(\chi)\delta(\mathcal{H}), \qquad (8.25)$$

with

$$D\bar{\mu} = \prod_x \prod_{i>j} \frac{\delta\tilde{\pi}^{ij}\delta\bar{q}_{ij}}{2\pi\hbar} \det[\{\chi_i, \mathcal{H}_j\}_{P.B.}]\delta(\chi_i)\delta(\mathcal{H}_j); \qquad (8.26)$$

consistent with the Faddeev-Popov prescription [102]. In functional integrals, the condition,

$$\chi_i = 0,$$

and the Faddeev-Popov determinant can also be realized by integrating over auxiliary fields $\tilde{b}^i$ and complex Grassmannian ghost fields $(\bar{c}^i, c^i)$ through

$$\int \prod_k \prod_{x\in\Sigma} \frac{d\tilde{b}^k}{2\pi} d\bar{c}^k dc^k \exp\left(i \int \tilde{b}^i \chi_i + \bar{c}^i \{\chi_i, \mathcal{H}_j\}_{P.B.} c^j d^3 x'\right).$$

8.4.2   *Addressing spatial diffeomorphisms*

In this subsection we address spatial diffeomorphisms and the resultant degrees of freedom of the theory. Supplementary to the

$$\mathcal{H}_i = 0$$

constraints and consistent with the gauge-invariance of transverse fluctuations, we introduce auxiliary conditions

$$\chi_i = \bar{q}^{*jl}\nabla^*_l\delta\bar{q}^{TT}_{ji} = q^{\frac{1}{3}}\nabla^{*j}\delta\bar{q}^{TT}_{ji} = 0. \qquad (8.27)$$

The physical infinitesimal perturbations are

$$\delta\bar{q}^{TT}_{ij} = \bar{q}_{ij} - \bar{q}^*_{ij}, \qquad (8.28)$$

wherein ∇^* denotes covariant derivative w.r.t. background $q^*_{ij} = q^{\frac{1}{3}}\bar{q}^*_{ij}$. It follows from

$$\bar{\pi}^{ij} := q^{\frac{1}{3}}\left[\tilde{\pi}^{ij} - \frac{q^{ij}}{3}\tilde{\pi}\right]$$

that the Poisson brackets for the barred variables are,

$$\{\bar{q}_{ij}(x), \bar{q}_{kl}(y)\}_{P.B.} = 0,$$

$$\{\bar{q}_{kl}(x), \bar{\pi}^{ij}(y)\}_{P.B.} = P^{ij}_{kl}\,\delta^3(x, y),$$

$$\{\bar{\pi}^{ij}(x), \bar{\pi}^{kl}(y)\}_{P.B.} = \frac{1}{3}(\bar{q}^{kl}\bar{\pi}^{ij} - \bar{q}^{ij}\bar{\pi}^{kl})\delta^3(x, y);$$

with

$$P^{ij}_{kl} := \frac{1}{2}(\delta^i_k \delta^j_l + \delta^i_l \delta^j_k) - \frac{1}{3}\bar{q}^{ij}\bar{q}_{kl}$$

denoting the traceless projection operator.

Under spatial diffeomorphisms, the unimodular metric variable changes by

$$\mathcal{L}_{\vec{\xi}}\bar{q}_{ij} = \xi^k \partial_k \bar{q}_{ij} + \bar{q}_{ik}\partial_j \xi^k + \bar{q}_{kj}\partial_i \xi^k - \frac{2}{3}\bar{q}_{ij}\partial_k \xi^k$$

$$= q^{-\frac{1}{3}}\left(\nabla_i \xi_j + \nabla_j \xi_i - \frac{2}{3}q_{ij}\nabla^k \xi_k\right)$$

$$= 2q^{-\frac{1}{3}}P^{lk}_{ij}\nabla_l \xi_k =: \mathcal{L}_{(ij)}{}^k \xi_k \tag{8.29}$$

wherein ∇ is the Levi–Civita connection of the spatial metric.

We may introduce the positive-definite block-diagonal metric

$$G^{IJ} = \begin{bmatrix} \bar{G}^{ijkl} & 0 \\ 0 & q^{mn} \end{bmatrix} \tag{8.30}$$

wherein $I = ((ij),m)$, $J = ((kl),n)$, and $\bar{G}^{ijkl} = \frac{1}{2}(\bar{q}^{ik}\bar{q}^{jl} + \bar{q}^{il}\bar{q}^{jk})$. This allows the definition a positive-definite inner product

$$\langle U, U' \rangle := \int \bar{G}^{IJ}U_I U'_J \sqrt{q}d^3 x$$

$$= \int \bar{G}^{ijkl}\delta\bar{q}^1_{ij}\delta\bar{q}^2_{kl}\sqrt{q}d^3 x + \int q^{mn}\xi^1_m \xi^2_n \sqrt{q}d^3 x \tag{8.31}$$

for any pair

$$U^{1,2}_I = \begin{bmatrix} \delta\bar{q}^{1,2}_{ij} \\ \xi^{1,2}_k \end{bmatrix}.$$

And associated with spatial diffeomorphisms is the operation

$$\mathcal{L}_I{}^J \xi_J = \begin{bmatrix} 0 & \mathcal{L}_{(ij)}{}^m \\ 1 & 0 \end{bmatrix} \cdot \begin{bmatrix} 0 \\ \xi_m \end{bmatrix} = \begin{bmatrix} \mathcal{L}_{\vec{\xi}}\bar{q}_{ij} \\ 0 \end{bmatrix}.$$

To demonstrate that the determinant of the relevant Poisson bracket is almost generically nonvanishing for an arbitrary background, we note that

$$\left\{ \int \zeta^m \chi_m \sqrt{q} d^3x, \int \xi^n \mathcal{H}_n d^3y \right\}_{P.B.}$$

$$= \left\{ \int \zeta^m \sqrt{q} q^{\frac{1}{3}} \nabla^{*n}(\bar{q}_{mn} - \bar{q}^*_{mn}) d^3x, \int (\mathcal{L}_{\xi}\bar{q}_{ij})^* \bar{\pi}^{ij} d^3y \right\}_{P.B.}$$

$$= -\frac{1}{2} \int \sqrt{q} q^{\frac{1}{3}} \left(\nabla^{*i}\zeta^j + \nabla^{*j}\zeta^i - \frac{2}{3} q^{ij} \nabla_k^* \zeta^k \right) \left(\mathcal{L}_{\xi}\bar{q}_{ij} \right)^* d^3y$$

$$= -\frac{1}{2} \int \bar{G}^{*ijkl} (\mathcal{L}_{(ij)}{}^m \zeta_m)^* (\mathcal{L}_{(kl)}{}^n \xi_n)^* \sqrt{q} d^3y$$

$$= -\frac{1}{2} \int \bar{G}^{*IJ} (\mathcal{L}_I^{*K}\zeta_K)(\mathcal{L}_J^{*L}\xi_L) \sqrt{q} d^3y = -\frac{1}{2} \langle \mathcal{L}^*\zeta, \mathcal{L}^*\xi \rangle$$

$$= -\frac{1}{2} \langle \zeta, \mathcal{L}^{*\dagger}\mathcal{L}^*\xi \rangle; \tag{8.32}$$

which means that the determinant of $\{\chi_m, \mathcal{H}_n\}$ is that of a negative semi-definite operator. It follows that a zero mode is present iff

$$\mathcal{L}_{(ij)}^* {}^k \xi_k = 0, \tag{8.33}$$

equivalently ξ^i must be a conformal Killing vector which obeys

$$\mathcal{L}_{\xi}q_{ij}^* = \frac{2}{3} q_{ij}^* \nabla_k^* \xi^k. \tag{8.34}$$

The orthogonal decomposition into physical and gauge changes

$$\delta\bar{q}_{ij} = \delta\bar{q}_{ij}^{TT} + \mathcal{L}_{\xi}\bar{q}_{ij} \tag{8.35}$$

does not fix $\vec{\xi}$ uniquely if one or more zero modes, $\vec{\xi}_{o\alpha}$, exist; but even then the physical $\delta\bar{q}_{ij}^{TT}$ is unaffected since these conformal Killing vectors satisfy

$$\mathcal{L}_{\vec{\xi}_{o\alpha}} \bar{q}_{ij} = 0. \tag{8.36}$$

This parallels the arguments given by York [103] for the decomposition of the momentum variable which we shall soon address.

In Electrodynamics, the physical degrees of freedom is the transverse projection

$$A_i^T = \left(\delta_i^j - \nabla_i \frac{1}{\nabla^2} \nabla^j \right) A_j \tag{8.37}$$

which is *gauge invariant* under

$$A_i \mapsto A_i + \nabla_i \eta \quad \forall \eta.$$

The GR analogy can be made concrete: Eq. (8.27) implies

$$\mathcal{L}^\dagger \cdot \begin{bmatrix} \delta\bar{q}_{ij}^{TT} \\ 0 \end{bmatrix}$$

vanishes, with · denoting matrix multiplication. By decomposing

$$\delta \bar{q}_{ij} = \delta \bar{q}_{ij}^{TT} + \mathcal{L}_{\bar{\eta}} \bar{q}_{ij},$$

or equivalently,

$$\begin{bmatrix} \delta \bar{q}_{ij} \\ 0 \end{bmatrix} = \begin{bmatrix} \delta \bar{q}_{ij}^{TT} \\ 0 \end{bmatrix} + \mathcal{L} \cdot \begin{bmatrix} 0 \\ \eta_k \end{bmatrix},$$

and acting with $\mathcal{L}^\dagger$ on the equation, it follows that

$$\begin{bmatrix} 0 \\ \eta_k \end{bmatrix} = (\mathcal{L}^\dagger \mathcal{L})^{-1} \cdot \mathcal{L}^\dagger \cdot \begin{bmatrix} \delta \bar{q}_{ij} \\ 0 \end{bmatrix}.$$

Substituting this back into the decomposition yields the physical[12]

$$\begin{bmatrix} \delta \bar{q}_{ij}^{TT} \\ 0 \end{bmatrix} = \left(I - \mathcal{L} \cdot (\mathcal{L}^\dagger \mathcal{L})^{-1} \cdot \mathcal{L}^\dagger\right) \cdot \begin{bmatrix} \delta \bar{q}_{kl} \\ 0 \end{bmatrix}, \tag{8.38}$$

which is *explicitly invariant* under spatial diffeomorphisms

$$\begin{bmatrix} \delta \bar{q}_{ij} \\ 0 \end{bmatrix} \mapsto \begin{bmatrix} \delta \bar{q}_{ij} \\ 0 \end{bmatrix} + \mathcal{L} \cdot \begin{bmatrix} 0 \\ \xi_k \end{bmatrix} \quad \forall \, \xi_k.$$

As will be addressed an analogous decomposition and projection Eq. (8.48) can be carried out for $\bar{\pi}_{TT}^{ij}$. These metric and momentum TT degrees of freedom are physical observables. In a system with second class Dirac constraints, *all variables have identically vanishing Dirac brackets with all constraints*. And, as was emphasized in [30], it is the promotion of Dirac, not Poisson, brackets to quantum commutators that is the prescription for quantizing a constrained classical theory.

8.4.3 *Note on the physical auxiliary or "gauge-fixing" conditions*

Instead of imposing the (three) weaker transversality conditions $\nabla^i \delta q_{ij} = q^{ik} \nabla_i \delta q_{kj} = 0$, a stronger set of (four) conditions, $\bar{q}^{ik} \nabla_i \delta \bar{q}_{kj} = 0$ *and* $\nabla_j \delta \ln q = 0$, constitute the full set of auxiliary conditions to complement

---

[12] As is often the practice in the physics literature, $(\mathcal{L}^\dagger \mathcal{L})^{-1}$ is the mnemonic for the inversion with the integral kernel or Green's function, $G(x, x')$, which satisfies $(\mathcal{L}^\dagger \mathcal{L})_x \cdot G(x, x') = \delta^3(x, x') I$. So Eq. (8.38) should read $\begin{bmatrix} \delta \bar{q}^{TT}(x) \\ 0 \end{bmatrix} = \begin{bmatrix} \delta \bar{q}(x) \\ 0 \end{bmatrix} +$ $\mathcal{L}_x \cdot \left(\int G(x, x') \cdot \left(\mathcal{L}_{x'}^\dagger \cdot \begin{bmatrix} \delta \bar{q}(x') \\ 0 \end{bmatrix} \right) d^3 x' \right)$. A positive-semidefinite self-adjoint operator has a complete set of eigenfunctions from which the Green's function can, in principle, be constructed as the Fredholm kernel.

the four constraints of GR and its extensions. To be precise, via the decomposition of $q_{ij} = q^{\frac{1}{3}}\bar{q}_{ij}$,

$$q^{ik}\nabla_i\delta q_{kj} = \nabla_i\bar{q}^{ik}\delta\bar{q}_{kj} + \nabla_j(q^{-\frac{1}{3}}\delta q^{\frac{1}{3}})$$
$$= \partial_i(\bar{q}^{ik}\delta\bar{q}_{kj}) + \Gamma^i_{il}\bar{q}^{lk}\delta\bar{q}_{kj} - \Gamma^l_{ij}\bar{q}^{ik}\delta\bar{q}_{kl} + \nabla_j\delta\ln q^{\frac{1}{3}}$$
$$= \bar{\nabla}_i\bar{q}^{ik}\delta\bar{q}_{kj} + \Delta^i_{il}\bar{q}^{lk}\delta\bar{q}_{kj} - \Delta^l_{ij}\bar{q}^{ik}\delta\bar{q}_{kl} + \nabla_j\delta\ln q^{\frac{1}{3}}, \quad (8.39)$$

wherein the decomposition $\Gamma^i_{jk} = \bar{\Gamma}^i_{jk} + \Delta^i_{jk}$ has been adopted, $\bar{\nabla}$ is the covariant derivative w.r.t. the connection $\bar{\Gamma}^i_{jk}$ which is the Christoffel symbol for the unimodular spatial metric, while, explicitly, $\Delta^i_{jk} = \frac{1}{2}(\delta^i_j\partial_k\ln q^{\frac{1}{3}} + \delta^i_k\partial_j\ln q^{\frac{1}{3}} - \bar{q}^{il}\bar{q}_{jk}\partial_l\ln q^{\frac{1}{3}})$. Not only does the stronger set imply $q^{ik}\nabla_i\delta q_{kj} = 0$, the conditions $\bar{q}^{ik}\nabla_i\delta\bar{q}_{kj} = 0$ are auxiliary to spatial diffeomorphism constraints $\mathcal{H}_i = -2\nabla_i\bar{\pi}^i_j = 0$; moreover, the extra $\nabla_j\delta\ln q = 0$ (equivalently, $q = q_0(x)e^{3(T-T_0)}$),[13] which is precisely the auxiliary condition $\chi = 0$ adopted for the Hamiltonian constraint $H = 0$. As shown, the four constraints supplemented by these precise auxiliary conditions (which are equivalent to the earlier set of $\{\chi_i = 0, \chi = 0\}$) form a second class Dirac system.

The auxiliary conditions are furthermore in agreement with considerations of orthogonality (with respect to a suitable positive-definite metric) of gauge and physical variations. The diffeomorphism generated by $\mathcal{H}_l = -2\nabla_k\bar{\pi}^k_l$ is

$$\delta_{gauge}\bar{q}_{ij} = \frac{1}{i\hbar}\left[\bar{q}_{ij}, -2\int\xi^l\nabla_k\bar{\pi}^k_l d^3x\right]$$
$$= q^{-\frac{1}{3}}(\nabla_i\xi_j + \nabla_j\xi_i) - \frac{2}{3}\bar{q}_{ij}\nabla_k\xi^k = \mathcal{L}_{\vec{\xi}}\bar{q}_{ij}. \quad (8.40)$$

This follows from the commutation relations of the momentric as the generator of $SL(3, R)$ transformations of the unimodular spatial metric; and $\xi_i = q_{ij}\xi^j$. Thus the requirement of

$$\int\sqrt{q}\bar{G}^{ijkl}(\delta_{phys}\bar{q}_{ij})(\delta_{gauge}\bar{q}_{kl})\,d^3x = 0 \quad \forall\,\xi_l, \quad (8.41)$$

leads to $\nabla_k\sqrt{q}\bar{G}^{ijkl}\delta_{phys}\bar{q}_{ij} = 0$, which is equivalent to

$$\chi_j = \bar{q}^{ik}\nabla_i\delta_{phys}\bar{q}_{kj} = 0. \quad (8.42)$$

Similarly,

$$\int\sqrt{q}(\delta_{phys}\ln q)(\mathcal{L}_{\vec{\xi}}\ln q)\,d^3x = \int\sqrt{q}(\delta_{phys}\ln q)\left(\frac{2}{3}\nabla_i\xi^i\right)d^3x = 0 \quad (8.43)$$

yields the aforementioned auxiliary condition $\nabla_i\delta_{phys}\ln q = 0$.

---

[13] This also renders $\partial_i\ln q = \partial_i\ln q_0$ in Δ^i_{jk}.

8.4.4 Solution of the diffeomorphism constraint

The key to a generic solution of

$$\mathcal{H}_i^T = -2\bar{q}_{ik}\nabla_j\bar{\pi}^{kj} = 0 \tag{8.44}$$

is to decompose the traceless momentum variable into

$$\bar{\pi}^{ij} = \bar{\pi}_{TT}^{ij} + \left(\nabla^i\bar{W}^j + \nabla^j\bar{W}^i - \frac{2}{3}q^{ij}\nabla_k\bar{W}^k\right)$$

$$=: \bar{\pi}_{TT}^{ij} + L_k^{ij}\bar{W}^k, \tag{8.45}$$

with $\bar{\pi}_{TT}^{ij}$ being transverse as well as traceless. In analogy with Eq. (8.30) and the subsequent steps, a corresponding positive-definite block diagonal metric (which defines the inner product between symmetric tensor densities of weight $q^{\frac{1}{3}}\sqrt{q}$ and also between vector densities of the same weight) yields the inner product relation

$$\langle\bar{\pi}|L\bar{W}\rangle := \int \frac{\bar{G}_{klij}}{\sqrt{q}}\bar{\pi}^{kl}(L_m^{ij}\bar{W}^m)d^3x$$

$$= -\int 2\left(\frac{\bar{G}_{klij}}{\sqrt{q}}\nabla^i\bar{\pi}^{kl}\right)\bar{W}^jd^3x$$

$$= \int \frac{\bar{q}_{ij}}{q^{\frac{1}{3}}\sqrt{q}}(L_{kl}^{\dagger i}\bar{\pi}^{kl})\bar{W}^jd^3x$$

$$= : \langle L^\dagger\bar{\pi}|\bar{W}\rangle, \tag{8.46}$$

with $\bar{G}_{ijkl}$ being the inverse of $\bar{G}^{mnop}$, and $L_{kl}^{\dagger i}\bar{\pi}^{kl} = -2q^{\frac{1}{3}}\bar{q}^{im}\bar{G}_{klmj}\nabla^j\bar{\pi}^{kl}$. The constraint $\mathcal{H}_i = 0 \Leftrightarrow L_{kl}^{\dagger m}\bar{\pi}^{kl} = 0$. Thus the physical momentum which is constrained to be TT satisfies $L_{jk}^{\dagger i}\bar{\pi}_{TT}^{jk} = 0$. Operating with $L^\dagger$ on Eq. (8.45) results in

$$L_{ij}^{\dagger k}\bar{\pi}^{ij} = 0 + L_{ij}^{\dagger k}L_l^{ij}\bar{W}^l, \tag{8.47}$$

which allows an inversion via the corresponding Green's function of $L^\dagger L$. Inversion of the positive semi-definite operator eliminates precisely three d.o.f. of $\bar{\pi}^{ij}$ which is contained in $\bar{W}^i$. So starting with the spatial metric and an arbitrary symmetric matrix p^{ij}, the densitized $\bar{\pi}^{ij} := q^{\frac{1}{3}}\sqrt{q}p^{ij}$ can be constructed, and with this, the projection defined by

$$\bar{\pi}_{TT}^{ij} := \bar{\pi}^{ij} - L_k^{ij}\bar{W}^k = \left(\left[I - L\frac{1}{L^\dagger L}L^\dagger\right]\bar{\pi}\right)^{ij}, \tag{8.48}$$

is TT and a physical solution of $\mathcal{H}_i = 0$.

The operator $L^\dagger L$ is positive semidefinite, with kernel consisting of those $\bar{W}_{o\alpha}^i$ which obey

$$(L\bar{W}_{o\alpha})^{ij} = 0.$$

The general solution,

$$\bar{W}^i = \bar{W}^i_p + \sum_\alpha c^\alpha \bar{W}^i_{o\alpha},$$

is a linear combination of the particular solution and elements of the kernel. But $(L\bar{W})^{ij} = (L\bar{W}_p)^{ij}$, so the presence of a non-trivial kernel does not affect the physical mode $\bar{\pi}^{ij}_{TT}$ and the uniqueness of the decomposition Eq. (8.45).

8.4.5   *Reduced symplectic potential*

Substituting the decomposition of $\tilde{\pi}^{ij}$ into the symplectic potential and integrating by parts terms with $\bar{W}^i$ reveal that

$$\int \tilde{\pi}^{ij} \delta q_{ij} d^3x = \int (\bar{\pi}^{ij}_{TT} \delta \bar{q}_{ij} - 2\bar{W}^j \nabla^i \delta \bar{q}_{ij} + \pi \delta \ln q^{\frac{1}{3}}) d^3x. \quad (8.49)$$

When restricted to the physical subspace with $\nabla^i \delta_{phys}.\bar{q}_{ij} = 0$ i.e. to traceless-transverse excitations $\delta \bar{q}^{TT}_{ij}$, the symplectic potential reduces to $\int (\bar{\pi}^{ij}_{TT} \delta \bar{q}^{TT}_{ij} + \pi \delta \ln q^{\frac{1}{3}})$ yielding two physical TT degrees of freedom, and an extra pair $(\ln q^{\frac{1}{3}}, \pi)$ to feature in intrinsic time and the Hamiltonian density. For perturbations about any background q^*_{ij}, the physical spatial metric excitations are traceless $(q^{*ij}\delta_{phys}\bar{q}_{ij} = 0)$ and transverse $(\nabla^{*i}\delta_{phys}\bar{q}_{ij} = 0)$ w.r.t. q^*_{ij}, correctly accounting for two perturbative graviton d.o.f. and two polarizations in classical gravitational waves (for instance, as explicitly discussed in Chapter 6, the $Y^{Klm}_{(4,5)ij}$ TT perturbations about S^3).

8.5   Physical Hamiltonian, and generalization beyond Einstein's theory

Reduction of the canonical functional integral to the *unconstrained* physical phase space can now be completed. Taking Eq. (8.8) and Eq. (8.10) into account, Eq. (8.23) leads, upon integrating over $D\mu$ of Eq. (8.25), to

$$\int D\bar{\mu} \exp\left[\frac{i}{\hbar} \int \int_\Sigma \left(\bar{\pi}^{ij} \frac{\partial \bar{q}_{ij}}{\partial T} - \frac{\bar{H}(q(T), \bar{q}_{ij}, \bar{\pi}^{ij})}{\beta}\right) d^3x dT\right], \quad (8.50)$$

with the *emergence of a Hamiltonian* generating T-translations. Further integration over $D\bar{\mu}$ of Eq. (8.26) yields the true physical functional integral, $\int D\bar{\mu}_{phys} \exp(\frac{i}{\hbar} S_{phys})$, with

$$S_{phys} = \int \left(\int_\Sigma \bar{\pi}^{ij}_{TT} \frac{\partial \bar{q}^{TT}_{ij}}{\partial T} d^3x\right) dT - \int H_{phys} dT, \quad (8.51)$$

$$D\bar{\mu}_{phys} = \prod_{x\in\Sigma} \prod_{i,j} \frac{\delta \bar{\pi}^{ij}_{TT}(x) \delta \bar{q}^{TT}_{ij}(x)}{2\pi\hbar}; \quad (8.52)$$

and corresponding emergent and nontrivial physical Hamiltonian,

$$H_{\text{phys}} = \frac{1}{\beta} \int_{\Sigma} \bar{H}(\ln q^{1/3}(T), \bar{q}_{ij}^{TT}, \bar{\pi}_{TT}^{ij}) \, d^3x. \tag{8.53}$$

A crucial point to note is that the precise form of X, Y_i, Z_i play no role in total determinant of the Poisson brackets of the constraints discussed earlier, and in the above derivation of the physical Hamiltonian; so changes in the precise form of $\mathcal{V}$ are in fact allowed within this framework. Einstein's GR (with $\beta = \frac{1}{\sqrt{6}}$ and $\mathcal{V} = -\frac{q}{(2\kappa)^2}[R - 2\Lambda_{\text{eff}}]$) *is thus a particular realization of this wider class of theories* which is *compatible with spatial diffeomorphism symmetry and an ever-expanding closed universe* [16, 72, 87]. Generalization of

$$\mathcal{H} = \beta^2 \tilde{\pi}^2 - \bar{H}^2$$

to other cases with

$$\bar{H} = \sqrt{\bar{G}_{ijkl}\bar{\pi}^{ij}\bar{\pi}^{kl} + \mathcal{V}[q_{ij}]}, \tag{8.54}$$

wherein $\mathcal{V}(q, \bar{q}_{ij})$ is a spatial scalar density of weight two is allowed [16]. In particular, the advantages of

$$\bar{H} = \sqrt{\hat{Q}_j^{\dagger i}\hat{Q}_i^j + q\mathcal{K}} \tag{8.55}$$

with positive coupling $\mathcal{K}$ have been elaborated in Chapter 7. Positivity of $\bar{H}$ in Eq. (8.55) also completes the earlier nonvanishing requirement of

$$\{\chi, \mathcal{H}\}_{P.B.} = 2\beta^2\tilde{\pi} = -2\beta\bar{H}.$$

8.5.1 *Ultraviolet-complete extension of Einstein's theory and the Yang-Mills analogy*

In the quantum context, this Hamiltonian with is Eq. (8.55) self-adjoint. Explicit spatial diffeomorphism invariance is achieved by introducing the interactions through

$$\hat{Q}_j^i := e^{W_T}\hat{\bar{\pi}}_j^i e^{-W_T} = \hat{\bar{\pi}}_j^i - ig\hbar\tilde{C}_j^i$$

with W_T being a combination of the Einstein-Hilbert action, W_{EH}, and the Chern-Simons functional, W_{CS}, in three spatial dimensions i.e.

$$\begin{aligned} W_T &= -gW_{CS} + \alpha W_{EH} \\ &= -\frac{g}{2}\int \tilde{\epsilon}^{ikj}\left(\Gamma_{im}^l\partial_j\Gamma_{kl}^m + \frac{2}{3}\Gamma_{im}^l\Gamma_{jn}^m\Gamma_{kl}^n\right)d^3x + \alpha\int\sqrt{q}R d^3x. \end{aligned} \tag{8.56}$$

As demonstrated in Ref. [92] and in Chapter 7, the final Hamiltonian in the limit of regulator removal is

$$H_{\text{phys}} = \int \frac{\bar{H}(x)}{\beta} d^3x,$$

$$\bar{H} = \sqrt{\hat{Q}_j^{\dagger i} \hat{Q}_i^j + q\mathcal{K}}$$

$$= \sqrt{\bar{\pi}_j^{\dagger i} \bar{\pi}_i^j + \hbar^2 g^2 \tilde{C}_j^i \tilde{C}_i^j - \frac{q}{(2\kappa)^2}(R - 2\Lambda_{\text{eff}})}, \qquad (8.57)$$

wherein $\tilde{C}^{ij} = \frac{\delta W_{CS}}{\delta q_{ij}}$ is the Cotton-York tensor (density). Associated with the dimensionless coupling constant g^2 in Eq. (8.57), the Cotton-York term (which contains up to six spatial derivatives) modifies the propagator of Einstein's theory and ensures power-counting ultraviolet convergence [31].

The Cotton-York extension in Eq. (8.57) is, remarkably, the analog of the magnetic field contribution in Yang-Mills Hamiltonian density,

$$\frac{q_{ij}}{2}(\hat{\bar{\pi}}_A^{\text{a}} \hat{\bar{\pi}}_A^{j\text{a}} + \hbar^2 \tilde{B}^{i\text{a}} \tilde{B}^{j\text{a}}) = \frac{q_{ij}}{2}\hat{Q}^{\dagger \, i\text{a}}\hat{Q}^{j\text{a}}, \qquad (8.58)$$

with

$$\hat{Q}^{i\text{a}} = e^{W_{CS}}\hat{\bar{\pi}}_A^{i\text{a}}e^{-W_{CS}} = \hat{\bar{\pi}}_A^{i\text{a}} + i\hbar\tilde{B}^{i\text{a}}; \qquad (8.59)$$

wherein $\hat{\bar{\pi}}_A^{i\text{a}} = \frac{\hbar}{i}\frac{\delta}{\delta A_{i\text{a}}}$ is the conjugate momentum to the gauge potential. The Yang-Mills magnetic field, $\tilde{B}^{i\text{a}} = \frac{\delta W_{CS}}{\delta A_{i\text{a}}}$, is the functional derivative of the Chern-Simons functional of the Yang-Mills connection $A_{i\text{a}}$; analogously the Cotton-York tensor density is metric functional derivative of the Chern-Simons functional W_{CS}. They both gave rise to dimensionless couplings and power-counting renormalizability. But for Yang-Mills there is no analog of the Einstein-Hilbert action term $\int \sqrt{q}R d^3x$, which can be added to W_T in Eq. (8.56). The contribution of W_{EH} to the Hamiltonian in the form of the spatial Ricci scalar potential in Eq. (8.57) is lower order in spatial derivatives than $\tilde{C}_j^i\tilde{C}_i^j$, so the ultraviolet behavior and renormalizability of the extended theory is now determined by the Cotton-York term.

8.6 Comparison of extrinsic, intrinsic, and scalar field time

8.6.1 *Extrinsic time with Hamilton implicitly dependent upon solution of Lichnerowicz-York equation, versus intrinsic time with explicit Hamiltonian*

In the extrinsic time formulation of York [17], constancy of

$$p := \frac{\tilde{\pi}}{\sqrt{q}} \qquad (8.60)$$

is invoked, and the Hamiltonian constraint $H = 0$ translates into the Lichnerowicz-York equation [61, 77],

$$p^2 - \frac{q}{\beta^2}\bar{H}^2(\bar{\pi}^{ij}, \bar{q}_{ij}, q) = 0. \tag{8.61}$$

This determines q uniquely, thus eliminating the degrees of freedom $(\ln q^{\frac{1}{3}}, \tilde{\pi})$ in terms of $(\bar{\pi}^{ij}, \bar{q}_{ij}, p)$. However, the reduced Hamiltonian, which can be seen from

$$\int \int \tilde{\pi} \frac{\partial \ln q^{\frac{1}{3}}}{\partial t} dt d^3x = -\frac{2}{3} \int \left(\int \sqrt{q} d^3x \right) dp \tag{8.62}$$

is then proportional to the spatial volume, wherein $q(\bar{\pi}^{ij}, \bar{q}_{ij}, p)$ is known only implicitly and non-locally from the Lichnerowicz-York equation.

In contradistinction, ITG adopts the $d \ln q^{\frac{1}{3}} = dT = \frac{2}{3} d \ln V$ as the universal cosmic time interval compatible with an expanding spatially closed universe, together with Hamiltonian constraint expressed as $\beta \tilde{\pi} = -\bar{H}$. This similarly eliminates the local d.o.f. in $(\ln q^{\frac{1}{3}}, \tilde{\pi})$. But the resultant Hamiltonian is now both *explicit*, and of the remarkable *relativistic square-root form*,

$$H_{phys.} = \frac{1}{\beta} \int \bar{H} d^3x = \frac{1}{\beta} \int \sqrt{\bar{\pi}^i_j \bar{\pi}^j_i + \mathcal{V}} d^3x.$$

Einstein's theory corresponds to $\beta = \frac{1}{\sqrt{6}}$ and $\mathcal{V} = -\frac{q}{(2\kappa)^2}[R - 2\Lambda_{\text{eff}}]$. But, as explained, this scheme permits *additional terms in the potential to improve ultraviolet convergence*, while *infrared divergence is completely curbed by spatial compactness*.

8.6.2 Scalar field time and problematic negative kinetic term

A scheme which uses a real scalar field as "time" would correspond to imposing the constraint

$$\chi = \phi - k = 0, \tag{8.63}$$

and the Hamiltonian constraint with an additional scalar field will be

$$\mathcal{H} = \beta^2 \tilde{\pi}^2 - \bar{H}^2 - \frac{1}{2\kappa} \left[\frac{\tilde{\pi}_\phi^2}{2} + \frac{1}{2} q q^{ij} \nabla_i \phi \nabla_j \phi + V(\phi) \right]$$

$$= 0. \tag{8.64}$$

The resultant Poisson bracket of these yields

$$\{\chi, \mathcal{H}\}_{P.B.} = -\frac{1}{2\kappa}\tilde{\pi}_\phi$$

$$= \mp \frac{1}{\sqrt{2\kappa}} \sqrt{2(\beta^2 \tilde{\pi}^2 - \bar{G}_{ijkl}\bar{\pi}^{ij}\bar{\pi}^{kl} - \mathcal{V}[q_{ij}]) - \frac{1}{2\kappa}V|_{\phi=k}} \tag{8.65}$$

on the constraint surface. Among other potential problems, the negative-(semi)definite entity $-\bar{G}_{ijkl}\bar{\pi}^{ij}\bar{\pi}^{kl}$ compromises[14] the reality of $\tilde{\pi}_\phi$, and this carries over to the reduced Hamiltonian term,

$$\int \int \tilde{\pi}_\phi \frac{d\phi}{dt} dt\, d^3x = \int \left(\int \tilde{\pi}_\phi d^3x \right) dk, \qquad (8.66)$$

in the action. It should be pointed out the origin of this difficulty arises from the relative sign difference, in Eq. (8.64), of the first term w.r.t. the rest; this very sign difference is a blessing in ITG which allows the apposite elimination of $\tilde{\pi}$ in terms of the rest of the variables. Moreover, unlike both the extrinsic and intrinsic time formulations above, it fails to eliminate the $(\ln q^{\frac{1}{3}}, \tilde{\pi})$ degrees of freedom, so while the total number of degrees of freedom modulo constraints and subsidiary conditions is preserved, there are nevertheless three (and not two) remaining *gravitational* degrees of freedom.

While consistent extension of the physical reduced Hamiltonian of GR is a significant aspect emphasised in this book, other approaches to resolve the problem of time and extract the physical degrees of freedom have been tried before. Reference [104] starts with a Baierlein-Sharp-Wheeler action [62] (in Ref. [16] this is also discussed within the context of intrinsic time gravity); and the main distinction is that instead of using $\ln q^{\frac{1}{3}}$ as time and solving the Hamiltonian constraint though the elimination of $\tilde{\pi}$, an extra λ time parameter is introduced with resultant constraint in the form of the Lichnerowicz-York equation. In Ref. [105], transverse-traceless physical decomposition was carried out with Minkowski background for the spatially non-compact case, while York's method was discussed for spatially compact manifolds. Functional path integral for the gravitational field has been developed earlier by Teitelboim by analogy with the quantum mechanics of covariant relativistic point particle [106], whereas our work formulates GR without general covariance. In Refs. [107] and [108] additional relativistic dust matter was invoked. Besides current observational conformity with cold, rather than relativistic, dark matter, the relative sign difference with the $\tilde{\pi}^2$ term is, as discussed in scalar field time, a potential problem in guaranteeing the reality of the square-root in the reduced Hamiltonian. Although may be curbed by imposing certain energy conditions; an alternative strategy discussed above is to exploit the sign difference and use York extrinsic time or our intrinsic time variable.

---

[14]Likewise, incorporation of Yang-Mills fields adds a problematic negative semi-definite term $\propto -H_{YM}$.

8.7 Incorporation of matter and Yang-Mills fields and solution of diffeomorphism constraint

Incorporating Yang-Mills and matter (both scalar and fermionic) fields into the theory changes the total Hamiltonian constraint to

$$\mathcal{H} = \beta^2 \tilde{\pi}^2 - \bar{H}^2 - H_{\text{matter+YM}} = 0. \tag{8.67}$$

But no couplings to $\tilde{\pi}$ appear in the usual $H_{\text{matter+YM}}$, so the Poisson bracket in Eq. (8.10) with the new $\mathcal{H}$ is unaffected. It follows that $\bar{H}$ is modified to $\sqrt{\bar{H}^2 + H_{\text{matter+YM}}}$ in the reduced physical Hamiltonian of Eq. (8.53). The momentum constraint will now include the generator of spatial diffeomorphisms $\tilde{D}_i$ for these additional fields i.e.

$$\mathcal{H}_i^{\text{GR+matter}} = -2\bar{q}_{ik}\nabla_j\tilde{\pi}^{jk} + \tilde{D}_i = 0; \tag{8.68}$$

but the decomposition Eq. (8.45) of the momentum variable $\tilde{\pi}^{ij}$ is still valid. The solution of the new diffeomorphism constraint is the momentum

$$\tilde{\pi}^{ij}_{phys} = \tilde{\pi}^{ij}_{TT} + (L\bar{W}_{phys})^{ij}, \tag{8.69}$$

with a certain $\bar{W}^i_{phys}$. Substituting this into Eq. (8.68) results in

$$- 2\bar{q}_{ik}\nabla_j(L\bar{W}_{phys})^{jk} + \tilde{D}_i = 0; \tag{8.70}$$

however,

$$L^{\dagger i}_{jk}L^{jk}_l\bar{W}^l_{phys} = -2\nabla_j(L\bar{W}_{phys})^{ij} = -\bar{q}^{ij}\tilde{D}_j, \tag{8.71}$$

so the particular solution $\bar{W}^i_{phys}$ is completely determined inverting $L^\dagger L$. This changes the explicit particular solution but does not disturb $\tilde{\pi}^{ij}_{TT}$ as free initial data for the momentum variable. In pure GR, $\tilde{D}_i$ vanishes, yielding $W^i_{phys} = 0$, and $\tilde{\pi}^{ij}_{phys} = \tilde{\pi}^{ij}_{TT}$ only. A general method for tackling constraints linear in the momentum and their solutions in the presence of sources will be presented in the next chapter; meanwhile, Electrodynamics is an apposite illustration. The analog in Electrodynamics is $E^i_{\text{phys}} = E^i_T - \nabla^i\phi_{\text{phys}}$, with ϕ_{phys} satisfying the (invertible) Poisson equation $\nabla^2\phi_{phys} = -4\pi\rho$; consequently, $\nabla_i E^i_{phys} = 4\pi\rho$. In the absence of sources the physical electric field is purely transverse.

Thus inclusion of matter and Yang-Mills gauge field content does not alter the salient fact the unconstrained or free gravitational initial data lie in the TT part of the momentum. In the reduction scheme of ITG, $\tilde{\pi}^{ij}_{phys}$ is remaining momentum d.o.f., while the trace part is eliminated by solving the Hamiltonian constraint as

$$\tilde{\pi} = -\frac{1}{\beta}\sqrt{\bar{H}^2 + H_{\text{matter+YM}}}$$

$$H_{phys} = \frac{1}{\beta}\int \sqrt{\bar{H}^2 + H_{\text{matter+YM}}}\, d^3x. \tag{8.72}$$

That the physical Hamiltonian, H_{phys}, is (through q) ultimately T-dependent may be disconcerting at first; but this feature also occurs in the York formulation wherein $\sqrt{q}$ depends on the extrinsic time parameter p, and k too appears in the potential of the scalar field Hamiltonian. This time-dependence of the Hamiltonian is the consequence of an internal clock which arose from a degree of freedom of the theory. In the most general formulation of classical and quantum mechanics, time-dependent Hamiltonians are allowed, and they can lead to perfectly well-defined theories. In the quantum context it is unitary, arising from self-adjoint but not necessarily time-independent Hamiltonians, which guarantees probability conservation. A time-dependent Hamiltonian comes with the additional features of time irreversibility and of its different terms dominating at different times; allowing our universe to have very disparate early and late eras, and to begin very differently from how it will end.

Chapter 9

Chiral fermions, gauging the Lorentz group without four-dimensional spacetime; and solving all the constraints

"Wherever anything lives, there is, open somewhere,
a register in which time is being inscribed."
— Henri Bergson

9.1 Are fermions fundamental when spacetime is not?

A main thesis of this book is the paradigm shift in Gravitation, from four-covariance to spatial diffeomorphism invariance. It has been shown, in earlier chapters, by solving the Hamiltonian constraint explicitly a physical reduced Hamiltonian which generates dynamical evolution in cosmic time emerges. This cosmic evolution is not only philosophically and intuitively appealing, but also observationally and physically compelling. A single 3dDI time-ordered quantum wave function governs alls; it is the embodiment of the shared past, present, and future of all that is our Universe. Quantum fields including quantum matter fields are the building blocks; yet the coexistence of quantum gravity and fermionic matter is not without challenges. For one, not all manifolds admit spin structures, and on such manifolds fermions cannot interact consistently with just gravitational fields. More importantly, is four-dimensional spacetime indispensable to the fundamental chiral fermions of our Standard Model and Grand Unified Theories; or is three-geometry enough? A fermion is a scalar under diffeomorphisms; however, chiral Weyl fermions also transform as spin $\frac{1}{2}$ entities under the Lorentz group of the tangent space; in the four-dimensional context they couple to the self- and anti-self-dual Lorentz spin connections. As "space-time is a concept of limited applicability", can fermions retain their fundamental nature when quantum gravity holds sway, or must they somehow be mere manifestations of even more basic entities?

189

A key to the resolution of these issues lies in the extraordinary nature of the Lorentz group itself. Often cited as the isometry group of the tangent space at each point, $SO(3,1)$ and its spinorial double cover seem to be predicated upon the existence of four-dimensional space-times. Yet, by a miracle, the isomorphism $SO(3,1) \cong SO(3,C)$ renders chiral fermions and triads as fundamental building blocks compatible and kindred with just three-geometry, rather four-geometry. $SL(2,C)$ is the double cover of both,[1] and 2-component Weyl fermions are the associated fundamental matter fields of particle physics. The physical Hamiltonian too is the spatial integral of three-dimensional entities, momenta are more fundamental than velocities, and Einstein's geometrodynamics is the dynamics of three-geometry, not four-geometry. Hamiltonian mechanics requires no *a priori* four-dimensional space-time, neither does quantum Schrödinger nor Heisenberg evolution with 3dDI time-ordering. Sections 9.2 and 9.3 reveal that in the coupling of a chiral fermion to gravitation, the fermion Hamiltonian is, without recourse to four-dimensional spacetime, dependent upon spatial $SO(3,C)$ triads,[2] Weyl fermions and their conjugate momentum as fundamental entities; and it is Lorentz group as $SO(3,C)$ for dreibein (the inverse of the triads), and as the double covering $SL(2,C)$ for Weyl fermions, that is effectively being gauged. Despite assertions in some QFT texts, *a priori* four dimensions is not needed to embrace the full Lorentz group and to define elementary chiral fermions [109]. In quantum gravity four-dimensional spacetime does not really exists, yet chiral fermions and particle physics theories constructed with them can remain fundamental.

9.1.1 *Inclusion of fermionic fields and solution of all the constraints: A brief overview*

A brief overview of this long chapter is as follows: Sections 9.2 and 9.3 present the conventional coupling of Weyl fermions to gravitation; first from the $(3+1)$-dimensional perspective of the action and its decomposition, and then without recourse to an *a priori* four-dimensional spacetime, from the Hamiltonian perspective and the gauging of the Lorentz group as $SO(3,C)$ and its universal and double cover $SL(2,C)$. The Hamiltonian density is derived, together with the constraints of spatial diffeomorphism and Lorentz symmetries. Several noteworthy features elaborated on include:

---

[1]Thus the Lorentz group is, amusingly, also the only known Lie group which possesses irreducible 2-, 3-, and 4-dimensional representations.

[2]See Eq. (9.13) and Eq. (9.36).

1) The absence of any coupling to $\tilde{\pi}$ in the hermitized fermion Hamiltonian; this means the introduction of fermions does not complicate the solution of the Hamiltonian constraint of expressing $\tilde{\pi}$ in terms of the rest of the variables. The solution of the general Hamiltonian constraint is presented in Sec. 9.6.

2) An independent spin connection with separate torsion d.o.f. is not necessary and not introduced, keeping the theory in its minimal form; rather, the total spin connection can be constructed, as in Eq. (9.39), from the combination of the torsionless spin connection compatible with the triad and the momentum conjugate to the triad. Axial torsion, as a composite rather than independent or extra d.o.f., and its effects can still be present in fermion coupling. The physical ramifications are addressed in Sec. 9.6.1.

3) Weight one-half fermion fields are introduced, with the goal of diffeomorphism-invariant fermion measure. Chiral fermion multiplets interacting with Yang-Mills and gravitational fields entail potential gauge and Lorentz anomalies. Both perturbative and global anomalies and the conditions for their absence are discussed.

4) Not all manifolds allow spin structures. A quantum origin of our universe from Euclidean instanton tunneling such as the scenario which will be presented in Chapter 10 would naturally need to ensure fermions and generic gravitation instantons can coexist consistently. Remarkably, the additional coupling of fermions to Yang-Mills connection of an apposite GUT group can lead to a generalized spin structure and the harmonious coexistence of chiral fermions, gravity and Yang-Mills fields. This resolution, and its implications for astroparticle physics, is elaborated on in Sec. 9.7.

All the constraints of the theory are solved exactly in this chapter in Secs. 9.4–9.6. A general scheme to solve exactly first class gauge constraints linear in the momentum, and to promote such first class systems to Dirac second class is presented in Sec. 9.5. Auxiliary (or "gauge-fixing") conditions for dreibein and spatial metric and their geometrical interpretation are discussed, together with the role they play in fixing the shift function.

It is also worth pointing out that all momenta can be written in terms dreibein and momentric; and all constraints, and also the Hamiltonian, can be expressed in terms of the latter which has well-defined $SL(3, R)$ group properties and group action on the spatial metric. In pure gravity, the spatial metric and momentric are the fundamental variables for ITG; when $SL(2, C)$ chiral fermions are introduced, $SO(3, C)$ dreibein becomes fundamental and the metric is now a composite field, but momentric still retains its cardinal role.

9.2   Weyl fermions, and gauging the Lorentz group: (3 + 1)-decomposition

Gauging of the tangential $SO(1,3)$ group in four-dimensional Lorentzian spacetime with standard 3+1 ADM decomposition has been worked out long ago; however, it will become apparent (as in the final Hamiltonian density $\tilde{\mu}_F$ in Eq. (9.40)), it is Lorentz group as $SO(3,C)$ for dreibein, and as the double covering $SL(2,C)$ for Weyl fermions, that is effectively being gauged. To relate to the perspective of four-covariant Lagrangian and action to the final fermion Hamiltonian, it suffices to consider just a single two-component left-handed Weyl fermion, ψ, coupled to gravity. In the chiral representation this has the Hermitian action

$$
\begin{aligned}
S_{Weyl} &= \frac{1}{2}\int d^4x (\det{}^4e)\psi^\dagger E^\mu_A \tau^A i D_\mu \psi + \text{hermitian conjugate}\\
&= \frac{1}{2}\int dt \int d^3x (Ne)\psi^\dagger (\mathcal{E}^\mu_A - n_A n^\mu)\tau^A i D_\mu \psi + h.c.\\
&= \frac{1}{2}\left[\int dt \int d^3x N e \psi^\dagger \mathcal{E}^\mu_A \tau^A i D_\mu \psi - \int dt \int d^3x e\psi^\dagger (n_A \tau^A)t^\mu i D_\mu \psi \right.\\
&\quad \left. + \int dt \int d^3x N^\mu e\psi^\dagger (n_A \tau^A)i D_\mu \psi\right] + h.c..
\end{aligned}
\tag{9.1}
$$

In Eq. (9.1), the definition for $\tau^{A=0,1,2,3}$ is $\tau^0 = I_2, \tau^{a=1,2,3}$ are the Pauli matrices, likewise $\bar{\tau}^A = (I_2, -\tau^a = (\tau^2\tau^a\tau^2)^T)$; and $\psi^\dagger = \psi^{*T}$. We have also decomposed the tetrad, in the tangential and normal directions, as

$$
E^\mu_A = (\mathcal{E}^\mu_A - n_A n^\mu),
\tag{9.2}
$$

with the normal vector n^μ to the ADM spatial hypersurface satisfying,

$$
n_\mu n^\mu = -1 \text{ and } n_A \equiv n_\mu E^\mu_A.
\tag{9.3}
$$

As usual, the tetrad and its inverse the vierbein realize the isomorphism between curved spacetime vectors and Minkowski tangent space vectors. It transforms as $E'^{\mu A} = (\Lambda)^A{}_B E^{\mu B}, E'^\mu{}_A = E^\mu{}_B (\Lambda^{-1})^B{}_A \ \forall \Lambda \in SO(1,3)$. In terms of the lapse and shift functions, the normal can be expressed as,

$$
n^\mu = \frac{(t^\mu - N^\mu)}{N}.
\tag{9.4}
$$

The conjugate momentum of the fermions field is[3]

$$
\tilde{\pi}_\psi = \frac{\partial L}{\partial(\mathcal{L}_t\tilde{\psi})} = -i\tilde{\psi}^\dagger (n_A \tau^A),
\tag{9.5}
$$

---

[3]See, for instance, (9.28).

wherein $\mathcal{L}_t$ is the Lie derivative with respect to the vector field

$$t^\mu \partial_\mu = (Nn^\mu + N^\mu)\partial_\mu. \tag{9.6}$$

Under the action of the Lorentz group, $\tilde{\pi}_\psi$ transforms as $\tilde{\pi}'_\psi = \tilde{\pi}_\psi u_L^{-1}$ i.e. inversely as the left-handed spinor $\tilde{\psi}' = u_L \tilde{\psi}$, $\forall u_L \in SL(2,C)$. (This property is required to maintain invariance of the fundamental anti-commutation relations under Lorentz transformations.) It also follows from Eq. (9.5), $u_R^\dagger = u_L^{-1}$, $n'_A = n_B(\Lambda^{-1})^B{}_A$, and the identities

$$(\Lambda^{-1})^A{}_B \tau^B = u_R \tau^A u_R^\dagger, \qquad (\Lambda^{-1})^A{}_B \bar{\tau}^B = u_L \bar{\tau}^A u_L^\dagger. \tag{9.7}$$

In fact one can view these identities as defining $\Lambda \in SO(1,3)$ from more fundamental $u_{L,R}$.

So the contribution to the Hamiltonian constraint, which has density weight one and comes with N as the Lagrange multiplier, is

$$\begin{aligned}
\tilde{\mu}_F &= -\frac{1}{2} e\psi^\dagger \mathcal{E}_A^\mu \tau^A D_\mu \psi + h.c. \\
&= -\frac{1}{2} e\psi^\dagger (n_B \tau^B)(n_C \bar{\tau}^C) \mathcal{E}_A^\mu \tau^A D_\mu \psi + h.c. \\
&= \frac{1}{2} \tilde{\pi}_\psi (n_C \mathcal{E}_A^\mu) \bar{\tau}^C \tau^A D_\mu \tilde{\psi} + h.c. .
\end{aligned} \tag{9.8}$$

In the second line of Eq. (9.8), we have inserted the identity matrix,

$$(n_B \tau^B)(n_C \bar{\tau}^C) = I_2. \tag{9.9}$$

Furthermore, it also works out that the entity $(n_C \mathcal{E}_A^\mu)\bar{\tau}^C \tau^A$ transforms according to the adjoint representation[4] of $SL(2,C)_L$ (ergo as an $SO(3,C)_L$ vector); and it is a spatial vector since

$$n_\mu (n_C \mathcal{E}_A^\mu)\bar{\tau}^C \tau^A = 0, \tag{9.10}$$

follows from Eq. (9.2) and Eq. (9.3).

As

$$n_A n^A = n_\mu n^\mu = -1, \tag{9.11}$$

the choice of

$$n_A = (-1, 0, 0, 0), \tag{9.12}$$

yields

$$(n_C \mathcal{E}_A^\mu)\bar{\tau}^C \tau^A = n_C(E_A^\mu + n_A E_B^\mu n^B)\bar{\tau}^C \tau^A = -E_a^\mu \tau^a. \tag{9.13}$$

Since,

$$n_\mu(E_a^\mu \tau^a) = 0, \tag{9.14}$$

---

[4]With the help of Eq. (9.7), it follows that $n'_C \mathcal{E}_A'^\mu \bar{\tau}^C \tau^A = u_L (n_C \mathcal{E}_A^\mu \bar{\tau}^C \tau^A)u_L^{-1}$.

a further choice of $n_\mu = -N\delta^0_\mu$ renders[5] $E^0_a = 0$, and $(n_C \mathcal{E}^i_A)\bar{\tau}^C \tau^A = -E^i_a \tau^a$; and E^i_a consistently transforms as an $SO(3, C)_L$ vector.

Physically n_μ and n^A are not dynamical fields; and the particular choices above do not break the respective fundamental $SO(3, C)$ and $SL(2, C)$ symmetry in $(n_C \mathcal{E}^\mu_A)\bar{\tau}^C \tau^A$ and ψ. The formalism relates E^μ_A which is compatible with the ADM four-metric $g_{\mu\nu}$ to the spatial triad E^i_a (or its inverse e_{ai}) which is assumed to be the basic variable in the Hamiltonian description of ITG with fermions. Conversely, given the spatial triad and N (which emerges *a posteriori* in ITG), one can reconstruct the classical four-dimensional vierbein and spacetime by reversing the steps above.

The fermion contribution to the density weight two Hamiltonian constraint reduces to

$$\tilde{\bar{\mu}}_F = e\tilde{\mu}_F = -\tilde{\pi}_\psi \tilde{E}^i_a \frac{\tau^a}{2} e^{\frac{1}{2}} D_i e^{-\frac{1}{2}} \tilde{\psi} + h.c.$$

$$= -\tilde{\pi}_\psi \tilde{E}^i_a \frac{\tau^a}{2} \left(e^{\frac{1}{2}} \partial_i e^{-\frac{1}{2}} \tilde{\psi} + \frac{1}{4} A_{iAB} \bar{\tau}^A \tau^B \tilde{\psi} \right) + h.c.$$

$$= -\tilde{\pi}_\psi \tilde{E}^i_a \frac{\tau^a}{2} \left[e^{\frac{1}{2}} \partial_i e^{-\frac{1}{2}} \tilde{\psi} + \left(A_{i0b} + \frac{i}{2} \epsilon_{0bcd} A^{cd}_i \right) \frac{\tau^b}{2} \tilde{\psi} \right] + h.c.. \quad (9.15)$$

This will be the same as in Eq. (9.40) with the identification of the specific left-handed spin connection gauging the $u_L \in SL(2, C)_L$ Lorentz freedom in ψ. Note that $\tilde{E}^i_a$ and $\tilde{\pi}_\psi$ transform respectively according to the $SO(3, C)_L$ and u_L^{-1} representations. In Eq. (9.1) it was the covariant derivative carries the full $SO(1, 3)$ spin connection, however, it is only the (anti)self-dual projection of the connection that is actually gauging u_L while in the hermitian conjugate part of the Hamiltonian the other projection of the connection gauges the u_R freedom; and $(E^i{}_a)^\dagger$ transforms as $(O_L)^* = (e^{(-\alpha_a + i\theta_a)T^a})^* = e^{(\alpha_a + i\theta_a)T^a} \in SO(3, C)_R$ since the generators are hermitian but anti-symmetric 3×3 matrices. It should also be noted that unlike the original Ashtekar theory which attempts to reproduce all of nature with spin connection of one chirality [46, 65, 110], in ITG hermiticity causes the accompanying right-handed spin connection to be present in the *h.c.* part of the Hamiltonian and Lagrangian. Performing a Legendre transformation, the corresponding fermionic contribution to the momentum constraint and Gauss Law constraint of Lorentz symmetry can be read off. With spatial N^i (note that $n_\mu N^\mu = 0$) and the $\mu = 0$ component of the spin connection as the respective Lagrange multipliers, these are associated with generators to be discussed in Eq. (9.45) and Eq. (9.50).

---

[5]This is consistent with $-1 = n_\mu n^\mu = g^{\mu\nu} n_\mu n_\nu = g^{00}(n_0)^2$ and $g^{00} = -1/N^2$ in ADM decomposition. And $-\delta^0_A = n_A = n_\mu E^\mu_A = -N E^0_A$ implies E^0_a must be trivial.

Finally to fledge out the four-dimensional picture, the torsion two-form is defined as $T_A := de_A + A_{AB} \wedge e^B$; with $T_A = \frac{1}{2}T_{A\mu\nu}dx^\mu dx^\nu$, veirbein $e^A = e^A_\mu dx^\mu$ and spin connection $A_{AB} = A_{AB\mu}dx^\mu$ one-forms. No matter what the torsion is, the generic spin connection can be decomposed as the torsionless spin connection $\omega_{\mu AB}$ plus torsion terms as

$$A_{AB\mu} = \omega_{AB\mu} - \frac{1}{2}[T_{A\sigma\mu}E^\sigma_B - T_{B\sigma\mu}E^\sigma_A - T_{C\rho\sigma}e^C_\mu E^\rho_A E^\sigma_B], \tag{9.16}$$

wherein the torsionless condition, $de_A + \omega_{AB} \wedge e^B = 0$, allows $\omega_{AB\mu}$ to be expressed totally in terms of the veirbein as

$$\omega_{AB\mu} = \frac{1}{2}[E^\nu_A(\partial_\mu e_{B\nu} - \partial_\nu e_{B\mu}) - E^\nu_B(\partial_\mu e_{A\nu} - \partial_\nu e_{A\mu})$$
$$- E^\alpha_A E^\beta_B(\partial_\alpha e_{C\beta} - \partial_\beta e_{C\alpha})e^C_\mu]. \tag{9.17}$$

A $GL(4,R)$ metric connection, $\Gamma^\alpha_{\beta\gamma}$, can be constructed from the vierbein through the tetrad (or metricity) condition,

$$\nabla^\Gamma_\mu E^\nu_A = \partial_\mu E^\nu_A + A^B_{\mu A}E^\nu_B + \Gamma^\nu_{\rho\mu}E^\rho_A = 0. \tag{9.18}$$

The spacetime metric $g_{\mu\nu}$ will then satisfy the metricity condition, $\nabla^\Gamma_\alpha g_{\mu\nu} = 0$, which preserves the norm of vectors and the orthogonality of vectors, under parallel transport with respect to the connection $\Gamma^\alpha_{\mu\nu}$. The symmetric part of $\Gamma^\alpha_{\mu\nu}$ is thus uniquely the Christoffel symbol constructed from the metric, and the tetrad condition relates torsion to its anti-symmetric part by

$$T_{A\mu\nu} = 2\Gamma^\alpha_{[\mu\nu]}e_{A\alpha}, \quad \Gamma^\alpha_{[\mu\nu]} = \frac{1}{2}(\Gamma^\alpha_{\mu\nu} - \Gamma^\alpha_{\nu\mu}). \tag{9.19}$$

In ITG, torsion degrees are not introduced as additional independent variables, nevertheless the spin connection can have fermionic torsion contributions.

9.3 Gauging local Lorentz symmetry without four-dimensional spacetime

At the fundamental level, without resorting to a four-dimensional spacetime, one can construct the quantum field theory of gravity with fermionic matter via the Hamiltonian description. On three-dimensional spatial Cauchy surfaces, the spatial metric q_{ij} is related to the dreibein, e_{ai} through $q_{ij} = e_{ai}e^a_i$ and it is invariant under local $SO(3,C)$ rotations of the dreibein. Under local Lorentz transformations, left and right-handed chiral Weyl fermions transform as

$$\psi'_{L,R} = u_{L,R}\psi_{L,R}, \tag{9.20}$$

where

$$u_{L,R} = e^{(\mp\alpha_a + i\theta_a)\frac{\tau^a}{2}} \in SL(2,C)_{L,R}, \tag{9.21}$$

with Pauli matrices τ^a and boost and rotation parameters α_a and θ_a respectively. The transformations are not unitary, however

$$u_R^\dagger = u_L^{-1} \text{ and } u_L^\dagger = u_R^{-1}. \tag{9.22}$$

The $SO(1,3)$ boost and rotational freedom of the spacetime vierbein is one-to-one with complex $SO(3,C)$ rotations of the dreibein above. The two groups are thus isomorphic. The left or right-handed $SL(2,C)_{L,R}$ above are the respective simply-connected double covers of $SO(3,C)_{L,R}$; and all of the Lorentz group can be captured with the $SO(3,C)$ freedom in the *spatial* dreibein (or its inverse, also called "triad") and the corresponding $SL(2,C)$ transformations of Weyl fermions. As a consequence, despite assertions in some QFT textbooks, four dimensions is not needed to embrace the full Lorentz group and to define elementary chiral fermions.

Under Lorentz transformations right and left-handed spin connections, $A_{ia}^\pm$, in the adjoint representation in three spatial dimensions transform as

$$A_{ia}'^\pm \frac{\tau^a}{2} = u_{R,L} A_{ia}^\pm \frac{\tau^a}{2} u_{R,L}^{-1} + u_{R,L} i\partial_i u_{R,L}^{-1}. \tag{9.23}$$

These (anti)self-dual connections are related by

$$A_{ia}^{\dagger\pm} = A_{ia}^\mp. \tag{9.24}$$

Left- and right-handed fermions can be reexpressed in terms of fields of the opposite chirality through

$$\psi_R = -i\tau^2 \chi_L^*. \tag{9.25}$$

This is useful in Grand Unified Theories wherein one expresses, say, all right-handed Weyl fermions in terms of left-handed ones, put all the fermions in a left-handed multiplet, and allow the components of the multiplet to transform into one another under a unifying Yang-Mills gauge group.

The gauging of the local Lorentz group with $SO(3,C)$ triads, $A_{ia}^\pm$ spin connections and chiral $SL(2,C)$ Weyl fermions will be taken up next. For simplicity, we consider only a single two-component Weyl fermion (as flavour and generation indices of fermions and non-gravitational Yang-Mills couplings are tangential to the discussion and can be tagged on without affecting the gauging of the local Lorentz group). Instead of a left-handed

Weyl fermion and its conjugate momentum (ψ, π_ψ), we may start with a left-handed density weight one-half Weyl fermion field,

$$\tilde{\psi} = e^{\frac{1}{2}}\psi, \qquad (9.26)$$

together with its density weight one-half conjugate momentum[6]

$$\tilde{\pi}_\psi = \pi_\psi e^{-\frac{1}{2}}. \qquad (9.27)$$

The motivation for this choice of tensor weights is covered in Sec. 9.3.3. It is interesting to observe that in the action of Eq. (9.1), the change of variables from (ψ, π_ψ) to $(\tilde{\psi} := e^{\frac{1}{2}}\psi, \tilde{\pi}_\psi := e^{-\frac{1}{2}}\pi_\psi)$ preserves the symplectic form since

$$ie\psi^\dagger(n_A\tau^A)\mathcal{L}_t\psi + h.c. = ie(\psi^\dagger(n_A\tau^A)\mathcal{L}_t\psi - (\mathcal{L}_t\psi^\dagger)(n_A\tau^A)\psi)$$
$$= i(\tilde{\psi}^\dagger(n_A\tau^A)\mathcal{L}_t\tilde{\psi} - (\mathcal{L}_t\tilde{\psi}^\dagger)(n_A\tau^A)\tilde{\psi}). \qquad (9.28)$$

Consequently, $\tilde{\pi}_\psi\mathcal{L}_t\tilde{\psi} + h.c. = \pi_\psi\mathcal{L}_t\psi + h.c.$, and the term containing $e^{-\frac{1}{2}}\mathcal{L}_t e^{\frac{1}{2}}$ and the complications it can bring are totally absent.

These fermion fields satisfy the fundamental anti-commutation relations,

$$\left[\tilde{\psi}^\alpha(x), (\tilde{\pi}_\psi)_\beta(x')\right]_+ = i\hbar\delta_\beta{}^\alpha\delta^3(x, x'), \qquad (9.29)$$

$$\left[\tilde{\psi}^\alpha(x), \tilde{\psi}^\beta(x')\right]_+ = [(\tilde{\pi}_\psi)_\alpha(x), (\tilde{\pi}_\psi)_\beta(x')]_+ = \left[(\tilde{\psi}^\dagger)_\alpha(x), (\tilde{\pi}_\psi)_\beta(x')\right]_+ = 0,$$

where α and β are the two-component spinor indices.

We can construct the corresponding covariant derivative $D_i^{A^-}$ acting on $\tilde{\psi}$ such that under local Lorentz transformations,

$$\tilde{\psi}' = u_L\tilde{\psi}, \quad \tilde{\pi}'_\psi = \tilde{\pi}_\psi u_L^{-1} \qquad \forall \, u_L \in SL(2, C), \qquad (9.30)$$

and Eq. (9.23), the Lorentz gauge-covariant derivative behaves as

$$(D_i^{A^-}\psi)' = u_L D_i^{A^-}\psi, \qquad iD_i^{A^-} := i\partial_i + A_{ia}^-\frac{\tau^a}{2}. \qquad (9.31)$$

(For notational simplicity and clarity, we use matrices and matrix multiplications instead of manipulating with spinorial indices; so a chiral Weyl fermion is a two-component single column matrix, $\tilde{\pi}_\psi$ and Pauli matrices are respectively 1×2 and 2×2.) In pure gravity, the anti-self-dual spin connection is also the Ashtekar connection [46]

$$A_{ia}^- = \Gamma_{ia} - iK_{ia} \qquad (9.32)$$

---

[6]π_ψ which is conjugate to ψ is of weight one.

wherein $e^a_{(i} K_{j)a}$ equals the extrinsic curvature K_{ij} on shell. K_{ia} is also conjugate to the densitized triad $\tilde{E}^{ia}$ of weight one ($\tilde{E}^{ia} = det(e)(e_{ai})^{-1}$); while Γ_{ia} is the torsionless spin connection compatible with the dreibein i.e. $de^a + \epsilon^{abc}\Gamma_b \wedge e_c = 0$, with one-forms $e^a = e^a_i dx^i$ and $\Gamma_a = \Gamma_{ia} dx^i$.

Under an arbitrary local Lorentz transformation Λ,

$$\tilde{E}'^{ia} = O(\Lambda^{-1})^a{}_b \tilde{E}^{ib}, \tag{9.33}$$

$$\tilde{E}'^i{}_a = \tilde{E}^i{}_b O(\Lambda)^b{}_a, \tag{9.34}$$

with $O \in SO(3, C)$. Since

$$O(\Lambda)^a{}_b \tau^b = u_L \tau^a u_L^{-1}, \qquad u_L \in SL(2, C), \tag{9.35}$$

the adjoint representation of $SL(2, C)$ is equivalent to the three-dimensional fundamental representation of $SO(3, C)$. This means

$$\tilde{E}'^i{}_a \frac{\tau^a}{2} = \tilde{E}^i{}_b O(\Lambda)^b{}_a \frac{\tau^a}{2}$$

$$= u_L \tilde{E}^i_a \frac{\tau^a}{2} u_L^{-1}. \tag{9.36}$$

The upshot is, with these ingredients, the *local Lorentz invariant* density weight two Hamiltonian density can be expressed as

$$\tilde{\tilde{\mu}}_F = -\tilde{\pi}_\psi \tilde{E}^i_a \frac{\tau^a}{2} e^{\frac{1}{2}} D_i e^{-\frac{1}{2}} \tilde{\psi} + h.c.. \tag{9.37}$$

A physical question arises as to whether or not one wishes to introduce more variables into the theory in the spin connection which is not torsionless in general. The Ashtekar spin connection offers a clue on how to gauge the Lorentz symmetry without introducing extraneous degrees of freedom. In pure GR, the EOM relates the extrinsic curvature to the momentum conjugate to the spatial metric. Similarly, for p_{ia} the conjugate momentum of the densitized triad,

$$K_{ia} = \kappa p_{ia}. \tag{9.38}$$

A well-founded off-shell prescription would be to define the left-handed spin connection as

$$A^-_{ia} := \Gamma_{ia} - i\kappa p_{ia} \tag{9.39}$$

with Γ_{ia} being uniquely the torsionless spin connection compatible with the dreibein e_{ai}. This means A^-_{ia} has the right transformation properties and is completely determined by just the dreibein and its conjugate momentum. There are then no extraneous degrees of freedom, even though A^-_{ia} can contain torsion. This is because, instead of Eq. (9.38), with fermions, the EOM for $\tilde{\pi}_{ai}$ can have fermionic contributions [44]. Off shell, the momentum p_{ia} is merely the entity conjugate to $\tilde{E}^{ia}$. With this choice of coupling, the final form of the Hamiltonian density,

$$\tilde{\tilde{\mu}}_F = -\tilde{\pi}_\psi \tilde{E}^i_a \frac{\tau^a}{2} e^{\frac{1}{2}} D^A_i e^{-\frac{1}{2}} \tilde{\psi} + h.c. \tag{9.40}$$

$$= -\tilde{\pi}_\psi \tilde{E}^i_a \frac{\tau^a}{2} e^{\frac{1}{2}} \left(\partial_i - i\Gamma_{ib} \frac{\tau^b}{2} \right) e^{-\frac{1}{2}} \tilde{\psi} + \frac{i}{2} \kappa \left(\tilde{E}^i_a p_{ib} \epsilon^{abc} \right) \tilde{\pi}_\psi \frac{\tau^c}{2} \tilde{\psi} + h.c.,$$

contains a further detail worth pointing out: the trace of the momentum conjugate to the spatial metric, $\tilde{\pi}$, is *totally absent* in the gravitational coupling to fermions. This can be deduced as follows. From, $\tilde{E}^{ia} = \frac{1}{2}\tilde{\epsilon}^{ijk}\epsilon^{abc}e_{bj}e_{ck}$, and the symplectic potential,

$$\int p_{ia}\delta\tilde{E}^{ia}d^3x = \int \tilde{\pi}^{ai}\delta e_{ai}d^3x, \qquad \tilde{\pi}^{ai} = \tilde{\epsilon}^{ijk}\epsilon^{abc}p_{jb}e_{ck}. \tag{9.41}$$

The trace term $\tilde{E}^{ia}p_{ia} = \tilde{\pi} = \frac{1}{2}e_{ai}\tilde{\pi}^{ai} = \tilde{\pi}$. With the help of $\tau^a\tau^b = \delta^{ab} + i\epsilon^{abc}\tau_c$, the κ dependent coupling to p_{ia} in A^-_{ia} can be expanded as

$$\tilde{\pi}_\psi \tilde{E}^i_a \frac{\tau^a}{2} \left(\kappa p_{ib} \frac{\tau^b}{2} \right) \tilde{\psi} = \frac{\kappa}{2} \left(\frac{\tilde{\pi}}{2} (\tilde{\pi}_\psi \psi) + i\tilde{E}^i_a p_{ib} \epsilon^{abc} \tilde{\pi}_\psi \frac{\tau^c}{2} \tilde{\psi} \right). \tag{9.42}$$

The resultant first term is purely anti-hermitian[7] and drops out by hermiticity of the Hamiltonian density, and the remaining[8] $(\tilde{E}^i_a p_{ib} \epsilon^{abc}) = -(\tilde{\pi}^i_a e_{bi} \epsilon^{abc})$ in Eq. (9.40) yields

$$\tilde{\pi}^i_a e_{bi} \epsilon^{abc} = \bar{\pi}^i_a \bar{e}_{bi} \epsilon^{abc}, \tag{9.43}$$

following the definitions Eqs. (2.86)–(2.88). Thus the fermion Hamiltonian density is just a function of $(\tilde{\psi}, \tilde{\pi}_\psi), (\bar{e}_{ai}, \bar{\pi}^{ai})$ and $e = \sqrt{q}$, but not of $\tilde{\pi}$. In ITG this is a crucial, because the absence of $\tilde{\pi}$ in $\tilde{\tilde{\mu}}_F$ means the inclusion of fermionic matter does not complicate the solution of the Hamiltonian constraint, which is achieved by expressing $\tilde{\pi}$ in terms of the rest of the variables. Since Yang-Mills and scalar fields couple to the metric but not the spin connection (ergo not to $\tilde{\pi}$), the absence of the possible fermionic complication is the only one that needs to be checked. Even though Γ_{ia} is in general an $SO(3,C)$ connection, it will be real-valued whenever the dreibein is. However, the (anti)self-dual spin connection must remain fully complex, and the p_{ia} contribution to the spin connection cannot be neglected. Expanding Eq. (9.37) in terms of the various contributions,

$$\tilde{\tilde{\mu}}_F = - \tilde{\pi}_\psi \tilde{E}^i_a \frac{\tau^a}{2} e^{\frac{1}{2}} \partial_i e^{-\frac{1}{2}} \tilde{\psi}$$

$$- \frac{i}{2} \tilde{\pi}_\psi \tilde{E}^i_a K_{ib} \epsilon^{abc} \frac{\tau^c}{2} \tilde{\psi} + i\tilde{\pi}_\psi \tilde{E}^i_a \frac{\Gamma^a_i}{4} \tilde{\psi} + h.c.. \tag{9.44}$$

---

[7]Recall that $\tilde{\pi}_\psi = -i\tilde{\psi}^\dagger n_A \tau^A$; while $\tilde{\pi}$ is $SO(3,C)$ invariant and hermitian.
[8]This is actually the generator of $SO(3,C)$ rotations of the variables.

Unlike the case of other variables, fermionic d.o.f. does not *a priori* yield a positive semidefinite energy density; however, the fermion Hamiltonian density is positive modulo a negative vacuum energy. The latter can be overcome by the bare positive cosmological constant (as was discussed in Chapter 7), so that the complete theory has resultant positive semi-definite Hamiltonian density.

9.3.1   *SO(3, C) and the unimodular spatial metric*

The determinant of the dreibein $e := det(e) = \sqrt{q}$ is a Lorentz invariant object, as is the unimodular spatial metric, $\bar{q}_{ij} := \bar{e}_{ai}\bar{e}^a{}_j$, which is unchanged under all $SO(3, C)$ rotations of $\bar{e}_{ai}$. Thus full Lorentz invariance is contained in the freedom of just the unimodular $\bar{e}_{ai}$. In general, any unimodular spatial metric, $\bar{q}_{ij}$, defines $\bar{e}_{ia}$ up to complex $SO(3, C)$ Lorentz rotations; however, it is gauge equivalent to a real $\bar{e}_{ia}$ upon imposing the reality condition for the spatial metric $\bar{q}_{ij} = \bar{q}^*_{ij}$ (and self-adjointness $q^\dagger_{ij} = q_{ij}$ in the quantum theory), and similarly for $e = \sqrt{q}$.

In the semiclassical context, spacetime is constructed from dreibein one-forms $e_{a=1,2,3}$ and the ITG lapse function, $N = \frac{\sqrt{q}(\partial_t \ln q^{1/3} - \frac{2}{3}\nabla_i N^i)}{4\beta\kappa\bar{H}}$, which defines the remaining piece of the veirbein via $e^0 = N dt$.

9.3.2   *Fermions, spatial diffeomorphisms, and*
fundamental anti-commutation relations

It still remains to construct the generator of spatial diffeomorphisms for the fermionic variables. In this respect ψ is a scalar field; while $\tilde{\psi}$ is a scalar density of weight one-half, ditto for its conjugate. In smeared form with shift vector N^i, the field theory generator is just

$$H^F_i[N^i] = \int \tilde{\pi}_\psi(\mathcal{L}_{\vec{N}}\tilde{\psi})d^3x, \qquad \mathcal{L}_{\vec{N}}\tilde{\psi} = N^i\partial_i\tilde{\psi} + \frac{1}{2}(\partial_i N^i)\tilde{\psi}. \tag{9.45}$$

$H^F_i[N^i]$ generates spatial diffeomorphism[9] of $\tilde{\psi}$ via commutation. To wit,

$$\delta_{Diff}\tilde{\psi} = \frac{1}{i\hbar}[\tilde{\psi}, H^F_i[N^i]] = \mathcal{L}_{\vec{N}}\tilde{\psi}, \qquad H^F_i = \tilde{\pi}_\psi\nabla_i\tilde{\psi}, \tag{9.46}$$

follows from the fundamental anti-commutation relation of the fermion and its conjugate and the use of

$$[A, BC] = [A, B]_+C - B[A, C]_+. \tag{9.47}$$

---

[9]Since $(\tilde{\pi}_\psi\tilde{\psi})$ is of weight one, over closed spatial manifolds $\int \mathcal{L}_{\vec{N}}(\tilde{\pi}_\psi\tilde{\psi})d^3x = 0$. It follows that $\int \tilde{\pi}_\psi(\mathcal{L}_{\vec{N}}\tilde{\psi})d^3x = -\int (\mathcal{L}_{\vec{N}}\tilde{\pi}_\psi)\tilde{\psi}\, d^3x$, which also correctly generates diffeomorphisms of $\tilde{\pi}_\psi$.

It should be noted that in QFT constraints and Hamiltonian evolution generate changes via commutator equations such as the Heisenberg EOM. Heisenberg and Schrödinger operators are related by $\hat{O}_H = U^\dagger(T, T_0)\hat{O}_S U(T, T_0)$, with time development operator $U(T, T_0)$ as in Eq. (3.17). They will evolve according to Heisenberg EOM,

$$\frac{d\hat{O}_H}{dT} = \frac{1}{i\hbar}[\hat{O}_H, H_H(T)] + \left(\frac{\partial\hat{O}_S}{\partial T}\right)_H, \tag{9.48}$$

$$H_H(T) = U^\dagger(T, T_0)\hat{H}_S(T)U(T, T_0), \left(\frac{\partial\hat{O}_S}{\partial T}\right)_H = U^\dagger(T, T_0)\left(\frac{\partial\hat{O}_S}{\partial T}\right)U(T, T_0),$$

$$[\hat{O}_H, H_H(T)] = U^\dagger(T, T_0)[\hat{O}_S, H_S(T)]U(T, T_0).$$

Commutation relations with the Hamiltonian rule. Nevertheless, *if the constraints and Hamiltonian involve only bispinor entities, the postulate of fundamental anti-commutation relations for fermionic variables suffices to compute all commutation relations* via the identity in Eq. (9.47). In this sense, *besides fundamental CR, Nature permits the option of fundamental anti-CR to realize Hamiltonian evolution of operators in QFT.*

9.3.3 *Diffeomorphism-invariant fermion measure and absence of anomalies*

With regard to spatial diffeomorphisms, the choice of weight one-half $\tilde{\psi}$ variable over scalar ψ confers a certain advantage. The Jacobian of the measure $\Pi_{\alpha,x}d\tilde{\psi}^\alpha(x)$ under infinitesimal diffeomorphisms is

$$J = \det\left(\frac{\delta\tilde{\psi}'^\alpha(x)}{\delta\tilde{\psi}^\beta(y)}\right) = 1 + tr\left(\delta_\beta^\alpha\left(N^i\partial_i + \frac{1}{2}(\partial_i N^i)\delta^3(x, y)\right)\right). \tag{9.49}$$

The trace is over all indices, including integration over spatial points i.e. $tr(M_\beta^\alpha(x, y)) = \int d^3x \int d^3y\, M_\beta^\alpha(x, y)\delta_\alpha^\beta\delta^3(x, y)$. The Jacobian of a field transformation however needs regularization. Following Fujikawa *et al.* [111] this can be achieved with mode cutoff. With a complete set of eigenmodes $\{\phi_I(x)\}$ of a self-adjoint operator (as will be apparent, the particular operator is not of the essence), and by substitution of $\delta(x, y) \rightarrow \sum_I \phi_I(x)\phi_I(y)$, the regulated expression for $tr(N^i(x)\partial_i^x + \frac{1}{2}(\partial_i^x N^i(x)))\delta^3(x, y)$ works out to be $\sum_{I=1}^N \int \phi_I(x)(N^i\partial_i + \frac{1}{2}(\partial_i N^i))\phi_I(x)d^3x$, with cutoff at large N. But this regulated expression is equal to $\sum_{I=1}^N \int \frac{1}{2}\partial_i(N^i(\phi_I)^2)d^3x$, which vanishes when the spatial integration is over a closed manifold (this is also consistent with the assumption of a discrete set of eigenmodes). Thus the Jacobian under spatial diffeomorphism transformations is unity and the fermion measure is invariant. The same result obtains for the measure of weight one-half conjugate momentum $\tilde{\pi}_\psi$. In canonical functional integral or path integral formalisms, this implies the absence of the corresponding anomaly [112].

Finally, the corresponding Gauss Law generator of $SL(2, C)$ Lorentz transformations of $\tilde{\psi}$ smeared with complex parameter ζ_a takes the form

$$G_F^a[\zeta_a] = \int d^3x \zeta_a i \tilde{\pi}_\psi \left(\frac{\tau^a}{2} \right) \tilde{\psi}; \qquad (9.50)$$

which produces the infinitesimal $SL(2, C)$ change,

$$\delta_G \tilde{\psi} = \frac{1}{i\hbar}[\tilde{\psi}, G_F^a[\zeta_a]] = i\zeta_a \left(\frac{\tau^a}{2} \right) \tilde{\psi}. \qquad (9.51)$$

9.3.4 *Perturbative and global gauge anomalies*

It is known that there can be perturbative QFT anomalies when chiral fermions are coupled to spin connections in $4n + 2$ spacetime dimensions [113], but not for four dimensions. Within the perspective presented in this chapter, this fact can be also be established by noting that for $SL(2, C)$, the Pauli matrices τ^a belong to a "safe representation" free of perturbative QFT gauge anomalies [114] i.e.

$$tr(\tau^a \{\tau^b, \tau^c\}) = 0; \quad \text{and also} \quad tr(\tau^a) = 0, \qquad (9.52)$$

signifying the absence of "mixed anomalies" in the coupling of chiral fermions to gravitational spin connections [110]. From the perspective of functional integral methods and fermion measures, if Lorentz transformations, in addition to spatial diffeomorphisms, were to be included, the extra contribution from the change in Eq. (9.51) is just the addition of the term $\zeta_a \frac{(\tau^a)_{\alpha\beta}}{2} \delta^3(x, y)$ to the trace term of the Jacobian in Eq. (9.49). This vanishes by the traceless property of Pauli matrices. In fact with the additional Lorentz symmetry the usual Lie derivative of fermion is not a totally covariant concept and the generator of covariant Lie derivative should be modified to

$$\mathcal{H}_i^F = \tilde{\pi}_\psi \left(\nabla_i - i\Gamma_{ia}\frac{\tau^a}{2} \right) \tilde{\psi}. \qquad (9.53)$$

Compared to (9.46) the difference in its action is just a Lorentz transformation.

When a chiral fermion multiplet is in also coupled to Yang-Mills connection one-form $A_a T^a$, the multiplet must belong to representations which are free of perturbative gauge anomalies; these representations satisfy $tr(T^a\{T^b, T^c\}) = 0$ and $tr(T^a) = 0$ [113, 114]. A regularization scheme for fermion loops which is explicitly invariant under all the local gauge symmetries of the theory, including local Lorentz invariance can be constructed for representations which satisfy these chiral gauge anomaly and mixed Lorentz-gauge anomaly cancellation conditions [115]. Anomalous theories on the other hand manifest themselves by having divergent fermion loops which remain unregularized by the scheme.

There is still the question of global anomaly. Witten [116] showed that in four dimensions, theories with an odd number of Weyl fermion doublets coupled to $SU(2)$ connection are inconsistent. Without gravity, the Standard Model has four $SU(2)_W$ doublets in each generation, and so it is not troubled by such a global anomaly. For Weyl fermions, the Lorentz group $SL(2, C)$ is the complexification of the $SU(2)$ associated with rotations; and an inconsistency can arise, due to the fact that on a Lorentz Weyl doublet $e^{i\pi\gamma_5}$ has the same action as $-1 \in SU(2)$. On the one hand a γ_5 transformation is associated with perturbative Adler-Bell-Jackiw chiral anomaly upon regularization, on the other $SU(2)$ is "safe" [110, 117, 118]. With generalized $Spin_G(4)$ structure (see Sec. 9.7), the $SO(10)$ GUT theory with 16 Weyl fermions per generation is the simplest and preeminent candidate which avoids such an inconsistency [110, 118].

9.3.5 *Diffeomorphism-invariant measures of other fields*

Appropriately weighted, the measures of scalars and Yang-Mills fields can all be made invariant under spatial diffeomorphisms. A scalar field densitized to weight one-half by multiplying with $e^{1/2}$ will have, as shown above, an invariant measure. A Yang-Mills gauge connection densitized as $\tilde{A}_{ia} := e^K A_{ia}$ transforms under both diffeomorphism and gauge variations as

$$\tilde{A}'_{ia} = \tilde{A}_{ia} + N^j\partial_j\tilde{A}_{ia} + K(\partial_j N^j)\tilde{A}_{ia} - N^j(\partial_i\tilde{A}_{ja}) + e^K(D_i^A\eta)_a. \quad (9.54)$$

This yields the Jacobian as

$$J = 1 + tr\left([(N^j\partial_j + K(\partial_j N^j))\delta_k^i\delta_b^a - N^k\partial_i\delta_b^a + f^{abc}\eta_c\delta_k^i]\delta^3(x, y)\right)$$
$$= 1 + 2\delta_a^a tr\left((N^j\partial_j + (3K/2)(\partial_j N^j))\delta^3(x, y)\right) \quad (9.55)$$

since the trace over antisymmetric Yang-Mills group structure constant f^{abc} is zero. Retracing the previous arguments a weight of $K = \frac{1}{3}$ will render the measure invariant. For the momentum conjugate to $\tilde{A}_{ia}$, a total weight of K yields the Jacobian of its measure as

$$J = 1 + 4\delta_a^a tr\Big((N^j\partial_j + (3K/4)(\partial_j N^j))\delta^3(x,y)\Big). \qquad (9.56)$$

As $\tilde{\pi}_A^{ia}$ the canonical conjugate momentum of a Yang-Mills connection is naturally already of weight one, multiplying it by $e^{-\frac{1}{3}}$ will result in total $K = \frac{2}{3}$ and an invariant measure. On the other hand with four-dimensional spacetime variable $A_{\mu a}$ the same steps lead to $K = \frac{3}{8}$ for an invariant measure [111].

For gravitational d.o.f. spatial diffeomorphism invariance is enforced in the canonical functional integral by introducing auxiliary conditions, Faddeev-Popov determinants, and integrating over TT excitations. This route was covered in the previous chapter. The generalization when fermionic and other d.o.f. are included will be addressed in the following sections.

9.3.6 *Total diffeomorphism and Lorentz Gauss Law constraints*

The total momentum constraint for spatial diffeomorphisms and Gauss Law constraint for Lorentz invariance are respectively

$$H_i^T = \mathcal{H}_i + \mathcal{H}_i^F \qquad (9.57)$$

$$= -2\nabla_j \bar{\pi}_i^j + \tilde{\pi}_\psi \left(\nabla_i - i\Gamma_{ia}\frac{\tau^a}{2}\right)\tilde{\psi} \qquad (9.58)$$

$$= 0;$$

and

$$G_T^a = G_{\bar{e}}^a + G_F^a$$

$$= -\epsilon^{abc}\bar{e}_{bi}\bar{\pi}_c^i + i\tilde{\pi}_\psi \left(\frac{\tau^a}{2}\right)\tilde{\psi} = 0. \qquad (9.59)$$

Although $\mathcal{H}_i = -2\nabla_j \bar{\pi}_i^j$ remains classically in same form as in pure metric theory, it is expressed in a novel way in terms of the momentric variable. Yet it is still capable of generating the correct symmetry. To wit, we first observe that the usual Lie derivative is

$$\mathcal{L}_{\vec{N}}\bar{e}_{ai} = N^k\nabla_k\bar{e}_{ai} + (\nabla_i N^k)\bar{e}_{ak} - \frac{1}{3}(\nabla_k N^k)\bar{e}_{ai}, \qquad (9.60)$$

with the first term expressible as

$$N^k\nabla_k\bar{e}_{ai} = N^k D_k^{\{\}+\Gamma}\bar{e}_{ai} - N^k\Gamma_{kb}\epsilon_a{}^{bc}e_{ci}. \qquad (9.61)$$

$D^{\{\}+\Gamma}$ denotes the total covariant derivative gauging both spatial and Lorentz indices. The former is gauged by the Christoffel symbol Γ^i_{jk} (which is symmetric under interchange of j and k) constructed from the spatial metric, and the latter by the torsionless connection compatible with e_{ai}. As a result the connections obey the metricity condition in three dimensions

$$D^{\{\}+\Gamma}_k \bar{e}_{ai} := \nabla_k \bar{e}_{ai} + \epsilon_a{}^{bc}\Gamma_{kb}\bar{e}_{ci} = 0. \tag{9.62}$$

Suitably modified by a Lorentz transformation with parameter $N^k\Gamma_{ka}$, the action of H_i^T is

$$\frac{1}{i\hbar}\left[\bar{e}_{ai}(x), H_k^T[N^k]_- + G_T^a[N^k\Gamma_{ka}]\right]_-$$

$$= \frac{1}{i\hbar}\left[\bar{e}_{ai}(x), \int N^k(-2\nabla_j\bar{\pi}_k^j)d^3y + G_T^a[N^k\Gamma_{ka}]\right]_-$$

$$= \frac{1}{i\hbar}\left[\bar{e}_{ai}(x), 2\int(\nabla_j N^k)\bar{\pi}_k^j d^3y + G_T^a[N^k\Gamma_{ka}]\right]_-$$

$$= (\nabla_i N^k)\bar{e}_{ak}(x) - \frac{1}{3}(\nabla_k N^k)\bar{e}_{ai}(x) - N^k\Gamma_{kb}\epsilon_a{}^{bc}\bar{e}_{ci}$$

$$= \mathcal{L}_{\vec{N}}\bar{e}_{ai} - N^k D^{\{\}+\Gamma}_k\bar{e}_{ai}, \tag{9.63}$$

which is, given the metricity condition, just the Lie derivative. In computing these results we may invoke the commutation relations between dreibein and the momentric, instead of its conjugate momentum. And for the fermion, the result, with the same gauge modification, is

$$\frac{1}{i\hbar}[\tilde{\psi}(x), H_k^T[N^k] + G_T^a[N^k\Gamma_{ka}]]_-$$

$$= \frac{1}{i\hbar}\left[\tilde{\psi}(x), \int N^k\tilde{\pi}_\psi\left(\nabla_k - i\Gamma_{ka}\frac{\tau^a}{2}\right)\tilde{\psi}\, d^3y\right]_-$$

$$= \mathcal{L}_{\vec{N}}\tilde{\psi}. \tag{9.64}$$

We may thus adopt the covariant total generator of spatial diffeomorpsims to be $\mathcal{H}_i^F$ which carries the bonus of commuting with the generator of Lorentz symmetry G_T^a; this is also true of the total H_i^T. And it suffices to adopt the set $\{H_i^T = 0, G_T^a = 0\}$ to encompass all the symmetries involved.

9.4 Solution of diffeomorphism and Lorentz Gauss Law constraints

The complete solution of the set of constraints turns out to be remarkably straightforward when the right decompositions are adopted. In Chapters 4 and 8 we have seen how transverse-traceless degrees emerge as the physical degrees of freedom for pure gravity. Incorporation of matter and other fields should not, and in fact do not, alter this conclusion. To wit, the solution of

$$H_i^T = -2\nabla_j \bar{\pi}_i^j + \tilde{\pi}_\psi \left(\nabla_i - i\Gamma_{ia}\frac{\tau^a}{2} \right) \tilde{\psi} = 0 \tag{9.65}$$

is to decompose the traceless momentric variable into

$$\bar{\pi}_i^j = \bar{\pi}_i^{Tj} + e \left(\nabla^j W_i - \frac{1}{3}\delta_j^i \nabla_k W^k \right) = \bar{\pi}_i^{Tj} + L_i^{jk} W_k, \tag{9.66}$$

with $\nabla_j \bar{\pi}_i^{Tj} = 0 \Leftrightarrow L_{jk}^{\dagger i} \bar{\pi}_i^{Tj} = 0$; and to reexpress the constraint as

$$\nabla_j \bar{\pi}_i^j = \frac{1}{2}\tilde{\pi}_\psi \left(\nabla_i - i\Gamma_{ia}\frac{\tau^a}{2} \right) \tilde{\psi}. \tag{9.67}$$

By substituting the above decomposition of the momentric, the constraint reduces to an operator equation on W_i which is similar to what has been achieved in the last chapter. This yields

$$(L^\dagger L)_i^j W_j = \frac{1}{2}\tilde{\pi}_\psi \left(\nabla_i - i\Gamma_{ia}\frac{\tau^a}{2} \right) \tilde{\psi}. \tag{9.68}$$

Inversion[10] of the positive semi-definite operator $(L^\dagger L)$ eliminates precisely three degrees of freedom in $\bar{\pi}_j^i$ which is contained in W^i. Moreover, this does not in any manner compromise the generality of the fermion content.

The Gauss Law is

$$G_T^a = -\epsilon^{abc}\bar{e}_{bi}\bar{\pi}_c^i + i\tilde{\pi}_\psi \left(\frac{\tau^a}{2} \right) \tilde{\psi} = -2\epsilon^{abc}\bar{e}_{bi}\bar{E}_c^k\bar{\pi}_k^i + \tilde{J}^a, \tag{9.69}$$

with $\tilde{J}^a := i\tilde{\pi}_\psi \frac{\tau^a}{2}\tilde{\psi}$. By contracting with invertible $\bar{e}_{aj}$, it is equivalently

$$\underset{\sim}{\epsilon}_{ijk}(\bar{q}^{jl}\bar{\pi}_l^i) = \frac{1}{2}\bar{e}_{ak}\tilde{J}^a. \tag{9.70}$$

This means the antisymmetric part of $(\bar{q}^{il}\bar{\pi}_l^j)$ is completely constrained by the fermion current (recall that in pure metric gravity, $\underset{\sim}{\epsilon}_{kij}(\bar{q}^{il}\bar{\pi}_l^j) = 0$ was just the requirement of the momentum $\tilde{\pi}^{ij}$ being symmetric, and also the origin of the Gauss Law when it is expressed terms of the dreibein and its conjugate momentum). The symmetric part, $\frac{1}{2}(\bar{q}^{ik}\bar{\pi}_k^j + \bar{q}^{jk}\bar{\pi}_k^i)$, remains completely free. Thus the Gauss constraint eliminates three further degrees leaving the "symmetric part of the TT momentric", $\frac{1}{2}(\bar{q}^{ik}\bar{\pi}_k^{Tj} + \bar{q}^{jk}\bar{\pi}_k^{Ti})$,

---

[10]It should be understood that "inversion" is carried out with the method of Green's function, as was explained in a footnote in the previous chapter.

with exactly two remaining degrees of freedom after all the constraints are solved. The TT momentric variable can in principle be extracted by projection, as given any traceless mixed index tensor density of weight one, $\bar{\alpha}^i_j$, the projection,

$$\bar{\pi}^{Ti}_j := \left[I - L\frac{1}{L^\dagger L}L^\dagger\right]^{ik}_{jl} \bar{\alpha}^l_k \tag{9.71}$$

automatically satisfies[11] $L^{\dagger i}_{jk}\bar{\pi}^{Tj}_i = 0$.

9.4.1 Auxiliary conditions for spatial dreibein

The auxiliary conditions on $\bar{e}_{ai}$ are that the physical excitations must obey a symmetrization and also a transversality condition. These requirements emerge naturally from the orthogonality of gauge and physical variations i.e.

$$\langle\delta_{phys}\bar{e}|\delta_{gauge}\bar{e}\rangle := \int \sqrt{q}\delta^{ab}\bar{q}^{ij}(\delta_{phys}\bar{e}_{ai})(\delta_{gauge}\bar{e}_{bj})d^3x = 0. \tag{9.72}$$

With metricity, the gauge variation is

$$\begin{aligned}
\delta_{gauge}\bar{e}_{ai} &= \mathcal{L}_{\vec{N}}e_{ai} + \epsilon_a{}^{bc}\eta_b\bar{e}_{ci} \\
&= (\nabla_i N^k)\bar{e}_{ak} - \frac{1}{3}(\nabla_k N^k)\bar{e}_{ai} - N^k\Gamma_{kb}\epsilon_a{}^{bc}e_{ci} + \epsilon_a{}^{bc}\eta_b\bar{e}_{ci} \\
&= (\nabla_i N^k)\bar{e}_{ak} - \frac{1}{3}(\nabla_k N^k)\bar{e}_{ai} + \epsilon_a{}^{bc}\eta'_b\bar{e}_{ci} \\
&= D^{\{\}+\Gamma}_i(N^k\bar{e}_{ak}) - N^k D^{\{\}+\Gamma}_i\bar{e}_{ak} - \frac{1}{3}(\nabla_k N^k)\bar{e}_{ai} + \epsilon_a{}^{bc}\eta'_b\bar{e}_{ci} \\
&= D^{\{\}+\Gamma}_i\bar{\phi}_a - \frac{1}{3}(\nabla_k N^k)\bar{e}_{ai} + \epsilon_a{}^{bc}\eta'_b\bar{e}_{ci}, \tag{9.73}
\end{aligned}$$

wherein $\bar{\phi}_a := N^k\bar{e}_{ak}, \eta'_b := \eta_b - N^k\Gamma_{kb}$. Thus the physical change must be determined by[12]

$$\int \sqrt{q}\delta^{ab}\bar{q}^{ij}(\delta_{phys}\bar{e}_{ai})(D^{\{\}+\Gamma}_j\bar{\phi}_b + \epsilon_b{}^{cd}\eta'_c\bar{e}_{dj})d^3x = 0, \quad \forall \eta'_b, \bar{\phi}_b. \tag{9.74}$$

This translates into $\epsilon_{abc}\bar{E}^{bi}\delta_{phys}\bar{e}^c_i = 0$, and through integration by parts, the transversality condition,

$$\bar{q}^{ij}D^{\{\}+\Gamma}_j\delta_{phys}\bar{e}_{ai} = 0. \tag{9.75}$$

---

[11] Again $\frac{1}{L^\dagger L}$ is the mnemonic for inversion with the Green function of $L^\dagger L$.

[12] The term with $\frac{1}{3}(\nabla_k N^k)\bar{e}_{bi}$ drops out due to the traceless condition $\bar{q}^{ij}(\delta_{phys}\bar{e}_{ai})\bar{e}_{aj} = \bar{E}^{ia}(\delta_{phys}\bar{e}_{ai}) = 0$.

These constitutes 6 conditions on the traceless variations, leaving two unconstrained degrees in $\delta_{phys}\bar{e}_{ai}$. The former condition, $\epsilon_{abc}\bar{E}^{ib}\delta_{phys}\bar{e}^c_i = 0$, is equivalent (by contraction with invertible $\bar{E}^{ia}$) to the requirement

$$\tilde{\epsilon}^{ijk}\bar{e}^a_i(\delta_{phys}\bar{e}_{aj}) = 0. \tag{9.76}$$

It follows that infinitesimal physical changes in the unimodular spatial metric satisfy

$$\delta^{phys}\bar{q}_{ij} = (\delta_{phys}\bar{e}_{ai})\bar{e}^a_j + \bar{e}_{ai}(\delta_{phys}\bar{e}^a_j) = 2(\delta_{phys}\bar{e}_{ai})\bar{e}^a_j; \tag{9.77}$$

and, as was the case in pure metric theory, transversality of physical metric excitations,

$$\begin{aligned}
\nabla^i(\delta^{phys}\bar{q}_{ij}) &= \bar{q}^{ik}\nabla_k(\delta^{phys}\bar{q}_{ij}) \\
&= 2\bar{q}^{ik}D^{\{\}+\Gamma}_k((\delta_{phys}\bar{e}_{ai})\bar{e}^a_j) \\
&= 2\bar{e}^a_j\bar{q}^{ik}D^{\{\}+\Gamma}_k\delta_{phys}\bar{e}_{ai} \\
&= 0, \tag{9.78}
\end{aligned}$$

remains reassuringly valid. Reality of spatial metric and its physical excitations $\delta^{phys}\bar{q}_{ij}$ ensures that $\bar{e}_{ai}$ and $\delta_{phys}\bar{e}^a_j$ are real up to $SO(3,C)$ Lorentz transformations.

9.5   A general scheme to solve first class gauge constraints exactly, and to promote to a Dirac second class system

The requirement of orthogonal decomposition of physical and gauge variations offers a systematic way to solve gauge constraints and identify physical excitations. For a generic field f_I with the index I denoting possibly coordinate and/or internal indices and a gauge variation linearly dependent on the gauge parameters $\eta_{J'}$ i.e. $\delta_{gauge}f_I = L^{J'}_I\eta_{J'}$, the scheme proceeds by demanding that gauge and physical changes are orthogonal with respect to a positive-definite metric G^{IJ}. The generalization to (9.72) reads

$$\langle\delta_{phys}f|\delta_{gauge}f\rangle := \int G^{IJ}(\delta_{phys}f_I)L^{K'}_J\eta_{K'}\,d^3x = 0. \tag{9.79}$$

It follows that[13]

$$0 = \langle\delta_{phys}f|\delta_{gauge}f\rangle = \langle\delta_{phys}f|L\eta\rangle = \langle L^\dagger\delta_{phys}f|\eta\rangle, \quad \forall\,\eta_{I'}. \tag{9.80}$$

[13]The number of independent components in unprimed and primed indices need not be equal, so a block diagonal total metric [26] with positive-definite G^{IJ} and $G'^{I'J'}$ defining the inner products between unprimed and primed entities should be adopted. The metrics are also assume to carry the appropriate tensor weights to render the inner product coordinate invariant.

The consequent "gauge-fixing", or rather "auxiliary", condition on the physical change is $(L^\dagger \delta_{phys} f)_{I'} = 0$. By decomposing the total change into physical and gauge variations,

$$\delta f_I = \delta_{phys} f_I + \delta_{gauge} f_I = \delta_{phys} f_I + L_I^{J'} \eta_{J'}, \qquad (9.81)$$

and applying the operator $L^\dagger$ to the equation, the result is

$$(L^\dagger \delta f)_I = (L^\dagger \delta_{phys} f)_I + (L^\dagger L \eta)_I = (L^\dagger L \eta)_I. \qquad (9.82)$$

This implies for an arbitrary change δf_I, the gauge parameter can be solved (or "uniquely fixed") by inversion of the positive-(semi)definite operator

$$\eta_{I'} = [(L^\dagger L)^{-1} \cdot L^\dagger \delta f]_{I'}; \qquad (9.83)$$

and the physical change can be extracted through projection as

$$\delta_{phys} f_I = \delta f_I - L_I^{J'} \eta_{J'} = \left[\left(I - L \frac{1}{L^\dagger L} L^\dagger \right) \delta f \right]_I. \qquad (9.84)$$

This automatically enforces $(L^\dagger \delta_{phys} f)_I = 0$; moreover, $\delta_{gauge} f_I = L_I^{J'} \eta_{J'}$ belongs to the kernel of the projection operator. In addition, any zero mode, ζ_0, of $L^\dagger L$ must satisfy $L\zeta_0 = 0$.

Let us assume the generator of the gauge transformation is linear in $p^I(x)$, the conjugate momentum to $f_I(x)$; and the corresponding "Gauss Law" constraint is $H^{I'} = D_J^{I'} p^J - S^{I'} = 0$, wherein $S^{I'}$ is a "source" term which may be present (for instance, in the total spatial diffeomorphism constraint, the generator of diffeomorphisms for the fermionic content plays this role). The analogous orthogonal decomposition of $p^I = p_{trans}^I + (D^\dagger W)^I$ with respect to a suitable positive definite metric leads to

$$0 = \langle p_{trans} | D^\dagger W \rangle = \langle D p_{trans} | W \rangle. \qquad (9.85)$$

This implies the "transverse" d.o.f. p_{trans}^I are required to satisfy $(Dp_{trans})^{I'} = 0$, and $(D^\dagger W)^I$ plays the role of the "longitudinal" complement. Thus

$$D_J^{I'} p^J = D_J^{I'} p_{trans}^J + (DD^\dagger W)^{I'} = (DD^\dagger W)^{I'} \qquad (9.86)$$

implies the elimination of $W^{I'}$ as $W^{I'} = [(DD^\dagger)^{-1} Dp]^{I'}$ and the extraction via projection of the transverse part

$$p_{trans}^I = \left[\left(I - D^\dagger \frac{1}{DD^\dagger} D \right) p \right]^I. \qquad (9.87)$$

The constraint

$$H^{I'} = D_J^{I'} p^J - S^{I'} = D_J^{I'} p_{trans}^J + (DD^\dagger W)^{I'} - S^{I'}$$
$$= (DD^\dagger W)^{I'} - S^{I'} = 0 \qquad (9.88)$$

is solved by inversion, and without compromising the value of $S^{I'}$, yielding

$$W^{I'} = [(DD^\dagger)^{-1} S]^{I'}. \tag{9.89}$$

In the absence of "sources", the longitudinal mode is trivial.

To recap, for the given operator D in (9.86), one adopts an *arbitrary* p^I, and obtain the transverse part, p^I_{trans}, through (9.87); and the corresponding longitudinal part through (9.89). The total physical $p^I_{phys} := p^I_{trans} + (D^\dagger W)^I$ will then satisfy $H^{I'}_{phys} = D^{I'}_J p^J_{phys} - S^{I'} = 0$.

9.5.1   *Electrodynamics as an example*

Electrodynamics, with $\tilde{E}^i = \tilde{E}^i_T - \sqrt{q} q^{ij} \nabla_j \phi$ conjugate to $\delta A_i = \delta A^T_i + \nabla_i \eta$, is the simplest illustration. The Gauss Law constraint, $\nabla_i \tilde{E}^i - \sqrt{q}\rho = 0$, is equivalent to

$$\nabla_i \tilde{E}^i = \nabla_i \tilde{E}^i_T - \sqrt{q} \nabla^2 \phi = \sqrt{q}\rho. \tag{9.90}$$

This yields the Poisson equation for the longitudinal mode, with solution $\phi = -\frac{1}{\nabla^2}\rho$. In the absence of sources the longitudinal mode is trivial. But with or without sources, the total electric field, via decomposition, satisfies $\nabla_i \tilde{E}^i = \nabla_i \tilde{E}^i_T - \sqrt{q} \nabla^2 \phi = -\sqrt{q} \nabla^2 \phi$; so the transverse d.o.f. can be obtained by projection as

$$\begin{aligned}
\tilde{E}^i_T &= \tilde{E}^i + \sqrt{q} q^{ij} \nabla_j \phi \\
&= \tilde{E}^i - \sqrt{q} q^{ij} \nabla_j \frac{1}{\nabla^2} \rho \\
&= \left(\delta^i_j - \nabla^i \frac{1}{\nabla^2} \nabla_j \right) \tilde{E}^j.
\end{aligned} \tag{9.91}$$

For the connection, the inner product $\langle \delta A_T | \nabla \eta \rangle = \int \sqrt{q} q^{ij} \delta A^T_i \nabla_j \eta \, d^3x = 0$ implies $q^{ij} \nabla_j \delta A^T_i = 0$. Since

$$q^{ij} \nabla_j \delta A_i = q^{ij} \nabla_j \delta A^T_i + \nabla^2 \eta = \nabla^2 \eta \quad \Rightarrow \eta = \frac{1}{\nabla^2} q^{ij} \nabla_j \delta A_i. \tag{9.92}$$

The transverse d.o.f. is expressible as $\delta A^T_i = (\delta^j_i - \nabla_i \frac{1}{\nabla^2} \nabla^j) \delta A_i$. The theory is abelian, and the relevant operator ∇_i is independent of the connection A_i; so the transverse change can be integrated to finite $A^T_i = (\delta^j_i - \nabla_i \frac{1}{\nabla^2} \nabla^j) A_j$.

9.5.2   *Relation to Dirac's second class system of constraints*

On the question of zero modes, w^α_o, of $DD^\dagger$, that are just homogeneous solutions of (9.88). It follows that $W^{I'} = W^{I'}_{particular} + \sum_\alpha w^\alpha_o$ is the general solution. As any zero mode must satisfy $D^\dagger w^\alpha_o = 0$, the zero modes do *not* contribute to p_{phys}^I since $(D^\dagger W)^I = (D^\dagger W_{particular})^I$. So the actual physical content can be obtained by ignoring the sector of zero modes; and the resultant invertible operator is really the one adopted in (9.89) to yield the particular solutions which contribute to p_{phys}^I. The same considerations apply to zero modes of $L^\dagger L$ and the inversion of the residual operator in (9.84).

The gauge symmetries of f_I are associated with the 1st class constraints. To wit,

$$\delta_{gauge} f_I(x) = \left\{ f_I(x), \int H^{J'}[\eta_{J'}] \right\}_{P.B.}$$
$$= \left\{ f_I(x), \int \eta_{J'} D_K^{J'} p^K \, d^3 y \right\}_{P.B.} = L_I^{J'} \eta_{J'}(x). \qquad (9.93)$$

Given G^{IJ} and $G''^{I'J'}$, there is a relation between L and D which can be worked out. In this particular scenario, the 1st class gauge constraints $H^{I'} = 0$ together with the auxiliary conditions $\chi_{I'} = 0$ form a 2nd class system provided $\det(\{\chi_{I'}(x), H^{J'}(y)\}_{P.B.})$ is non-vanishing. As the constraints generate only *infinitesimal gauge transformations*, the separation of physical from gauge changes can be expressed only through auxiliary conditions which are restrictions on infinitesimal changes of the fields, $\chi_{I'} := (L^\dagger \delta f)_{I'} = 0$, rather than as conditions on f_I. For non-abelian theories, L_f and $L_f^\dagger$ (as indicated by the subscripts) can also depend on f_I. However, the requirement of the non-vanishing determinant for a 2nd class system has a clear physical picture. Since $H^{J'}$ generates gauge symmetries of f_I, it follows that $\{\chi_{I'}, H^{J'}[\eta_{J'}]\}_{P.B.} = \delta_{gauge} \chi_{I'}$. Thus the non-vanishing determinant criterion is the same as saying the auxiliary condition must *not* possess any invariance under gauge transformations i.e. it must "fix the gauge freedom completely". With infinitesimal $\delta f = f_{phys} - f$, one must thus decide whether $\chi_{I'} := (L_f^\dagger(f_{phys} - f))_{I'} = 0$ (which determines the desired f_{phys}) has any residual gauge invariance. To wit, the change in the gauge-fixing condition induced by an infinitesimal gauge variation is

$$\delta_g \chi_{I'} = (L_{f+\delta_g f}^\dagger(f_{phys} - (f + \delta_g f)))_{I'} - (L_f^\dagger(f_{phys} - f))_{I'}$$
$$= (L_f^\dagger \delta_g f)_{I'} + ...; \qquad \text{with } \delta_g f := \delta_{gauge} f. \qquad (9.94)$$

Thus the infinitesimal change is

$$\delta_{gauge} \chi_{I'} = (L_f^\dagger \delta_{gauge} f)_{I'} = (L_f^\dagger L_f \eta)_{I'}. \qquad (9.95)$$

It follows that the determinant in question vanishes iff $L_f^\dagger L_f$ has one or more zero modes. Given the previous understanding that zero modes do not affect $\delta_{phys} f$, it is the operator $\bar{L}$ with domain complementary to the kernel of L that is of physical relevance. By subtraction of zero modes, the modified auxiliary condition $(\bar{L}^\dagger \delta f)_{I'} = 0$ comes with invertible $\bar{L}^\dagger \bar{L}$ and the realization of a system of Dirac 2nd class constraints. Among the upshots of a 2nd class system are weak inequalities are replaced by strong ones, and all phase space variables have identically vanishing Dirac brackets with *all* constraints (so subject to the constraints and auxiliary conditions all the d.o.f. are "invariant observables" Dirac-commuting with the constraints). And it is fundamental Dirac, not Poisson, brackets which are promoted to quantum CR. The theorem of Maskawa-Nakajima [101] states that a true reduced phase space can, in principle, be found in which Dirac brackets equal Poisson brackets in the reduced phase space. Note also that complete solution of the constraints and auxiliary conditions only projects out the "transverse" d.o.f. of the relevant Yang-Mills, or gravitational fields as the case may be, without imposing any further restrictions on the matter d.o.f.

9.5.3 *Auxiliary conditions and the shift function*

The conditions $\chi_j := \nabla^i \delta \bar{q}_{ij} = 0$ and $\chi := \ln q - \ln(q_0(x)e^{3T}) = 0$ fix N^i uniquely. This can be argued as follows: The shift vector can be decomposed into transverse and longitudinal parts as $N_i = N_i^T + \nabla_i \zeta$. Thus

$$\{\chi, [N^k H_k]\}_{P.B.} = \left\{\chi, -\frac{2}{3}\int N^k (\nabla_k \tilde{\pi}) \, d^3 x\right\}_{P.B.}$$
$$= \mathcal{L}_{\vec{N}} \ln q = 2\nabla^k N_k = 2\nabla^2 \zeta, \qquad (9.96)$$

can vanish only if ζ is trivially constant, which does not contribute to N_i. Under diffeomorphisms

$$\delta \bar{q}_{ij} = \{\bar{q}_{ij}, [N^k H_k]\}_{P.B.} = \mathcal{L}_{\vec{N}} \bar{q}_{ij} = \nabla_i N_j + \nabla_j N_i - \frac{2}{3} q_{ij} \nabla^k N_k$$
$$= \nabla_i N_j^T + \nabla_j N_i^T + \nabla_i \nabla_j \zeta + \nabla_j \nabla_i \zeta - \frac{2}{3} q_{ij} \nabla^2 \zeta. \qquad (9.97)$$

Imposing $\chi_j = 0$ implies

$$\nabla^i (\nabla_i N_j^T + \nabla_j N_i^T) = -\nabla^i \left(\nabla_i \nabla_j \zeta + \nabla_j \nabla_i \zeta - \frac{2}{3} q_{ij} \nabla^2 \zeta\right). \qquad (9.98)$$

Equivalently,

$$\nabla^2 N_j^T + [\nabla^i, \nabla_j] N_i^T + \nabla_j \nabla^i N_i^T = \nabla^2 N_j^T + R_j^i N_i^T$$
$$= -\nabla^i \left(\nabla_i \nabla_j \zeta + \nabla_j \nabla_i \zeta - \frac{2}{3} q_{ij} \nabla^2 \zeta\right). \qquad (9.99)$$

Since $\{\chi, [N^k H_k]\}_{P.B.}$ vanishes iff $\nabla_i \zeta = 0$, this leads to the requirement that $\{\bar{q}_{ij}, [N^k H_k]\}_{P.B.}$ must satisfy

$$(\nabla^2 \delta^i_j + R^i_j) N^T_i = 0 \tag{9.100}$$

if there were any residual freedom. But N^T_i must be trivial since $(\nabla^2 \delta^i_j + R^i_j)$ is an elliptic operator.[14] Thus $\chi = 0$ and $\chi_i = 0$ fix N^i uniquely.

9.6 Solving the total Hamiltonian constraint with fermion, Yang-Mills and scalar fields

It was explained in earlier chapters that the total Hamiltonian constraint in ADM formalism can be written as

$$\mathcal{H} = \beta^2 \tilde{\pi}^2 - \bar{H}^2 - \frac{1}{2\kappa} H_{non-GR} = 0. \tag{9.101}$$

H_{non-GR} is just the sum of standard fermionic, Yang-Mills and scalar field Hamiltonian densities, but with each of these being of weight two. For instance, in the case of a simple real scalar field, we have the usual

$$H_\phi(x) = \frac{1}{2}(\tilde{\pi}_\phi \tilde{\pi}_\phi + qq^{ij}\nabla_i\phi\nabla_j\phi) + qV(\phi), \tag{9.102}$$

written as a tensor density of weight two, while for a simple Weyl fermion, $\tilde{\tilde{\mu}}_F$ was the result arrived at earlier. Generalizations to multiple and complex scalar fields, Standard Model of particle physics with 15 Weyl fermions per generation and other GUT fermion multiplets with generic Yang-Mills gauge coupling are all straightforward. The weight two Hamiltonian density for Yang-Mills field is

$$H_{YM} = \frac{q_{ij}}{2}(\tilde{\pi}^{ia}_A \tilde{\pi}^{ja}_A + \hbar^2 \tilde{B}^{ia} \tilde{B}^{ja}), \tag{9.103}$$

with weight one entities $\tilde{\pi}^{ia}_A$ being the conjugate momentum to, and $\tilde{B}^{ia}$ the magnetic field of, the Yang-Mills connection A_{ia}. Since none of these contributions contain $\tilde{\pi}$, the solution of the Hamiltonian constraint is algebraic and easy. It again factorizes as

$$\mathcal{H} = \left(\beta\pi - \sqrt{\bar{H}^2 + \frac{1}{2\kappa}H_{non-GR}}\right)\left(\beta\pi + \sqrt{\bar{H}^2 + \frac{1}{2\kappa}H_{non-GR}}\right) = 0 \tag{9.104}$$

with $H_{non-GR} = H_\phi + \tilde{\tilde{\mu}}_F + H_{YM}$. For an expanding universe, $\tilde{\pi} = -\frac{1}{\beta}\tilde{H}_T$ with

$$\tilde{H}_T = \frac{1}{\beta}\sqrt{\bar{H}^2 + \frac{1}{2\kappa}H_{non-GR}} \tag{9.105}$$

---

[14]In a normal coordinate system based at any point p, $R_{ij}(p) = -\frac{3}{2}\nabla^2_p q_{ij}$.

is taken as the solution. This implies the Hamiltonian which generates cosmic time evolution (or T-translation) is $H_{Phys.} = \frac{1}{\beta} \int \tilde{H}_T d^3 x$. With reference to earlier derivations, this merely amounts to $\frac{1}{2\kappa}(H_{non-GR})$ being added under the square root to the generic $\mathcal{V}$.

9.6.1 *Torsion*

With the incorporation of fermions, and the presence of $\bar{\pi}^{ia}$ conjugate to unimodular $\bar{e}_{ai}$ in Eq. (9.44), a modification to the EOM occurs. Since only $\tilde{\bar{\mu}}_F$ contains this conjugate momentum, it suffices to consider just the fermionc addition to GR. Focusing on changes generated by the physical Hamiltonian (inclusion of spatial diffeomorphisms and local Lorentz transformations effects is straightforward), the physical Hamiltonian for GR with fermion is

$$H_{Phys} = \frac{1}{\beta} \int \tilde{H}_T d^3 x = \frac{1}{\beta} \int \sqrt{\bar{\pi}^i_j \bar{\pi}^j_i + \mathcal{V}_{GR}(q_{ij}) + \frac{1}{2\kappa}\tilde{\bar{\mu}}_F} \; d^3 x. \quad (9.106)$$

The equation of motion for the unimodular dreibein is (using the commutation relation between momentric and dreibein in Eq. (5.18)),

$$\frac{d\bar{e}_{ai}}{dT} = \{\bar{e}_{ai}, H_{Phys}\}_{PB}$$

$$= \frac{1}{4\beta\tilde{H}_T}(2\bar{e}_{aj}\bar{\pi}^j_i - \epsilon_a{}^{bc}\bar{e}_{bi}\tilde{J}_c). \quad (9.107)$$

In ITG the effective or emergent lapse function will (modulo spatial diffeomorphisms) now be $Ndt = \frac{\sqrt{\tilde{q}}dT}{4\beta\kappa\tilde{H}_T}$. As $\bar{q}_{ij}\bar{\pi}^{ja} = 2\bar{e}^a_j\bar{\pi}^j_i$, instead of the usual simple relation between the momentum, $\bar{\pi}^{ia}$, conjugate to $\bar{e}_{ai}$ and its derivative w.r.t. ADM coordinate time, this is now

$$\bar{q}_{ij}\bar{\pi}^j_a = \frac{e}{\kappa N}\left(\frac{d\bar{e}_{ai}}{dt}\right) + \epsilon_a{}^{bc}\bar{e}_{bi}\tilde{J}_c. \quad (9.108)$$

The additional term implies the total spin connection from Eq. (9.16) has a fermionic torsion contribution, even though no extraneous or explicit independent torsion degrees of freedom have been introduced in the theory. This gives rise to an effective four-fermion interaction in the Hamiltonian density that is of the form $\kappa \tilde{J}^a \tilde{J}_a$. In the case of the Standard Model or in GUT the fermion current must sum over all fermion flavors, and as all particles couple to gravity, this can contribute to extra channels of interactions and increase scattering cross sections, for instance to the $e^+ e^- \to \mu^+ \mu^-$ process. Measurements of this and related cross sections put an empirical bound of the Fermi four-fermion interaction mass scale weaker than several hundred GeVs [119]. The corresponding coupling that emerges in our torsion above is governed by κ, with the Fermi constant $\sim M_{Planck}^{-2}$, which is too weak to lead to currently detectable effects, and the current empirical bound is reassuringly not violated.

Introduction of independent and full-fledged torsion degrees of freedom in the spin connection causes the fermion current to couple only the trace and axial part of torsion, with the anti-hermitian trace torsion-fermion coupling canceling out when hermiticity of the Hamiltonian is imposed. The remaining axial torsion field is associated with the axial anomaly in QFT due to the breaking of chiral symmetry by regularization. As a result, the axial torsion field has a non-transverse vacuum polarization amplitude, causing it to be manifested as a massive torsion mode. This mass is naturally of the cut-off scale, which is also much higher than the current empirical lower bound of a few hundred GeV [120].

9.7 Generalized spin structure; GUT, sterile neutrino, and cold dark matter

The initial Chern-Simons state of the universe which will be discussed in Chapter 10 arises from the Hartle-Hawking proposal of summing over Euclidean manifolds. If this idea is pursued to its logical conclusion, the sum ought to be over all possible manifolds. However, it is known that not all Euclidean four-manifolds support the ordinary spin structure which permit the usual coupling of fermion fields to gravity. With respect to the Euclidean $SO(4)$ Einstein-Cartan connection, parallel transport of the veirbein $e_{A\mu}$ around a closed loop gives rise to a holonomy element of the group. When fermions which transform according to the double covering of $SO(4)$ are introduced and parallel transported, the spin connection must be consistently lifted to the double cover $Spin(4)$ group. However, an obstruction to the lifting can happen for certain Euclidean manifolds which do not permit this spin structure. Technically, this happen whenever the second Stiefel-Whitney class is non-trivial. These manifolds contain noncontractible closed two surfaces. To see how such spaces can fail to have an ordinary spin structure consider parallel transporting the veirbein around a one-parameter family of closed curves l_α, with $0 \leq \alpha \leq 1$, each of which begins and ends at the same point x_{ref}. The family of curves l_α, starting and ending with l_0 and l_1 which are trivial loops consisting of just a point x_{ref}, are to map out a topological two-sphere. For the veirbein, the holonomy element for each curve will belong to $SO(4)$, with identity element corresponding to the trivial curves l_0 and l_1. Thus the holonomy elements described a closed path in $SO(4)$ (which is however doubly-connected i.e. $\Pi_1(SO(4)) = Z_2$). Parallely propagating a spinor field on the other hand should give rise to a holonomy element in the double cover $Spin(4)$ group. Suppose the holonomies from the parallel transport of the veirbein as above leads to a non-trivial element of $\Pi_1(SO(4))$ i.e. a loop which cannot be deformed to a point in $SO(4)$, then upon lifting to its double cover this holonomy must start and end at different elements of $Spin(4)$. This leads to a contradiction since l_0 and l_1 are trivial curves which should correspond to trivial holonomy, even for the spinor field. When such a situation arises for certain manifolds, ordinary spin structure cannot be defined consistently.

Remarkably, the introduction of Yang-Mills gauge fields and generalized spin structures [121] can allow fermions and these gravitational manifolds to coexist. A *raison d'etre* for Yang-Mills fields! Given a general grand unification (simple) simply-connected gauge group G with a $Z_2 = \{I, c\}$ center, generalized spin structures with group $Spin_G(4) := \{Spin(4) \times G\}/Z_2$ can be constructed, wherein the Z_2 equivalence relation is defined by $(s, g) \equiv (-s, c \cdot g)$, $\forall (c, g) \in Spin(4) \times G$. Note that $Spin_G(4)$ is the double cover of $SO(4) \times (G/Z_2)$, and the coupling of spinors to both gravitational and Yang-Mills connections causes then to transform according to double cover $Spin_G(4)$ group. Parallel transport of these fermions does not give rise to any inconsistency. The aforementioned problematic lift of the non-trivial holonomy from l_1 in $SO(4)$ is canceled by the extra holonomy from the Yang-Mills connection so that the combined holonomy starts and end on the identity element of $Spin_G(4)$ for both trivial curves l_0 and l_1.

In the event that these spin structures are defined by Yang-Mills GUT groups, the simplest choice would be $SO(10)$ (more correctly, $G = Spin(10)$). It is easy to check that the 16 Weyl fermions indeed belong to the 16-dimensional representation of $Spin(10)$ and satisfy the generalized spin structure equivalence relation for $\{Spin(4) \times Spin(10)\}/Z_2$ (in this context "$SO(10)$ GUT" is a misnomer since its practitioners attempt to unify the 16 Weyl spinors of each generation in a GUT gauge group and $SO(10) = Spin(10)/Z_2$ does not have a 16-dimensional representation). It is worth emphasizing that the generalized spin structure implies an additional "isospin-spin" relation in that fermions must belong to $Spin(10)$ representations, while bosons must belong to $SO(10)$ representations of the GUT. This has implications for spontaneous symmetry breaking via fundamental bosonic Higgs, which cannot belong to the spinorial representations of $SO(10)$. On the other hand $SU(5)$ GUT (with 15 Weyls in the $\bar{5} \oplus 10$

$\bar{5} \oplus 10$ representations) fails to yield such a generalized spin structure, not to mention being empirically ruled out by both its shorter proton decay times and prediction of electroweak Weinberg angle [42, 122]. It also lacks the extra "sterile neutrino" of "$SO(10)$ GUT" which can participate in neutrino oscillations, even though the evidence for an eV-scale sterile neutrino remains inconclusive [123] (the natural sterile neutrino mass would more naturally be of GUT symmetry breaking scale). As this particle carries no standard model $SU(3)_C \times SU(2)_W \times U(1)_Y$ charges, it would be detectable only through its gravitational interaction. It is an excellent candidate for cold dark matter; current critical density of the universe is equivalent to the mass of 5 hydrogen atoms per cubic meter and all baryonic matter constitutes $\sim 5\%$ of this, so even if the "missing dark matter" density is required to be 5 times the baryonic density, it would only take one massive $\sim 10^9$ GeV sterile neutrino per $10^9 m^3$ to make up for all the missing cold dark matter; whereas a keV scale neutrino may be more compatible with the structure of the universe observed today.

As $Spin_G(4)$ is not globally the direct product group $SO(4) \times (G/2)$, this actually represents a "unification" of Lorentz and Yang-Mills gauge groups; well-known "no-go theorem" of Supersymmetry notwithstanding. The theorem asserts that spacetime and internal symmetries can be combined only in a trivial way in a Lie algebra; this however does not contradict the fact that the Lie algebra of the generalized spin structure is trivially $so(4) \oplus so(10)$ while gravitational and Yang-Mills symmetries are united *globally* as $Spin_G(4)$ with $G = Spin(10)$ [124].

Chapter 10

Initial state of the universe

道無常名，聖無常體 ... 。
詞無繁説，理有忘筌，濟物利人，宜行天下。[1]

"The right way has no invariable name, its sages are different
persons at different times, ...
Free of complicated sayings, its principles will outlast memories
of its procedures, it benefits mankind, let it propagate to all
under heaven."

10.1 The dog that did not bark: Weyl curvature and Einstein's theory

Gregory (Scotland Yard detective): "Is there any other point to which you would wish to draw my attention?"
Holmes: "To the curious incident of the dog in the night-time."
Gregory: "The dog did nothing in the night-time."
Holmes: "That was the curious incident."
The excerpt, from the *The Adventure of Silver Blaze*, in *The Memoirs of Sherlock Holmes* by Arthur Conan Doyle, epitomized a nonhappening which divulged a crucial piece of information.

In four or more dimensions, the Riemann curvature tensor takes its full-fledged complexity beyond mere Ricci curvatures. It can be decomposed as

$$R_{\mu\nu\alpha\beta} = C_{\mu\nu\alpha\beta} + \frac{1}{2}(g_{\mu\alpha}R_{\nu\beta} + g_{\nu\beta}R_{\mu\alpha} - g_{\nu\alpha}R_{\mu\beta} - g_{\mu\beta}R_{\nu\alpha})$$
$$- \frac{1}{6}(g_{\mu\alpha}g_{\nu\beta} - g_{\nu\alpha}g_{\mu\beta})R. \tag{10.1}$$

---

[1] Proclamation of the Tang Emperor Taizong on the Nestorian Stele.

In four dimensions there are 10 components to the Ricci tensor, and to the Weyl tensor $C_{\mu\nu\alpha\beta}$ the same number. Einstein's field equations are

$$R_{\mu\nu} - \frac{1}{2}g_{\mu\nu}R = \frac{8\pi G}{c^4}T_{\mu\nu}, \tag{10.2}$$

or, equivalently in the form Einstein first wrote it correctly,

$$R_{\mu\nu} = \frac{8\pi G}{c^4}\left(T_{\mu\nu} - \frac{1}{2}g_{\mu\nu}T^{\alpha}{}_{\alpha}\right). \tag{10.3}$$

Thus Einstein's equations directly correlate the Ricci curvature to the energy-momentum tensor; whereas the ten independent components of the Weyl tensor are totally, and one may add, unnaturally or inexplicably, absent from the scene. Yet roping in the Weyl curvature in modifications of Einstein's theory can lead to better quantum field theory convergence. It is known through the works of many [125] that the Einstein-Hilbert action supplemented by higher 4-curvature invariants (including the square of Weyl conformal tensor),

$$S_{renor.} = \int \left[\frac{1}{16\pi G}(2\Lambda - R) + c_1 C_{\mu\nu\alpha\beta}C^{\mu\nu\alpha\beta} + c_2 R^2 + c_3 E\right]\sqrt{|g|}d^4x, \tag{10.4}$$

leads to a renormalizable, albeit non-unitary, theory. In the above, unlike the negative mass dimension of the coupling constant in the Einstein-Hilbert term, the coupling constants $c_{1,2,3}$ are dimensionless signifying perturbative power-counting renormalizability (actual renormalizability of the total theory has indeed been demonstrated). $E = R^2 - 4R_{\mu\nu}R^{\mu\nu} + R_{\mu\nu\alpha\beta}R^{\mu\nu\alpha\beta}$ is the Gauss-Bonnet topological integrand (which allows the trading of $R_{\mu\nu}R^{\mu\nu}$ term for $C_{\mu\nu\alpha\beta}C^{\mu\nu\alpha\beta}$ contribution in $S_{renor.}$). This theory which is four-covariant comes with higher time derivatives and physical ghost (state of negative norm), implying a violation of unitarity. Probability in the scattering matrix is not conserved, and the classical theory is also unstable. The renormalizability from better convergence with higher derivatives comes at the expense of higher time derivatives and unitarity. Horava's bold proposal was to forgo four-covariance in favor of supplementing Einstein's theory with higher spatial but not time derivatives and retaining just spatial diffeomorphism symmetry. In this context, it is no surprise that the Cotton-York tensor, which is the analog of the Weyl tensor specific to three dimensions, should feature prominently in our extension of Einstein's theory. Some of the ramifications have already been addressed in this tome. Others will be taken up in the following discussions.

10.1.1 *Bekenstein-Hawking black hole entropy*

A Schwarzschild black hole is associated with Bekenstein-Hawking entropy [55]

$$S_{bh} = \frac{1}{4} \left(\frac{A}{L_{Planck}^2} \right) k_B, \qquad (10.5)$$

wherein $A = 4\pi r_s^2$ is the area of the Schwarzschild event horizon at $r_s = \frac{2GM}{c^2}$. Compared to other objects, a black hole carries an inordinate amount of entropy. Consider an object of mass M containing $\frac{M}{m_B}$ number of baryons. If it were to collapse into a black hole, the entropy per baryon would be $\frac{1}{4} [\frac{4\pi r_s^2}{L_{Planck}^2 (M/m_B)}] k_B$. This is roughly an entropy of $10^{20} k_B$ per baryon for $M = 10 M_\odot$. All other known forms of entropy are by comparison negligible. If the more than 3×10^{80} baryons in our universe were totally made up of such black holes their entropy (in units of Boltzmann constant k_B) would exceed 10^{100}, a googol! Since the horizon area of black holes or entropy increases during mergers, Penrose estimates that the total entropy of a black hole dominated universe could reach $S = 10^{123} k_B$. Thermodynamic postulate of equal a priori probabilities for all the accessible microstates leads to density of state $\rho = 1/\Omega$ that maximizes the statistical von Neumann entropy $S = -k_B Tr(\rho \ln \rho)$. This yields the Boltzmann entropy relation $S = k_B \ln \Omega$ for total number of microstates Ω. Under such conditions, a staggering number of microstates $\Omega \sim e^{10^{123}}$, which is much greater than a googolplex (10^{googol}), would correspond to black hole initial data. This carries the profound implication that black hole states would swamp all other, and extreme fine tuning would be necessary for the universe to steer away from the preponderance of being born a black hole or from being black hole dominated at the beginning. Thus isotropic Robertson-Walker Big Bang without black hole singularities would be, in Penrose's words, "extraordinarily special", even though formation of structure through the clumping of matter to form stars, black holes, galaxies and clusters of galaxies can entail. Penrose also emphasized the argument does not need to rely on strict thermodynamic adherence. What can reasonably be argued on general physical grounds is a phase space that includes gravitational degrees of freedom would quite certainly be predominantly black hole initial data. To pick out a Big Bang without black hole singularity would require imposing an extreme form of selection rule. This would require the Weyl curvature to vanish, a statement extraneous to Einstein's field equations.

10.1.2   *Penrose's Weyl Curvature Hypothesis*

Compared to what it theoretically could have been in a black hole dominated phase space, the entropy content of the cosmological gravitational field at the Big Bang was extremely low. Robertson-Walker isotropic singularity is one in which the Ricci tensor dominates over the Weyl tensor, whereas black holes are associated with Weyl singularities. Penrose links the initial low entropy content of the universe with the effective vanishing of the Weyl curvature tensor near the Big Bang. He proposes the Weyl Curvature Hypothesis of vanishing Weyl tensor as a selection rule for the initial data of our universe. This is a four-covariant statement. Furthermore, the Hypothesis does imply from the initial low entropy condition to formation of structure through clumping of matter into stars, black holes, galaxies and gigantic black holes, the overall entropy increases as the universe ages as black hole entropy by far outweighs all other forms of entropy.[2] Such an initial condition would lead to an entropy arrow of time in accordance with the Second Law of Thermodynamics. In ITG, in an ever expanding universe the cosmological arrow of time points in the direction of increasing volume, this gravitational-cosmological arrow of time and the thermodynamic arrow of time are aligned if the initial quantum state of the universe favors the Penrose selection rule. This links the quantum arrow of time to the thermodynamic arrow within the context of a time-asymmetric Hamiltonian in ITG.

10.2   Quantum origin of our universe from a confluence of circumstances in ITG

While the Weyl Curvature Hypothesis proffers a tantalizing resolution of fundamental problems, *the means to achieving and realizing it in a natural context within semiclassical and quantum Einstein theory remain elusive.* Penrose advocates the theory of Cyclic Conformal Cosmology wherein the final conformally flat (hence vanishing Weyl tensor) stage of one universe is mapped onto the beginning of another universe together with the associated remnants of the past universe [24, 126]. But this has its difficulties [127]; furthermore, Penrose's Weyl Curvature Hypothesis is a classical statement which is prone to quantum and semiclassical corrections, especially at the

---

[2]Even if all black holes were to eventually evaporate away after eons, it is possible a very small cosmological constant, ergo very large cosmological horizon and entropy, remains. In a de Sitter manifold the area of the cosmological horizon is inversely proportional to the value of cosmological constant.

beginning of the universe wherein quantum effects are expected to be significant for Planck size beginnings. In addition, and as emphasized by Penrose himself, the selection of initial data of vanishing Weyl curvature at the Big Bang *is exceedingly unlikely within the context of a gravitational theory described entirely by Einstein's GR.* In Penrose's hypothesis, the strict vanishing of Weyl curvature is also problematic to the existence of primordial gravitational waves characterized by nontrivial Weyl invariants.

A compelling initial quantum state of our universe which is compatible with initial semiclassical vanishing of Weyl curvature, but which goes beyond Penrose's classical hypothesis and Einstein's GR, is brought about by a remarkable confluence of circumstances within the framework of ITG. The contributing factors are

1) in ITG the Hamiltonian is inherently asymmetrical in intrinsic time, and an initial era of Cotton-York domination at early intrinsic times when our universe was small holds sway before Einstein's theory rule at later epochs;

2) the natural inclusion of a Cotton-York term to Einstein's theory as advocated in ITG which leads to the Chern-Simons [21] ground state;

3) quantum origin of the universe from Hartle-Hawking proposal [22, 71, 92] and instanton tunneling which produce this Chern-Simons state;

4) critical points of this exact Hartle-Hawking wavefunction are spatially conformally flat; furthermore, vanishing expectation values of extrinsic curvature and Cotton-York tensor lead to the Penrose's Weyl Curvature Hypothesis being satisfied semiclassically;

5) this physical condition of effective vanishing Weyl curvature is a solution of the initial data problem of Einstein's theory *with Cotton-York addition* (it solves the diffeomorphism constraints and the generalized Lichnerowicz-York equation with Cotton-York term [61, 72, 73, 75, 76] as in Chapter 4); and

6) the exact Chern-Simons Hartle-Hawking state exhibits quantum fluctuations about the critical points which produce correlations functions that carry characteristic physical signals which are scale-invariant at the lowest approximation, and linearized classical perturbations of the theory produces gravitational waves [52] with two polarizations.

These ideas and their physical manifestations will be focus of the following sections.

10.2.1 *Euclidean-Lorentzian tunneling*

A noteworthy feature of the choice of spatial volume as time is that it is common to Euclidean and Lorentzian manifolds which can interface at a particular instant of intrinsic time, provided junction conditions for signature change are met. A requirement of vanishing momentum suffices [56]. As the commutator $[\bar{\pi}^i_j, \tilde{C}^j_i]$ vanishes identically, the Hamiltonian can be reexpressed as

$$H = \frac{1}{\beta} \int \sqrt{\bar{\pi}^i_j \bar{\pi}^j_i - \frac{q}{(2\kappa)^2}(R - 2\Lambda) + g^2 \hbar^2 \tilde{C}^i_j \tilde{C}^j_i} \, d^3x$$

$$= \frac{1}{\beta} \int \sqrt{(Q^\dagger_{CS})^i_j (Q_{CS})^j_i - \frac{q}{(2\kappa)^2}(R - 2\Lambda)} \, d^3x;$$

$$(Q_{CS})^i_j := e^{-gW_{CS}} \bar{\pi}^i_j e^{gW_{CS}} = \bar{\pi}^i_j - ig\hbar\tilde{C}^i_j. \tag{10.6}$$

Null classical value of the complex combination $(Q_{CS})^i_j$ implies the vanishing of both the traceless momentric and the Cotton-York tensor. The latter is just the statement of spatial conformal flatness. Since the Hamiltonian density in ITG is proportional to $\tilde{\pi}$, the junction condition needed is that spatial Ricci scalar R cancels the effective cosmological constant term on the conformally flat spatial interface with zero momentric.[3] For pure gravity this Lorentzian-Euclidean interface happens precisely at the throat of the Lorentzian de Sitter solution with $k = +1$ conformally flat S^3 slicings; continuation to purely Euclidean signature occurs at smaller volumes or earlier intrinsic times. The Euclidean de Sitter continuation constitutes a hemisphere of S^4; at the interface which is the equator of the Euclidean hemisphere as well as the Lorentzian throat, $R = 6k/a^2 = 2\Lambda$. At even smaller volumes H with $(Q_{CS})^i_j = 0$ becomes imaginary, and Eq. (6.22) leads to dynamics with respect to Euclidean interval $d\tau' := -idt'$ wherein

$$\frac{da}{d\tau'} = -\sqrt{1 - \frac{\Lambda}{3}a^2} \quad \Rightarrow \quad a = \sqrt{\frac{3}{\Lambda}} \cos\left(\sqrt{\frac{\Lambda}{3}}(\tau' - \tau'_0)\right);$$

$$ds^2_{Euclidean} = d\tau'^2 + a^2(\tau')(d\chi^2 + \sin^2 \chi d\Omega^2). \tag{10.7}$$

Remarkably, this Euclidean-Lorentzian continuation is an exact and explicit instanton tunneling solution of the emergence of a Lorentzian signature universe from a Euclidean origin at a point of zero volume.

---

[3]Only the gravitational contribution has so far been addressed; but Yang-Mills and fermionic additions will not affect the argument if the initial state is also the vacuum of Yang-Mills theory (with gauge instantons this nonperturbative vacuum can be a Yang-Mills Chern-Simons state) with the fermions being zero modes of the matter Hamiltonian.

Hawking had presented such an intuitive picture in his popular discourse [54]; ITG bolsters this imagery, it finds correspondence and happy union between the Hartle-Hawking Proposal and Penrose's Weyl Curvature Hypothesis. This classical picture is the semiclassical result of an exact quantum state which is the exponent of the (relative) Chern-Simons functional of the affine connection. This state satisfies $(Q_{CS})^i_j|\Psi_{CS}\rangle = 0$; its critical point $\frac{\delta W_{CS}}{\delta q_{ij}} = \tilde{C}^{ij} = 0$ is conformal flatness; and the semiclassical momentum $\tilde{\pi}^{ij} \propto \frac{\delta W_{CS}}{\delta q_{ij}} = \tilde{C}^{ij}$ vanishes at this critical value. The correspondence goes beyond critical point analysis. For this state, $\langle\Psi_{CS}|(Q_{CS})^i_j|\Psi_{CS}\rangle = 0$ implies $\langle\Psi_{CS}|\tilde{\pi}^i_j|\Psi_{CS}\rangle = 0$ and $\langle\Psi_{CS}|\tilde{C}^i_j|\Psi_{CS}\rangle = 0$ actually hold at the level of expectation values (this is analogous to the situation in the simple harmonic oscillator: the action of the complex annihilation operator on the vacuum state results in $\langle 0|a|0\rangle = 0$, consequently $\langle 0|x|0\rangle = \langle 0|p|0\rangle = 0$). For the explicit classical de Sitter solution, Penrose's criterion of vanishing Weyl curvature actually holds for the initial data of vanishing momentum and Cotton-York tensor at the Lorentzian de Sitter throat at the beginning of the universe (the de Sitter solution is a special case of Robertson-Walker spacetimes, all of which come with vanishing Weyl tensor).

10.2.2 *A hot beginning*

The current CMB radiation with temperature variations of one part in 10^5 is to very good approximation a black body spectrum, and it is even closer to Planck spectrum at earlier times before structure formation. The Planck formula for energy density as a function of wavelength at the current Roberston-Walker scale factor is

$$du_\lambda|_{a_0} = \frac{8\pi hc}{\lambda_{a_0}^5} \frac{d\lambda_{a_0}}{[e^{hc/(k_B\lambda_{a_0}\mathcal{T})} - 1]}$$

$$\Rightarrow du_\lambda|_a = \frac{(\frac{h\nu|_a}{a^3})}{(\frac{h\nu|_{a_0}}{a_0^3})} du_\lambda|_{a_0} = f^4 du_\lambda|_{a_0}$$

$$= \frac{8\pi hc}{\lambda_a^5} \frac{d\lambda_a}{[e^{hc/(k_B\lambda_a(f\mathcal{T}))} - 1]};$$

$$f := \frac{a_0}{a} = \frac{\lambda_{a_0}}{\lambda_a}. \tag{10.8}$$

Thus the Planck distribution retains its form during Robertson-Walker expansion, while the temperature $\mathcal{T}$ is shifted to $f\mathcal{T}$. Our universe must have cooled from a very hot beginning (even an infinitely hot one if one entertains a singular starting point of vanishing scale factor and infinite scalar curvature) to its current 2.73K as it follows the flow in intrinsic time interval $dT = 2d\ln a$. Emergence from Euclidean tunneling offers a tantalizing explanation of a hot beginning. The Big Bang happens at the Euclidean-Lorentzian interface which has a Euclidean time variable that is periodic. The explicit Lorentzian de Sitter solution is joined at the throat by the Euclidean de Sitter manifold with periodic time variable governed by the effective value of the cosmological constant. The scale factor at the throat is $a = \sqrt{\frac{3}{\Lambda}}$, and the periodicity of the Euclidean time τ' read off from Eq. (10.7) is $\mathcal{P} = \frac{2\pi}{c}\sqrt{\frac{3}{\Lambda}}$. The upshot is correlation functions of fields should obey the Kubo-Martin-Schwinger condition [58] which is a manifestation in statistical mechanics of quantum mechanical systems and quantum field theory of a system in thermal equilibrium, with the partition function of a thermal system at temperature $\mathcal{T}_{initial} = \frac{\hbar}{k_B \mathcal{P}}$. A Planck scale interface with $a \sim L_p$ implies $\mathcal{T} = \frac{\hbar c}{2\pi k_B}\sqrt{\frac{\Lambda}{3}} = \frac{\hbar c}{2\pi k_B L_P} = \frac{1}{2\pi}\mathcal{T}_P \sim 10^{31}$K. A very hot beginning is associated with the narrowness of throat and the high value of effective cosmological constant at that instant. If we retrace, within the context of the standard model of cosmology and particle physics, the current CMB temperature to the Planck scale the same temperature of $\sim 10^{31}$K would emerge. This is not really a happenstance since the Planck scale is the only one that is invoked in both cases; however, retracing the CMB does not really explain why the universe has a *very hot beginning* that was in *pure* thermal equilibrium which lead to the very near Planck CMB that we observe today. The origin of the universe from Euclidean tunneling offers a resolution.

It is also worth pointing out if our universe started out in a very hot Big Bang in thermal equilibrium, Penrose's argument may need to be adjusted if both spatial compactness and strict thermal equilibrium between primordial black holes and the universe are enforced. If we require black holes formed at the beginning of the universe to be in thermal equilibrium with an initial Planck scale universe ($a(t) \sim L_P$) at Planck temperature $\mathcal{T}_P = \frac{M_P c^2}{k_B}$, setting the black hole Hawking temperature $\mathcal{T}_{Hawking} = \frac{1}{8\pi}(\frac{M_P}{M})\mathcal{T}_P \sim \mathcal{T}_P$ implies these primordial black holes are of mass $M \sim \frac{1}{8\pi}M_P$, and each black hole will only contribute $S_{bh} = 4\pi(\frac{M}{M_P})^2 k_B < 1k_B$ unit of entropy. Whether or not the universe is spatially compact is of importance. In spatially non-compact $k = 0, -1$ Robertson-walker manifiolds for instance, the spatial volume is always infinite regardless of how small the scale factor

$a(t)$ is, whereas for the $k = +1$ compact S^3 case the spatial volume is $2\pi^2 a^3$. In the non-compact cases, there is no limit on the number of Planck scale black holes the universe can contain and on the total entropy of these Planck size black holes, whereas in the spatially compact case there is only room for single black hole of $\sim 1k_B$ entropy to be in thermal equilibrium with a universe of Planck size at Planck scale temperature. All these considerations will however be superseded by the exact nature of the quantum origin of the universe which will be presented in the following sections.

10.3 Gravitational Chern-Simons functional and the Cotton-York tensor

The Chern-Simon functional is [128, 129]

$$W_{CS} = \frac{1}{2} \int \tilde{\epsilon}^{ikj} \left(\Gamma^m_{in} \partial_j \Gamma^n_{km} + \frac{2}{3} \Gamma^m_{in} \Gamma^n_{js} \Gamma^s_{km} \right) d^3x. \tag{10.9}$$

Its the variation leads, upon integration by parts over closed spatial manifold, to

$$\delta W_{CS} = \frac{1}{2} \int \tilde{\epsilon}^{ikj} R^n{}_{mjk} \delta \Gamma^m_{in} d^3x. \tag{10.10}$$

If GR were just a pure Yang-Mills theory of connection, the functional derivative of the Chern-Simons would merely yield the analogous curvature term, $\frac{\delta W_{CS}}{\delta \Gamma^m_{in}} = \frac{1}{2} \tilde{\epsilon}^{ikj} R^n{}_{mjk}$, with Riemann curvature tensor

$$R^n{}_{mjk} = \partial_j \Gamma^n_{km} - \partial_k \Gamma^n_{jm} + \Gamma^n_{js} \Gamma^s_{km} - \Gamma^n_{ks} \Gamma^s_{jm}. \tag{10.11}$$

Geometrodynamics being a theory with metric, rather than connection, as the fundamental variable allows a further step. In three dimensions the Riemann curvature tensor can be expressed completely in terms of the Ricci tensor through

$$R^n{}_{mjk} = q_{mk} R^n_j + \delta^n_j R_{mk} - q_{mj} R^n_k - \delta^n_k R_{mj} + \frac{R}{2} (\delta^n_k q_{mj} - \delta^n_j q_{mk}). \tag{10.12}$$

Variation of the connection leads to $\delta \Gamma^m_{in} = \frac{1}{2} q^{mr} (\nabla_i \delta q_{rn} + \nabla_n \delta q_{ri} - \nabla_r \delta q_{in})$. These two results conspire to yield

$$\begin{aligned} \delta W_{CS} &= \frac{1}{2} \int \frac{1}{2} \tilde{\epsilon}^{ikj} (-4 R^n_j \nabla_k \delta q_{in} + R \nabla_k \delta q_{ij}) d^3x \\ &= \int \tilde{\epsilon}^{ikj} \left(\delta q_{in} \nabla_k R^n_j - \frac{1}{4} \delta q_{ij} \nabla_k R \right) d^3x \\ &= \int \tilde{\epsilon}^{imn} \nabla_m (R^j_n - \frac{1}{4} \delta^j_n R) \delta q_{ij} d^3x. \end{aligned} \tag{10.13}$$

The end result is an identity neither present nor foreseen in Yang-Mills theory with fundamental connection variables: the functional derivative of the Chern-Simons functional W_{CS} with respect to the spatial metric yields the Cotton-York tensor density [20, 44] i.e.

$$\frac{\delta W_{CS}}{\delta q_{ij}} = \tilde{C}^{ij} := \tilde{\epsilon}^{ikl}\nabla_k \left(R^j{}_l - \frac{1}{4}\delta^j{}_l R \right).$$ (10.14)

This entity differs saliently in being a third order derivative of the spatial metric, whereas Yang-Mills and Riemann curvatures are, respectively, first order and second order derivatives of the connection and the metric. Thus $\tilde{C}^{ij}$ has physical dimension $(\text{length})^{-3}$ instead of $(\text{length})^{-2}$.

As a functional solely of the Christoffel symbol which is uniquely defined by the spatial metric, W_{CS} possesses the additional property of being dependent only on the unimodular part, $\bar{q}_{ij}$, of the spatial metric ($q_{ij} = q^{\frac{1}{3}}\bar{q}_{ij}$) and independent of q i.e. $W_{CS}[\Gamma^i_{jk}(q_{kl})] = W_{CS}[\bar{\Gamma}^i_{jk}(\bar{q}_{kl})]$. This can be demonstrated by explicit computations and also through

$$\frac{\delta W_{CS}}{\delta q(x)} = \int \frac{\delta W_{CS}}{\delta q_{ij}(y)}\frac{\delta q_{ij}(y)}{\delta q(x)}d^3y$$

$$= \tilde{C}^{ij}(x)\frac{1}{3q(x)}q_{ij}(x)$$

$$= 0,$$ (10.15)

as the Cotton-York tensor is traceless. The antisymmetric part of $\tilde{C}^{ij}$ vanishes via the Bianchi identity i.e.

$$\epsilon_{ijk}\tilde{C}^{ij} = \epsilon_{ijk}\tilde{\epsilon}^{imn}\nabla_m \left(R^j{}_n - \frac{1}{4}\delta^j{}_n R \right)$$

$$= \nabla_m \left(R^m{}_k - \frac{1}{2}\delta^m{}_k R \right)$$

$$= 0;$$ (10.16)

so in explicitly symmetric form,

$$\tilde{C}^{ij} = \tilde{\epsilon}^{imn}\nabla_m \left(R^j{}_n - \frac{1}{4}\delta^j{}_n R \right)$$

$$= \frac{1}{2}(\tilde{\epsilon}^{imn}\nabla_m R^j{}_n + \tilde{\epsilon}^{jmn}\nabla_m R^i{}_n).$$ (10.17)

Equation (10.13) also proves that W_{CS} is invariant under spatial diffeomorphisms of the metric, as $\delta W_{CS} = \int \tilde{C}^{ij}\delta q_{ij}\, d^3x$ vanishes (due to the transversality and symmetry of the Cotton-York tensor, and upon integration by parts) when $\delta q_{ij} = \mathcal{L}_{\vec{N}}q_{ij} = \nabla_i N_j + \nabla_j N_i$.

The incorporation in ITG of the Cotton-York tensor which is powerful tool unique to three dimensions brings surprising new features which could not be foreseen within the four-covariant paradigm of Einstein's theory. The effect of a Cotton-York term in the extended Lichnerowicz-York Equation in narrowing down the initial data of the Universe has been discussed in Chapter 4. Chapters 7 and 8 sketched how the presence of the Cotton-York term in the Hamiltonian of ITG leads to the Chern-Simons ground state, and described the analogy to the magnetic field term in the Hamiltonian of renormalizable Yang-Mills theories. The upcoming sections will address in detail the Hartle-Hawking Chern-Simons state as the origin of the Universe, its semiclassical correspondence, its quantum gravity fluctuations and correlations functions in the early universe, and other novel physical consequences.

10.3.1 One-to-one correspondence of Cotton-York and transverse-traceless metric excitations

It can be deduced that variations of the Cotton-York tensor are in fact one-to-one with $\delta \bar{q}_{ij}^{TT}$ regardless of the background. The Hessian H^{ijkl} is defined by

$$\frac{\delta^2 W_{CS}}{\delta q_{kl}(y)\delta q_{ij}(x)} = \frac{\delta \widetilde{C}^{ij}(x)}{\delta q_{kl}(y)} =: H^{ijkl}(x,y). \tag{10.18}$$

The action of the Hessian on the space of transverse traceless metric variations is

$$\int H^{ijkl}(x,y)\delta q_{kl}^{TT}(y)d^3y =$$

$$\frac{1}{4}\left\{ \tilde{\epsilon}^{inp}q^{jo}R_n^c[\delta_o^m\delta_p^l\delta_q^k - \delta_o^l\delta_p^k\delta_q^m - \delta_o^k\delta_p^m\delta_q^l]\nabla_m + \tilde{\epsilon}^{imp}\nabla_m \right.$$

$$\left. \left[q^{jk}R_p^l + 3\delta_p^l\left(R^{jk} - \frac{1}{3}q^{jk}R\right) - \delta_p^lq^{jk}\nabla^2 \right] + (i \leftrightarrow j) \right\}\delta q_{kl}^{TT}(x). \tag{10.19}$$

Despite its complicated appearance, the operator acting on δq_{kl}^{TT} is in fact invertible in general. To demonstrate this consider local Riemann normal coordinates at an arbitrary point P where $\partial_i q_{jk} = 0$. It follows that at that point $\Gamma_{jk}^i = 0$, and its derivative satisfies

$$\Gamma_{jk,l}^i + \Gamma_{lj,k}^i + \Gamma_{kl,j}^i = 0. \tag{10.20}$$

Consequently the identity

$$\partial_k\Gamma_{jl}^i = -\frac{1}{3}(R_{jlk}^i + R_{ljk}^i) \tag{10.21}$$

holds at P. From these $\partial_k\partial_l q_{ij} = -\frac{1}{3}(R_{ikjl} + R_{iljk})$ obtains. Thus at the point P,

$$(\nabla_P)_k(\nabla_P)_l(q_P)_{ij} = -\frac{1}{3}[(R_P)_{ikjl} + (R_P)_{iljk}]; \qquad (10.22)$$

$$(R_P)_{ij} = -\frac{3}{2}(\nabla_P)^2(q_P)_{ij}. \qquad (10.23)$$

The R.H.S. of Eq. (10.19) at P then reduces to

$$O_P^{ijkl}(\delta q_P)_{kl}^{TT} := \frac{5}{64}[(q_P)^{jk}\tilde{\epsilon}^{irl} + (q_P)^{ik}\tilde{\epsilon}^{jrl} + (k \leftrightarrow l)](\nabla_P)^2(\nabla_P)_r(\delta q_P)_{kl}^{TT}. \qquad (10.24)$$

This has an explicit inverse operator

$$(O_P^{-1})_{mnij} = \frac{16}{5}\frac{1}{(\nabla_P)^2}(\nabla_P)^s\frac{1}{(\nabla_P)^2}[(q_P)_{mi}\underline{\epsilon}_{jsn} + (q_P)_{mj}\underline{\epsilon}_{isn} + (m \leftrightarrow n)], \qquad (10.25)$$

with $(O_P^{-1})_{mnij}(O_P)^{ijkl}(\delta q_P)_{kl}^{TT} = \frac{1}{2}(\delta_m^k\delta_n^l + \delta_m^l\delta_n^k)(\delta q_P)_{kl}^{TT}$. This implies the action of the Hessian on transverse traceless metric variations is invertible, and Cotton-York variations are one-to-one with $\delta\bar{q}_{ij}^{TT}$ in general. Both have two physical d.o.f.; and the Cotton-York tensor which is also TT, has two independent 2 eigenvalues at each point. It may be regarded as the non-perturbative manifestation of the physical d.o.f. in geometrodynamics.

10.3.2 *Conformal flatness and the Cotton-York tensor*

The decomposition $q_{ij} = q^{\frac{1}{3}}\bar{q}_{ij}$ yields $\Gamma_{jk}^i = \bar{\Gamma}_{jk}^i + \Delta_{jk}^i$, wherein

$$\bar{\Gamma}_{jk}^i = \frac{1}{2}\bar{q}^{il}\left(\partial_j\bar{q}_{kl} + \partial_k\bar{q}_{jl} - \partial_l\bar{q}_{jk}\right),$$

$$\Delta_{jk}^i = \frac{1}{2}\left(\delta_j^i\partial_k\ln q^{1/3} + \delta_k^i\partial_j\ln q^{1/3} - \bar{q}^{il}\bar{q}_{jk}\partial_l\ln q^{1/3}\right).$$

The respective Ricci tensors of the barred and unbarred variables are then related by

$$R_{ij} = \bar{R}_{ij} - \frac{1}{2}\bar{\nabla}_i\partial_j\ln q^{1/3} - \frac{1}{2}\bar{q}_{ij}\bar{q}^{kl}\bar{\nabla}_k\partial_l\ln q^{1/3}$$
$$+ \frac{1}{4}(\partial_i\ln q^{1/3})(\partial_j\ln q^{1/3}) - \frac{1}{4}\bar{q}_{ij}\bar{q}^{kl}(\partial_k\ln q^{1/3})(\partial_l\ln q^{1/3}). \qquad (10.26)$$

A spatial 3-manifold is locally conformally flat when it admits coordinate systems in which $q_{ij} = q^{1/3}\delta_{ij}$ i.e. when $\bar{R}_{ij} = 0$. It follows that

$$R_{ij} = -\frac{1}{2}\bar{\nabla}_i w_j - \frac{1}{2}\bar{q}_{ij}\bar{q}^{kl}\bar{\nabla}_k w_l + \frac{1}{4}w_i w_j - \frac{1}{4}\bar{q}_{ij}\bar{q}^{kl}w_k w_l, \qquad (10.27)$$

wherein $w_i := \partial_i \ln q^{1/3}$. In three spatial dimensions the Weyl tensor vanishes identically, and curvature is completely determined by the Ricci tensor which generically has six components. Conformal flatness means R_{ij} is completely parametrized as above by just one variable $\ln q^{1/3}$. There must therefore be five constraints; these arise from the integrability conditions of w_i that R_{ij} can be expressed as in Eq. (10.27). It follows that the covariant derivative of the Schouten tensor, $R_{ij} - \frac{1}{4}Rq_{ij}$, is

$$\nabla_k \left(R_{ij} - \frac{1}{4}Rq_{ij}\right) = -\frac{1}{2}\bar{\nabla}_k \bar{\nabla}_j w_i - \frac{1}{2}w_i w_j w_k - \frac{1}{2}[w_j \bar{\nabla}_k w_i + w_i \bar{\nabla}_k w_j$$

$$+ w_k \bar{\nabla}_j w_i] - \frac{1}{4}\bar{q}_{ij}\bar{q}^{lm}w_l \bar{\nabla}_k w_m - \frac{1}{4}\bar{q}_{ki}\bar{q}^{lm}w_l \bar{\nabla}_j w_m -$$

$$\frac{1}{4}\bar{q}_{jk}\bar{q}^{lm}w_l \bar{\nabla}_i w_m + \frac{1}{8}\bar{q}^{lm}w_l w_m[\bar{q}_{jk}w_i + \bar{q}_{ij}w_k + \bar{q}_{ik}w_j]. \qquad (10.28)$$

And the integrability condition is that the Cotton tensor C_{ijk} must obey

$$C_{ijk} = \nabla_k \left(R_{ij} - \frac{1}{4}Rq_{ij}\right) - \nabla_j \left(R_{ik} - \frac{1}{4}Rq_{ik}\right)$$

$$= \frac{1}{2}[\bar{\nabla}_j, \bar{\nabla}_k]w_i = \frac{1}{2}\bar{R}^l_{\ ikj}w_l = 0 \qquad (10.29)$$

due to the vanishing of $\bar{R}_{ij}$; in three dimensions this also implies the existence of coordinates in which we may replace $\bar{\nabla}_i$ by ∂_i. Thus $[\partial_j, \partial_k]w_i = 0$, the necessary and sufficient condition of Frobenius' theorem that the vector field w_i gives rise to integral curves, is precisely equivalent to $C_{ijk} = 0$. This is the same as saying the vanishing of the Cotton-York tensor (density) $\tilde{C}^i_j := \tilde{\epsilon}^{ikl}\nabla_k(R_{jl} - \frac{1}{4}Rq_{jl}) = \frac{1}{2}\tilde{\epsilon}^{ikl}C_{jlk}$ is the necessary and sufficient condition for conformal flatness in three spatial dimensions. The entity $\tilde{C}^i_j$ is conformal invariant (to be explicit, $\bar{\tilde{C}}^i_j$ of $\bar{q}_{ij}$ and $\tilde{C}^i_j$ of q_{ij} obeys $\bar{\tilde{C}}^i_j = \tilde{C}^i_j$). Since $\tilde{C}^{ij}$ is symmetric and traceless, its vanishing imposes precisely the five constraints to restrict R_{ij} to the form of Eq. (10.27).

10.3.3 On conformal flatness, Poincaré Conjecture and the three-sphere

It is known that every compact simply connected conformally flat manifold is conformally diffeomorphic to the n-sphere [80]. Topologically, the Poincaré Conjecture says that every simply connected, closed three-manifold is homeomorphic to the three-sphere. The proof was first presented by Grigori Perelman in a series of three articles posted to arXiv:math.DG [81]. Perelman saw no need to submit these articles to any journal; he also declined the Fields Medal, and the Millennium Prize, despite the fact that the Poincaré Conjecture is the only solved Millennium problem. A fact of differential topology is the notion of topological and differentiable isomorphism is the same in dimension three and below, so the generalized smooth Poincaré Conjecture that a manifold which is a homotopy three-sphere *is in fact diffeomorphic* to the standard three-sphere holds.

10.3.4 *Standard three-sphere and $k = +1$ Robertson-Walker metric*

The $k = +1$ Robertson-Walker metric with S^3 spatial slicing has the standard form,

$$ds^2 = -dt^2 + a^2(t)(dl)^2_{S^3}; \tag{10.30}$$

and the spatial metric can be made manifestly conformally flat as

$$
\begin{aligned}
(dl)^2_{S^3} &= d\chi^2 + \sin^2\chi(d\theta^2 + \sin^2\theta d\phi^2) \\
&= \Phi^2(dr^2 + r^2(d\theta^2 + \sin^2\theta d\phi^2)) \\
&= \Phi^2(dx^2 + dy^2 + dz^2),
\end{aligned} \tag{10.31}
$$

with $r := \frac{1-\cos\chi}{\sin\chi}$, $\Phi^2 = \frac{4}{(1+r^2)^2} = [\frac{\sin^2\chi}{1-\cos\chi}]^2$, and $x = r\sin\theta\cos\phi, y = r\sin\theta\sin\phi, z = r\cos\theta$ being the usual correspondence between Cartesian and spherical coordinates.

All Robertson-Walker spacetimes have vanishing Weyl tensors i.e. they also exhibit four-dimensional conformal flatness. For the closed $k = +1$ case, an explicit conformally flat description is

$$
\begin{aligned}
ds^2 &= a^2(\eta)(-d\eta^2 + d\chi^2 + \sin^2\chi(d\theta^2 + \sin^2\theta d\phi^2)) \\
&= a^2(\zeta, R)A^2(\zeta, R)(-d\zeta^2 + dR^2 + R^2 d\Omega^2),
\end{aligned} \tag{10.32}
$$

with $d\eta := \frac{dt}{a(t)}, \zeta := \frac{2\sin\eta}{\cos\eta + \cos\chi}, R := \frac{2\sin\chi}{\cos\eta + \cos\chi}$ and $A^2(\zeta, R)R^2 = \sin^2\chi$.

10.3.5 *Chern-Simons functional and the Pontryagin invariant*

In four dimensional manifolds with Euclidean signature, the Pontryagin invariant is $P_4 = \frac{1}{8\pi^2}\int_M tr(R \wedge R)$ with $R = d\Gamma + \Gamma \wedge \Gamma$ being the curvature 2-form. Since

$$\int_M tr(R \wedge R) = \int_M d\left(tr\left(\Gamma d\Gamma + \frac{2}{3}\Gamma \wedge \Gamma \wedge \Gamma\right)\right). \tag{10.33}$$

If a three manifold of interest, Σ, is the boundary of M i.e. $\Sigma = \partial M$, it follows, by Stokes' theorem,

$$
\begin{aligned}
-\frac{1}{2} \int_M tr(R \wedge R) &= -\frac{1}{2} \int_{\partial M = \Sigma} tr\left(\Gamma d\Gamma + \frac{2}{3} \Gamma \wedge \Gamma \wedge \Gamma\right) \\
&= -\frac{1}{2} \int_\Sigma \Gamma^i_j d\Gamma^j_i + \frac{2}{3} \Gamma^i_j \wedge \Gamma^j_k \wedge \Gamma^k_i \\
&= \frac{1}{2} \int_\Sigma \tilde{\epsilon}^{ikj} \left(\Gamma^m_{in} \partial_j \Gamma^n_{km} + \frac{2}{3} \Gamma^m_{in} \Gamma^n_{js} \Gamma^s_{km}\right) d^3x \\
&= W_{CS}[\Gamma|_\Sigma],
\end{aligned}
\tag{10.34}
$$

wherein $\Gamma^i_j = \Gamma^i_{jk} dx^k$ denotes the connection one-form on the spatial manifold Σ. However $Tr(R \wedge R)$ can only be locally exact in general; its associated Chern-Simons 3-form can be related by a "large gauge transformation" with non-trivial winding number. Two locally defined "gauge related" connections, Γ and Γ', can produce the same Pontryagin density. If Σ is the common boundary of two orientable manifolds M' and M i.e. $\Sigma = \partial M' = \partial M$, gluing one to the other with reversed orientation yields a closed four manifold $M' \sqcup \overline{M}$ (the overline denotes reversed orientation). The Pontryagin index of the resultant closed four-manifold need not be trivial since

$$
\begin{aligned}
P_4 &= \frac{1}{8\pi^2} \int_{M' \sqcup \overline{M}} tr(R \wedge R) = \frac{1}{8\pi^2} \left(\int_{M'} tr(R \wedge R) - \int_M tr(R \wedge R)\right) \\
&= \frac{1}{8\pi^2} \left(\int_{\partial M' = \Sigma} tr\left(\Gamma' d\Gamma' + \frac{2}{3} \Gamma' \wedge \Gamma' \wedge \Gamma'\right) \right. \\
&\quad \left. - \int_{\partial M_4 = \Sigma} tr\left(\Gamma d\Gamma + \frac{2}{3} \Gamma \wedge \Gamma \wedge \Gamma\right)\right) \\
&= -\frac{1}{24\pi^2} \int_\Sigma tr(U dU^{-1} \wedge U dU^{-1} \wedge U dU^{-1}),
\end{aligned}
\tag{10.35}
$$

which is also the winding number of the transition function in the overlap region, $U : \Sigma \to SL(3, R)$, with the two connections related by $\Gamma' = U\Gamma U^{-1} + U dU^{-1}$. If Σ is also simply connected besides being closed, it is homeomorphic to S^3 by Perelman's proof of the Poincaré Conjecture;[4] thus the mapping U is classified by $\Pi_3(SL(3, R))$. As $SL(3, R)$ has the same homotopy type as its deformation retract $SO(3)$ (which is diffeomorphic to

---

[4]These weak assumptions on Σ are enough to pin down the initial q_0 at the origin of the universe to be proportional to the determinant of the standard S^3 metric, and the volume element is $\sqrt{q_0(x)} d^3x = a^3 \sin^2 \chi \sin \theta d\chi d\theta d\phi$. As discussed, the initial a is correlated with the temperature and effective value of the cosmological constant at that time.

the real projective space RP^3), it follows that $\Pi_3(SL(3,R)) = \Pi_3(RP^3) = Z$ [130].

10.4   How did the universe begin?

The Schrödinger equation is first-order in time, and so are the Heisenberg equations of motion for quantum operators; thus neither Schrödinger nor Heisenberg evolution can uniquely determine the initial state. The beginning needs an extra physical input. Without first solving the Hamiltonian constraint and having a nontrivial reduced physical Hamiltonian, in conventional canonical quantum gravity there is not even a notion of ground state as the state of lowest energy in the quantum mechanics of closed universes. Without boundary terms, the total energy for a closed universe is always "zero" as the Hamiltonian is totally made up of constraints. Hartle and Hawking proposed it is still reasonable, however, to expect to be able to define a state of minimum excitation corresponding to the classical notion of a geometry of high symmetry. It is suggested that the universe began with Euclidean four-geometries and made a quantum transition to a Lorentzian geometry with the three space and one time dimension that we observe today. The explicit input for the beginning of our physical universe is the Hartle-Hawking no-boundary proposal [22].

In contradistinction, with intrinsic cosmic time and a reduced physical Hamiltonian, the notion of ground state is well-defined even for a spatially closed universe. It was advocated in Ref. [92] that this corresponds to the Chern-Simons state $\sim e^{-gW_{CS}}$ (see Eq. (10.44) for the precise definition). Unlike the abstract Hartle-Hawking wave function, it is an *explicit* functional of metric three-geometries. What is noteworthy is that this same state has an analogous Hartle-Hawking no-boundary interpretation involving Euclidean four-manifolds, instantons and tunneling.

10.4.1   *The Hartle-Hawking no-boundary proposal*

The Hartle-Hawking state for a spatially closed universe is

$$\Psi[q_{ij}^\Sigma] = N \int Dg_{\mu\nu}^M \, e^{-S_E[g_{\mu\nu}^M]}, \tag{10.36}$$

where S_E is the Euclidean action for gravity including a cosmological constant; and Σ, on which q_{ij} is the induced spatial metric, is the only boundary ($\partial M = \Sigma$) of all the compact Euclidean four-geometries summed over.

The L.H.S. is the wave function of the universe, so the functional integral over all compact four-geometries bounded by a given three-geometry is the amplitude for that three-geometry to arise from a zero three-geometry i.e. a single point of zero volume, which is also the beginning of intrinsic time. "In other words, the ground state is the amplitude for the universe to appear from nothing", and "the boundary condition of the universe is that it has no boundary" [22]. Hawking went as far as to say "it seems we don't need string theory even for the beginning of the universe" [54].

10.4.2 *Chern-Simons functional and the Hartle-Hawking state*

As the exterior derivative of the Chern-Simons three-form is the Pontryagin integrand, $W_{CS}[\Gamma|_{\Sigma=\partial M}] = -\frac{1}{2}\int_M tr(R \wedge R)$, a Chern-Simons state $e^{-gW_{CS}[\Gamma_\Sigma]} = e^{\frac{g}{2}\int_M tr(R \wedge R)}$ is in fact expressible as the boundary contribution of the exponential of a Euclidean action which is the Pontryagin integral. However, the Chern-Simons functional is not invariant under large gauge transformations, so two generic compact Euclidean manifolds, M and M', with the same boundary Σ may yield the same Chern-Simons contribution only up to the winding number of a large gauge transformation. As the Chern-Simons state is a real exponential factor, rather than a mere phase, the winding number discrepancy cannot be mod out by quantizing the coupling constant g (as was done in theories with three-dimensional Chern-Simons *action*). Furthermore, the real exponential factor will be unbounded as it can be arbitrarily stepped up or down with the number of windings. However, there is a device to define the Chern-Simons functional modulo large gauge transformations. This can be done by subtracting a topological invariant that is the Chern-Simons functional, of a flat connection in Yang-Mills gauge theory [131], or, of a conformally flat connection, $\Gamma_o|_\Sigma$, for the metric gravitational case at hand. Since Chern-Simons functional $W_{CS}[\Gamma_o|_\Sigma]$ changes by the same winding number, the difference

$$W_{CS}[\Gamma'|_\Sigma] - W_{CS}[\Gamma'_o|_\Sigma] = W_{CS}[\Gamma|_\Sigma] - W_{CS}[\Gamma_o|_\Sigma], \qquad (10.37)$$

is invariant under U for $\Gamma' = U\Gamma U^{-1} + UdU^{-1}$, $\Gamma'_o = U\Gamma_o U^{-1} + UdU^{-1}$ on $\Sigma = \partial M = \partial M'$, regardless of the winding number of U. Fortuitously this equality extends to different compact Euclidean manifolds M and M', so long as they have the same boundary Σ although their respective boundary connections, $\Gamma|_\Sigma$ and $\Gamma'|_\Sigma$, can differ by a large gauge transformation to account for the possible non-triviality of $P_4 = \frac{1}{8\pi^2}\int_{M' \sqcup \overline{M}} tr(R \wedge R)$.

Explicitly

$$\int_{M'\sqcup\overline{M}} tr(R \wedge R) = \int_{M'} tr(R \wedge R) - \int_{M} tr(R \wedge R)$$

$$= -\frac{1}{3}\int_{\Sigma} tr(U\,dU^{-1} \wedge U\,dU^{-1} \wedge U\,dU^{-1})$$

$$= -2(W_{CS}[\Gamma'_o|_{\Sigma}] - W_{CS}[\Gamma_o|_{\Sigma}]), \qquad (10.38)$$

and Eq. (10.37) leads to

$$-\frac{1}{2}\int_{M'} tr(R \wedge R) - W_{CS}[\Gamma'_o|_{\Sigma=\partial M'}] = -\frac{1}{2}\int_{M} tr(R \wedge R) - W_{CS}[\Gamma_o|_{\Sigma=\partial M}]$$

$$= W_{CS}[\Gamma|_{\Sigma}] - W_{CS}[\Gamma_o|_{\Sigma}]. \qquad (10.39)$$

So the no-boundary proposal with Pontryagin Euclidean action leads to state

$$\Psi_{\text{no boundary}}[\Gamma|_{\Sigma}] = N \int D\Gamma_M \, e^{-g(-\frac{1}{2}\int_{M:\partial M=\Sigma} tr(R\wedge R) - W_{CS}[\Gamma_o|_{\Sigma=\partial M}])}$$

$$(10.40)$$

wherein we may sum over *all* compact four-manifolds M with the criterion the affine connection reduces to gauge-equivalent class of Γ on $\Sigma = \partial M$. The corresponding Hartle-Hawking wave function is therefore,

$$\Psi_{\text{no boundary}}[\Gamma|_{\Sigma}] = \mathcal{N}e^{-g(W_{CS}[\Gamma(q_{ij})|_{\Sigma}] - W_{CS}[\Gamma_o|_{\Sigma}])}. \qquad (10.41)$$

This too has the interpretation as the amplitude for a three-geometry to arise from a single point. Since Σ is the boundary of Euclidean signature manifolds, it follows that Σ will naturally and advantageously be Riemannian, rather than pseudo-Riemannian; and the wave function of the universe will a functional of three-geometry. Furthermore Σ being the boundary of a (four) manifold must be closed.

10.4.3 *Relative Chern-Simons functional*

The wave function of the initial state can be stated in the more rigorous mathematical language of relative Chern-Simons forms [132] $CS[\Gamma_2, \Gamma_1]$. Using a connection $\Gamma(s) = (1-s)\Gamma_1 + s\Gamma_2 =: \Gamma_1 + s\Delta\Gamma$ (for $0 \le s \le 1$) which interpolates between Γ_1 and Γ_2 and the Bianchi identity $D^{\Gamma(s)}R(s) = 0$, a particular application of Chern-Weil theory yields the difference

$$tr(R_2 \wedge R_2) - tr(R_1 \wedge R_1) = \int_{s=0}^{1} ds \frac{d}{ds} tr(R(s) \wedge R(s))$$

$$= \int_{s=0}^{1} ds \, tr(2(D^{\Gamma(s)} \Delta\Gamma) \wedge R(s))$$

$$= d\left(2\int_{s=0}^{1} ds \, tr(\Delta\Gamma \wedge R(s))\right)$$

$$= d\left(2\int_{s=0}^{1} ds \, tr(\Delta\Gamma \wedge (R_1 + sD^{\Gamma_1}\Delta\Gamma + s^2\Delta\Gamma \wedge \Delta\Gamma))\right)$$

$$= dCS(\Gamma_2, \Gamma_1), \tag{10.42}$$

$$CS(\Gamma_2, \Gamma_1) := tr\left(2\Delta\Gamma \wedge R_1 + \Delta\Gamma \wedge D^{\Gamma_1}\Delta\Gamma + \frac{2}{3}\Delta\Gamma \wedge \Delta\Gamma \wedge \Delta\Gamma\right). \tag{10.43}$$

As $CS(\Gamma_2, \Gamma_1)$ involves only tensorial quantities, this difference is *globally exact* and well defined (it also demonstrates the integral of $tr(R \wedge R)$ over a closed manifold is independent of the connection). It should be kept in mind that $\Delta\Gamma = \Gamma - \Gamma_o$ transforms covariantly as the difference between two connections (rather than inhomogenously as a connection). The relative Chern-Simons form is invariant, $CS(\Gamma_2^U, \Gamma_1^U) = CS(\Gamma_2, \Gamma_1)$, under $\Gamma_{1,2}^U = U\Gamma_{1,2}U^{-1} + UdU^{-1}$ for all $U \in GL(3, R)$, including "large gauge transformations" with non-trivial winding numbers. This is true of the Chern-Simons functional $W_{CS}[\Gamma, \Gamma_o] := -\frac{1}{2}\int_{\Sigma} CS[\Gamma, \Gamma_o]$ of Γ relative to a conformally flat connection Γ_o.

The Hartle-Hawking wave function discussed earlier in Eq. (10.41) can thus be phrased in the precise context of relative Chern-Simons functional; this Chern-Simons Hartle-Hawking state is

$$\Psi_{CS} = \Psi_{no\,boundary} = \mathcal{N}\exp(-gW_{CS}[\Gamma, \Gamma_o]); \tag{10.44}$$

$$W_{CS}[\Gamma, \Gamma_o] := -\frac{1}{2}\int_{\Sigma} CS[\Gamma, \Gamma_o].$$

The subtraction of the Chern-Simons functional of a conformally flat connection does not alter the fact that this relative Chern-Simons state satisfies $(Q_{CS})_j^i \Psi_{CS} = 0$. This is because $[\bar{\pi}_j^i, W_{CS}[\Gamma_o]] = (\tilde{C}|_o)_j^i = 0$, with $(\tilde{C}|_o)_j^i$ denoting the Cotton-York tensor density of a conformally flat connection (as discussed, the Cotton-York tensor vanishes iff there is conformal flatness).

S^3 can support a non-trivial conformally flat connection (see for instance the explicit Robertson-Walker $dl_{S^3}^2$ displayed earlier). W_{CS} is however independent of q as discussed, and a conformally flat Γ_o therein can be replaced by its equivalent flat connection. If Σ is topologically a Poincaré sphere, a pure gauge connection can carry non-trivial $\Pi_3(SL(3,R))$ winding number.

10.4.4   *Chern-Simons state and the superficial degree of divergence of associated Feynman graphs*

The Chern-Simons Hartle-Hawking state has a norm which is essentially just the partition function of a relative Chern-Simons action,

$$\langle \Psi_{CS} | \Psi_{CS} \rangle \propto Z = \int \mathfrak{D}\bar{q} \, e^{-2g(W_{CS}[\Gamma] - W_{CS}[\Gamma_o])}; \qquad (10.45)$$

wherein

$$-2(W_{CS}[\Gamma] - W_{CS}[\Gamma_o])$$
$$= \int tr \left(2\Delta\Gamma \wedge R_o + \Delta\Gamma \wedge D^{\Gamma_o} \Delta\Gamma + \frac{2}{3} \Delta\Gamma \wedge \Delta\Gamma \wedge \Delta\Gamma \right). \qquad (10.46)$$

Note that it is the *relative Chern-Simons functional*, after subtraction of $W_{CS}[\Gamma_o]$, which is adopted as the physical state. The upshot is that it is invariant under large gauge transformations which would have rendered the exponent unbounded (there is no factor of i factor in the exponent, so quantization of the coupling constant will not make it invariant under transformations with nontrivial winding numbers).

The Taylor expansion is exact when we take into account *finite* (not just infinitesimal) $\Delta\Gamma$ to third order. As the Chern-Simons functionals depend only on $\bar{q}_{ij}$ but not on the conformal factor q, we may replace the $GL(3,R)$ connection Γ by its $SL(3,R)$ counterpart $\bar{\Gamma}$ which is the Christoffel symbol of $\bar{q}_{ij}$. Since we are expanding about a conformally flat background $\bar{\Gamma}_o = 0$, this simplifies the finite change to $\Delta\bar{\Gamma} = \bar{\Gamma} - \bar{\Gamma}_o = \bar{\Gamma}$ which is fortuitously polynomial in unimodular $\bar{q}_{ij}$, because its inverse is $\bar{q}^{ij} = \frac{1}{2}\tilde{\epsilon}^{ikl}\tilde{\epsilon}^{jmn}\bar{q}_{km}\bar{q}_{ln}$. Moreover we may take $\bar{q}^o_{ij}$ to be flat since q^o_{ij} is conformally flat. To wit we substitute $\Delta\bar{\Gamma}(\bar{q}_{ij}) = \bar{\Gamma}(\bar{q}_{ij})$ wherein $\bar{q}_{ij} = \delta_{ij} + \bar{h}_{ij}$, and consider the perturbative QFT by expanding all interactions in the action $W_{CS}[\Gamma] - W_{CS}[\Gamma_o]$ in powers of $\bar{h}_{ij}$ and its spatial derivatives.

There are no infrared divergences due to the cutoff from spatial compactness. For a general diagram the ultraviolet behaviour at large momentum k can be computed as follows. The propagator behaves as k^{-3}. The superficial degree of divergence for a general diagram for this theory in 3 dimensions is

$$D = 3L - 3P + \sum_i n_i V_i, \qquad (10.47)$$

where L is number of loops, P the number of propagators, V_i is the number of interaction vertices with i factors of $\bar{h}_{ij}$ and n_i denotes its k^{n_i} dependence. Each type of vertex is cubic in spatial derivative (this can also be observed from dimensional analysis since the metric variables and coupling constant are dimensionless and the only dimensional dependence come from spatial derivatives), thus $n_i = 3\,\forall i$. With the help of $L = P - \sum_i V_i + 1$ which is an identity relating the numbers of loops, vertices and propagators, the superficial degree of divergence for any particular diagram is thus

$$D = 3L - 3P + 3\sum_i V_i$$

$$= 3P - 3\sum_i V_i + 3 - 3P + 3\sum_i V_i$$

$$= 3; \tag{10.48}$$

which is a constant(!) regardless of which diagram is involved. This implies with a regularization or ultraviolet cutoff set at K, each diagram diverges as K^3. The divergences encountered is rather trivial because all normalized expectation values must be divided by the yet to be normalized partition function and, as D is constant, all physical *ratios* of matrix amplitudes will be preserved irrespective of the cutoff K. Interestingly, there is no renormalization of the coupling constant g, which cannot be required since the number of vertices and factors of g vary over different diagrams whereas D does not. The conformal invariance and independence of scale factor q in the action may be the underlying reason for the absence of a running coupling constant that is dependent upon cutoff scale.

10.5 Quantum metric fluctuations and correlation functions of the Chern-Simons Hartle-Hawking state

With the relative Chern-Simons functional, the quadratic term of an expansion about a background results in

$$\int d^3x \int d^3y\, \Delta\bar{q}_{ij}(x)\, \frac{\delta^2 W_{CS}}{\delta\bar{q}_{ij}(x)\delta\bar{q}_{kl}(y)}\bigg|_{\text{background}}\, \Delta\bar{q}_{kl}(y)$$

$$=: \int d^3x \int d^3y\, \Delta\bar{q}_{ij}(x) H^{ijkl}(x,y)\Delta\bar{q}_{kl}(y). \tag{10.49}$$

The critical point of the Chern-Simons functional is precisely at the vanishing the Cotton-York tensor i.e. at conformal flatness. For the simply connected spatially closed case this is uniquely the standard three-sphere.

About S^3, unimodular metric variations restricted to physical 3dDI transverse traceless infinitesimal changes $\delta\bar{q}_{ij}^{TT}$ yields the Hessian as

$$H^{ijkl}(x,y)\big|_{S^3} =$$
$$\tfrac{1}{8}((\bar{q}^{ik}\tilde{\epsilon}^{jlm} + \bar{q}^{il}\tilde{\epsilon}^{jkm} + \bar{q}^{jk}\tilde{\epsilon}^{ilm} + \bar{q}^{jl}\tilde{\epsilon}^{ikm})\bar{\nabla}_m\bar{\nabla}^2)\big|_{S^3}\,\delta^3(x,y). \quad (10.50)$$

The inverse operator is

$$H_{ijkl}^{-1}(x,y)\big|_{S^3} =$$
$$\tfrac{1}{2}((\bar{q}_{ik}\underline{\epsilon}_{jrl} + \bar{q}_{il}\underline{\epsilon}_{jrk} + \bar{q}_{jk}\underline{\epsilon}_{irl} + \bar{q}_{jl}\underline{\epsilon}_{irk})\tfrac{1}{\bar{\nabla}^2}\bar{\nabla}^r\tfrac{1}{\bar{\nabla}^2})\Big|_{S^3}\,\delta^3(x,y). \quad (10.51)$$

The two-point function is proportional to this operator evaluated at conformally flat S^3. The two physical TT excitations about S^3 and their eigenvalues are known explicitly and introduced in Chapter 6. On the basis of these TT tensor harmonics, $\bar{Y}_{(4)ij}^{Klm}$ and $\bar{Y}_{(5)ij}^{Klm}$, the Hessian has the following action:

$$H^{ijkl}\big|_{S^3}\begin{bmatrix}\bar{Y}_{(4)kl}^{\{K\}}\\[4pt]\bar{Y}_{(5)kl}^{\{K\}}\end{bmatrix}$$
$$= \bar{G}^{ijkl}\big|_{S^3}\begin{bmatrix} 0 & \frac{[K(K+2)-2](K+1)}{2a^3}\\[8pt] \frac{[K(K+2)-2]^2}{2(K+1)a^3} & 0 \end{bmatrix}\begin{bmatrix}\bar{Y}_{(4)kl}^{\{K\}}\\[4pt]\bar{Y}_{(5)kl}^{\{K\}}\end{bmatrix}, \quad (10.52)$$

wherein the notation for the set of quantum numbers Klm has been simplified to $\{K\}$ to eliminate some clutter; and a is the $k = +1$ Roberston-Walker scale factor of the three-sphere.

Expectation values of operators with respect to the Chern-Simons Hartle Hawking state can be evaluated explicitly. The two-point correlation function for TT modes is defined by

$$\langle\delta\bar{q}_{ij}^{TT}(x)\delta\bar{q}_{kl}^{TT}(y)\rangle = \frac{1}{Z}\int\mathcal{D}\bar{q}\,\delta\bar{q}_{ij}^{TT}(x)\delta\bar{q}_{kl}^{TT}(y)e^{-2g(W_{CS}[\Gamma]-W_{CS}[\Gamma_o])}.$$
$$(10.53)$$

Expanding the exponent in terms of the coupling constant,[5] and noting that the inverse of the above Hessian is essentially the two-point function, to lowest order approximation,

$$\langle\delta\bar{q}_{ij}^{TT}(x)\delta\bar{q}_{kl}^{TT}(y)\rangle = \frac{1}{2g}H_{ijkl}^{-1}(x,y)\big|_{S^3}. \quad (10.54)$$

---

[5]In usual practice, g is absorbed in the quadratic "propagator" term by redefinition, and $\frac{1}{\sqrt{g}}$ is considered as "the coupling constant" in the interaction terms. So in usual QFT language, a large value for g implies weak coupling, and vice versa.

In flat space, or with flat coordinates for $\bar{\nabla}_i$, Eq. (10.51) just works out in momentum space, or "k-space", to be "$\frac{1}{k^3}$ in character". This is the well-known signature of a power spectrum of scale-invariant perturbations.[6]

Unlike R^3, the S^3 spatial manifold is closed; fortuitously the explicit TT modes $\bar{Y}_{(4,5)ij}^{\{K\}}$ and their eigenvalues are known exactly even when flat space Fourier decomposition is not applicable. Every physical $\delta\bar{q}_{ij}^{TT}$ can be expanded in terms of orthonormal modes as

$$\delta\bar{q}_{ij}^{TT} = \sum_{\{K\}} c_{\{K\}}^{(4)} \bar{Y}_{(4)ij}^{\{K\}} + c_{\{K\}}^{(5)} \bar{Y}_{(5)ij}^{\{K\}}$$

$$= \sum_{\{K\}} a_{\{K\}}^{+} \bar{Y}_{(+)ij}^{\{K\}} + a_{\{K\}}^{-} \bar{Y}_{(-)ij}^{\{K\}}; \qquad (10.55)$$

wherein $\bar{Y}_{(\pm)ij}^{\{K\}}$ are chosen to be orthonormal linear combinations of $\bar{Y}_{(4,5)ij}^{\{K\}}$ which diagonalize the Hessian. Explicitly,

$$\bar{Y}_{(\pm)ij}^{\{K\}} = \sqrt{\frac{K(K+2)-2}{3}}\bar{Y}_{(4)ij}^{\{K\}} \pm \frac{(K+1)}{\sqrt{3}}\bar{Y}_{(5)ij}^{\{K\}}; \qquad (10.56)$$

and $(a_{KLM}^{\pm})^* = (-1)^M a_{KL-M}^{\pm}$ ensures the reality of the metric. In the basis of the two orthonormal modes $\bar{Y}_{(\pm)ij}^{\{K\}}$, the traceless Hessian is diagonal and its eigenvalues are equal and opposite (respectively positive and negative for $\bar{Y}_{(\pm)ij}^{\{K\}}$). The S^3 background is actually the saddle point of the Chern-Simons functional of the affine connection. To wit, the quadratic term yields

$$\int d^3x \int d^3y \delta\bar{q}_{ij}(x)H^{ijkl}(x,y)\delta\bar{q}_{kl}(y)$$

$$= \sum_{\{K\}} \frac{[K(K+2)-2]^{\frac{3}{2}}}{2a^3}\left(\left(a_{\{K\}}^+\right)^* a_{\{K\}}^+ - \left(a_{\{K\}}^-\right)^* a_{\{K\}}^-\right). \qquad (10.57)$$

The correlation function $\langle \delta\bar{q}_{ij}^{TT}(x)\delta\bar{q}_{kl}^{TT}(y)\rangle$ can be expressed through the fundamental two-point functions of these eigenmodes. At this level of approximation, the $\{K\}$-space behavior of the correlation functions are given

---

[6]Explicitly with flat background $\bar{q}_{ij} = \delta_{ij}$, under $\mathbf{x} \to \alpha\mathbf{x}, \mathbf{y} \to \alpha\mathbf{y}$ scaling, $\langle \delta\bar{q}_{ij}^{TT}(\mathbf{x})\delta\bar{q}_{kl}^{TT}(\mathbf{y})\rangle \propto \int d^3k(\bar{q}_{ik}\varepsilon_{jrl} + \bar{q}_{il}\varepsilon_{jrk} + \bar{q}_{jk}\varepsilon_{irl} + \bar{q}_{jl}\varepsilon_{irk})\frac{k^r}{k^4}e^{i\mathbf{k}\cdot(\mathbf{x}-\mathbf{y})}$ remains unchanged, by redefining the dummy variable $\mathbf{k} \to \frac{1}{\alpha}\mathbf{k}$ in the integration.

by[7]

$$\langle (a_{\{K\}}^{\pm})^* a_{\{K'\}}^{\pm} \rangle = \frac{2a^3}{g[K(K+2)-2]^{\frac{3}{2}}} \delta_{\{K\}\{K'\}};$$

$$\langle (a_{\{K\}}^{\pm})^* a_{\{K'\}}^{\mp} \rangle = 0. \tag{10.58}$$

For large values of K the nontrivial correlation functions of $a_{\{K\}}^{\pm}$ do behave as $\sim \frac{1}{K^3}$ as

$$\frac{1}{[K(K+2)-2]^{\frac{3}{2}}} = \frac{1}{K^3} \left(1 - \frac{3}{K} + \frac{21}{2K^2} - \frac{65}{2K^3} + \frac{795}{8K^4} + \ldots \right). \tag{10.59}$$

This is a consequence both of the discrete spectrum in a closed manifold, and of the Chern-Simons state. Also noteworthy is K is a dimensionless number, so its large value limit is intrinsically meaningful without the need to compare with another scale (c.f. "large momentum limit" in Fourier transform space). Since K is the same label as in $\bar{Y}_{(4,5)ij}^{Klm}(x)$ and $\bar{Y}_{(\pm)ij}^{Klm}(x)$, the dependence of the correlation functions can be stated more succinctly in terms of the eigenvalues of the Laplacian operator. These modes satisfy [51]

$$\bar{\nabla}^2 \bar{Y}_{(4,5)ij}^{Klm} = -(\lambda_K)^2 \bar{Y}_{(4,5)ij}^{Klm},$$

$$(\lambda_K)^2 = \frac{K(K+2)-2}{a^2} \quad \text{with K} \geq 2, \tag{10.60}$$

and $\bar{Y}_{(\pm)ij}^{Klm}$ have the same eigenvalues. Therefore $\langle (a_{\{K\}}^{\pm})^* a_{\{K'\}}^{\pm} \rangle$ in Eq. (10.58) varies precisely as $(\lambda_K)^{-3}$. This is in complete agreement with the scale-invariant k^{-3} dependence in the earlier analysis with flat coordinate analysis. Therein the eigenvalues of $\bar{\nabla}^2$ for Fourier modes are $-k^2$. After all, the dependence is expressed in terms of coordinate independent eigenvalues of the Laplacian. These two-point and other further correlation functions are all explicitly computable.

The functional derivative of the Chern-Simons functional is the Cotton-York tensor density $q^{\frac{1}{3}} \tilde{C}^{ij} = \frac{\delta W_{CS}}{\delta \bar{q}_{ij}}$. And the Hessian $H_{ijkl}(x,y) = \frac{\delta^2 W_{CS}}{\delta \bar{q}_{kl}(y) \delta \bar{q}_{ij}(x)} = q^{\frac{1}{3}} \frac{\delta \tilde{C}^{ij}(x)}{\delta \bar{q}_{kl}(y)}$. Variations of the Cotton-York tensor are thus one-to-one with variations of $\delta \bar{q}_{ij}$ when the Hessian is invertible. This is explicitly true of Eq. (10.50) and Eq. (10.51).[8] So the physical

---

[7]As one of the eigenvalues is negative, usual QFT Gaussian integration has to be generalized to the method of steepest descents in the integration over complex variables $a_{\{K\}}^{\pm}$ and $(a_{\{K\}}^{\pm})^*$ with S^3 metric as the saddle point.

[8]It is also true in general of the relation between TT metric variations and $\tilde{C}_j^i$ (this mixed tensor density is independent of q). The Cotton-York tensor which is also TT may be regarded as the non-perturbative manifestation of the two physical d.o.f. in geometrodynamics.

excitations $\delta\bar{q}_{ij}^{TT}$ *translates directly into Cotton-York tensor curvature per-turbations* $\delta\tilde{C}_j^i(x) = \int H^i{}_j{}^{kl}(x,y)\big|_{S^3} \delta\bar{q}_{kl}^{TT}(y)d^3y$ about conformally flat S^3. From this and Eq. (10.54) the two-point correlation function of the Chern-Simons state for Cotton-York fluctuations about the conformally flat S^3 saddle point is

$$\langle\delta\tilde{C}_j^i(x)\delta\tilde{C}_l^k(y)\rangle = \frac{1}{2g} H^i{}_j{}^k{}_l\big|_{S^3} \delta^3(x,y). \qquad (10.61)$$

In the correlation functions, no less than *metric and curvature fluctuations in quantum gravity* are being manifested. The defining characteristics of these correlations emerge from the culmination and confluence of the Hartle-Hawking proposal, the Chern-Simons state, and the framework of Intrinsic Time Geometrodynamics.

Bibliography

[1] W. Pauli, *General Principle of Quantum Mechanics*, translated by P. Achuthan and K. Venkatesan (Springer-Verlag, 1980), p. 63.

[2] K. Gödel, *An Example of a New Type of Cosmological Solutions of Einstein's Field Equations of Gravitation*, Rev. Mod. Phys. **21** (1949) 447.

[3] K. Gödel, *A Remark About the Relationship between Relativity Theory and Idealistic Philosophy*, in Albert Einstein: Philosopher - Scientist, ed. P. A. Schilpp (New York, Harper & Row, 1959). See also, for instance, P. Yourgrau, *A World Without Time: The Forgotten Legacy of Gödel and Einstein* (Basic Books, 2005), and references therein.

[4] S. W. Hawking, *Chronology protection conjecture*, Phys. Rev. D. **46** (1992) 603.

[5] A resource page on the Planck observatory is `https://www.cosmos.esa.int/web/planck`.

[6] Resources and references on WMAP can be found at `https://map.gsfc.nasa.gov/`.

[7] Details and results of the *Supernova Cosmology Project* (`https://supernova.lbl.gov/`) and the *High-z Supernova Search* (`https://lweb.cfa.harvard.edu/supernova/home.html`) can be found at their respective websites.

[8] For observational evidence of our compact universe, see, for instance, E. Di Valentino, A. Melchiorri and J. Silk, *Planck evidence for a closed Universe and a possible crisis for cosmology*, Nat. Astron. **4** (2020) 196.

[9] J. A. Wheeler, *Superspace and the nature of Quantum Geometrodynamics*, in Battelle Rencontres, eds. C. M. DeWitt and J. A. Wheeler (NY: W. A. Benjamin, 1968).

[10] R. L. Arnowitt, S. Deser and C. W. Misner, *Dynamical structure and definition of energy in General Relativity*, Phys. Rev. **116** (1959) 1322; in Gravitation: an introduction to current research, ed. L. Witten (Wiley, N.Y., 1962).

[11] See, for instance, C. Isham, *Canonical quantum gravity and the problem of time*, (Preprint, gr-qc/9210011v1); K. V. Kuchar, *The Problem of Time in Canonical Quantization of Relativistic Systems*, in *Conceptual Problems*

of Quantum Gravity, eds. A. Ashtekar and J. Stachel (Boston, Birkhauser, 1991); E. Anderson, *Problem of Time in Quantum Gravity*, annalen der Physik, **524** (2012) 757, and references therein.

[12] Bryce S. DeWitt, *Quantum Theory of Gravity. I. The Canonical Theory*, Phys. Rev. **160** (1967) 1113.

[13] S. Tomonaga, *On a Relativistically Invariant Formulation of the Quantum Theory of Wave Fields*, Prog. Theor. Phys. **1** (1946) 27; J. Schwinger, *Quantum Electrodynamics. I. A Covariant Formulation*, Phys. Rev. **74** (1948) 1449.

[14] J. Wheeler with K. Ford, *Geons, Black Holes, and Quantum Foam: A Life in Physics*, (W.W. Norton & Company, New York, 1998); https://www.manhattanprojectvoices.org/oral-histories/john-wheelers-interview-1965.

[15] P. A. M. Dirac, *The theory of gravitation in Hamiltonian form*, Proc. Roy. Soc. Lond. A**246** (1958) 333.

[16] C. Soo and H. L. Yu, *General Relativity without the paradigm of spacetime covariance and resolution of the Problem of Time*, Prog. Theor. Phys. (2014) 013E01.

[17] J. W. York Jr., *Gravitational degrees of freedom and the initial-value problem*, Phys. Rev. Lett. **26** (1971) 1656.

[18] C. N. Yang and R. Mills, *Conservation of Isotopic Spin and Isotopic Gauge Invariance*, Phys. Rev. **96** (1954) 191.

[19] L. Smolin and R. M. Unger, *The Singular Universe and the Reality of Time*, (Cambridge University Press, 2014).

[20] E. Cotton, *Sur les varits a trois dimensions*, Annales de la Facult des Sciences de Toulouse. II. **1** no. 4, (1899) 385; for a review of the properties of the Cotton tensor, see, for instance, A. Garcia, F. W. Hehl, C. Heinicke and A. Macias, *The Cotton tensor in Riemannian spacetimes*, Class. Quantum Grav. **21** (2004) 1099.

[21] S. S. Chern and J. Simons, *Characteristic forms and geometric invariants*, Annals of Mathematics **99** (1974) 48.

[22] J. Hartle and S. W. Hawking, *Wave function of the Universe*, Phys. Rev. D**28** (1983) 2960.

[23] R. Penrose, *The Emperor's New Mind* (Oxford University Press, 1989) and *The Road to Reality* (Alfred A. Knopf, Publisher, New York, 2005).

[24] R. Penrose, *Cycles of Time - An Extraordinary New View of the Universe* (The Bodley Head, London, 2010).

[25] R. Penrose, *Difficulties with Inflationary Cosmology*, in Proceedings of the 14th Texas Symposium on Relativistic Astrophysics, ed. E. J. Fergus (New York Academy of Sciences) pp. 249–264.

[26] E. E. Ita III, C. Soo and H. L. Yu, *Cosmic time and reduced phase space of General Relativity*, Phys. Rev. D**97** (2018) 104021.

[27] M. D. Srinivas and R. Vijayalakshmi, *The 'time of occurrence' in quantum mechanics*, Pramāna **16**, No. 3 (1981) 173.

[28] L. Mandelstam and I. Tamm, *The Uncertainty Relation Between Energy and Time in Non-relativistic Quantum Mechanics*, In: Bolotovskii, B.M.,

Frenkel, V.Y., Peierls, R. (eds.) Selected Papers. (Springer, Berlin, Heidelberg, 1991).

[29] P. A. M. Dirac, *Relativity Quantum Mechanics with an Application to Compton Scattering*, Proc. Roy. Soc. A**111** (1926) 405; *The Quantum Theory of the Electron*, Proc. Roy. Soc. A**117** (1928) 610.

[30] P. A. M. Dirac, *Lectures in Quantum Mechanics* (New York, Yeshiva University Press, 1964).

[31] P. Horava, *Quantum Gravity at a Lifshitz Point*, Phys. Rev. D**79** (2009) 084008.

[32] D. Blas, O. Pujolas and S. Sibiryakov, *On the extra mode and inconsistency of Horava gravity*, JHEP **10** (2009) 029.

[33] S. Hojman, K. Kuchar and C. Teitelboim, *New approach to General Relativity*, Nature Phys. Sci. **245** (1973) 97; *Geometrodynamics Regained*, Ann. Phys. **96** (1976) 88; C. Teitelboim, *The Hamiltonian structure of spacetime*, in General Relativity and Gravitation Vol. 1, ed. A. Held (Plenum, New York, 1980).

[34] M. Li and Y. Pang, *A trouble with Horava-Lifshitz gravity*, JHEP **0908** (2009) 015.

[35] M. Henneaux, A. Kleinschmidt and G. LucenaGomez, *A dynamical inconsistency of Horava gravity*, Phys. Rev. D**81** (2010) 064002.

[36] C. Soo, J. Yang and H. L. Yu, *New formulation of Horava-Lifshitz quantum gravity as a master constraint theory*, Phys. Lett. B**701** (2011) 275.

[37] See, for instance, T. Thiemann, *Modern Canonical Quantum General Relativity* (Cambridge University Press, 2008).

[38] T. Thiemann, *The Phoenix Project: Master Constraint Programme for Loop Quantum Gravity*, Class. Quantum Grav. **23** (2006) 2211.

[39] J. R. Klauder, *Overview of Affine Quantum Gravity*, Int. J. Geom. Meth. Mod. Phys. **3** (2006) 81, and references therein.

[40] E. P. Wigner, *Göttinger Nachr.* **31** (1932) 546; E. P. Wigner, *Group Theory and its Application to the Quantum Mechanics of Atomic Spectra* (Academic Press, New York, 1959).

[41] S. Weinberg, *On the Development of Effective Field Theory*, EPJ H *46* (2021) 6.

[42] See, for instance, R. E. Marshak, *Conceptual Foundations of Modern Particle Physics* (World Scientific Publishing Co., 1993), and references therein.

[43] H. Weyl, *Gravitation und Elektrizität*, Sitzungsber. Preuss. Akad. Wiss., Berlin (1918) 465; H. Weyl, *Space, Time, Matter* (Dover, New York, 1952); Weyl's ideas and refutation of his gravitational theory are also discussed in W. Pauli, *Theory of Relativity* (Dover, New York, 1981).

[44] M. Blagojevic and F. W. Hehl, *Gauge Theories of Gravitation: A Reader with Commentaries* (World Scientific, 2013).

[45] Information on gravitational wave detections and a resource website for the LIGO-Virgo-KAGRA collaboration can be found at https://www.ligo.org/.

[46] A. Ashtekar, *Lectures on non-perturbative canonical gravity* (World Scientific, 1991).

[47] A. Einstein, *Prinzipielles zur allgemeinen Relativitätstheorie*, Annalen der Physik, **55** (1918) 241.

[48] J. Barbour, *The definition of Mach's Principle*, Found. Phys. **40** (2010) 1263.

[49] J. A. Isenberg and J. M. Nester, *Canonical gravity*, in General Relativity and Gravitation. One Hundred Years After the Birth of Albert Einstein, ed. A. Held (Plenum Press, NewYork and London, 1980).

[50] C. Misner, *Mixmaster Universe*, Phys. Rev. Lett. **22** (1969) 1071; *Quantum Cosmology. I*, Phys. Rev. **186** (1969) 1319.

[51] L. Lindblom, N. W. Taylor and F. Zhang, *Scalar, vector and tensor harmonics on the three-sphere*, Gen. Rel. Grav. **49** (2017) 139.

[52] E. E. Ita III, C. Soo and H. L. Yu, *Gravitational waves in Intrinsic Time Geometrodynamics*, Eur. Phys. J. C**78** (2018) 723.

[53] R. Penrose, *Singularities and Time-Asymmetry*, in General Relativity: An Einstein Centenary Survey, eds. S. W. Hawking and W. Israel (Cambridge University Press, 1979) pp. 581–638.

[54] S. Hawking, in S. Hawking and R. Penrose, *The Nature of Space and Time* (Princeton University Press, New Jersey, 1996).

[55] J. D. Bekenstein, *Black Holes and the Second Law*, Lettere al Nuovo Cimento **4** (1972) 737; *Black holes and entropy*, Phys. Rev. D**7** (1973) 2333; S. W. Hawking, *Particle creation by black holes*, Commun. Math. Phys. **43** (1975) 199.

[56] S. A. Hayward, *Signature Change in General Relativity*, Class. Quantum Grav. **9** (1992) 1851; erratum: Class. Quantum Grav. **9** (1992) 2543.

[57] S. Weinberg, *The cosmological constant problem*, Rev. Mod. Phys. **61** (1989) 1.

[58] R. Kubo, *Statistical-Mechanical Theory of Irreversible Processes. I. General Theory and Simple Applications to Magnetic and Conduction Problems*, J. Phys. Soc. Japan, **12** (1959) 570; P. C. Martin and J. Schwinger, *Theory of Many-Particle Systems. I*, Phys. Rev. **115** (1959) 1342. See also, for instance, Chapter 17 of G. Parisi, *Statistical field theory* (Addison-Wesley Pub. Co., 1988).

[59] R. Courant and D. Hilbert, *Methods of Mathematical Physics* (Wiley Classics Edition, 1989); see also, for instance, the discussion in L. D. Landau and E. M. Lifshitz, *Mechanics (Course of theoretical physics, Vol. 1)* (Pergamon Press, 1960).

[60] U. H. Gerlach, *Derivation of the Ten Einstein Field Equations from the Semiclassical Approximation to Quantum Geometrodynamics*, Phys. Rev. **177** (1968) 1929.

[61] J. W. York, *Role of Conformal Three-Geometry in the Dynamics of Gravitation*, Phys. Rev. Lett. **28** (1972) 1082; J. W. York, *Conformally invariant orthogonal decomposition of symmetric tensors on Riemannian manifolds and the initial-value problem of General Relativity*, J. Math. Phys. **14** (1973) 456.

[62] R. F. Baierlein, D. H. Sharp, and J. A. Wheeler, *Three-Dimensional Geometry as Carrier of Information about Time*, Phys. Rev. **126** (1962) 1864.

[63] A. Peres, *On Cauchy's problem in general relativity - II*, Nuovo Cimento, **26** (1962) 53.

[64] K. V. Kuchar, *Canonical Quantum Gravity*, in General Relativity and Gravitation, eds. R. J. Gleiser, C. N. Kozameh and O. M. Moreschi (Bristol, Institute of Physics Pub., 1992).

[65] A. Ashtekar, J. D. Romano and R. S. Tate, *New variables for gravity: Inclusion of matter*, Phys. Rev. D**40** (1989) 2572; L. N. Chang and C. Soo, *Standard Model with Gravity Couplings*, Phys. Rev. D**53** (1996) 5682.

[66] J. Barbour, B. Foster and N. Ó Murchadha, *Relativity without relativity*, Class. Quantum Grav. **19** (2002) 3217; E. Anderson and J. Barbour, *Interacting vector fields in relativity without relativity*, Class. Quantum Grav. **19** (2002) 3249.

[67] H. C. Lin and C. Soo, *Intrinsic Time Geometrodynamics: explicit examples*, Chin. J. Phys. **53** (2015) 110102 (Chinese Journal of Physics Special Issue On the occasion of 100 years since the birth of Einstein's General Relativity).

[68] W. C. Chen, *Dynamical basis of relativistic time effects in generic curved spacetime*, National Cheng Kung University Master Thesis, 2018.

[69] M. A. Hohensee, B. Estey, F. Monsalve, G. Kim, P. C. Kuan, S. Y. Lan, N. Yu, A. Peters, S. Chu, and H. Mueller, *Equivalence Principle, and Matter Waves*, J. Phys. Conf. Ser. **264** (2011) 012009.

[70] C. Soo and H. L. Yu, *New commutation relations for Quantum Gravity*, Chin. J. Phys. **53**, (2015) 110106 (Chinese Journal of Physics, Special Issue On the occasion of 100 years since the birth of Einstein's General Relativity).

[71] E. E. Ita III, C. Soo and H. L. Yu, *Intrinsic Time Quantum Geometrodynamics*, Prog. Theor. Exp. Phys. (2015) 083E01.

[72] N. Ó Murchadha, C. Soo and H. L. Yu, *Intrinsic time gravity and the Lichnerowicz-York equation*, Class. Quantum Grav. **30** (2013) 095016.

[73] A. Lichnerowicz, *L'integration des equations de la gravitation relativiste et le probleme des n corps*, J. Math. Pures Appl. **23** (1944) 37, reprinted in A. Lichnerowicz: *Choix d'oeuvres mathematiques* (Hermann, Paris, 1982), p. 4.

[74] Y. Choquet-Bruhat, *General relativity and the Einstein equations* (Oxford Univ. Press, 2009).

[75] N. Ó Murchadha and J. W. York, *Existence and uniqueness of solutions of the Hamiltonian constraint of General Relativity on compact manifolds*, J. Math. Phys. **14** (1973) 1551.

[76] N. Ó Murchadha, *Readings of the Lichnerowicz-York equation*, Acta Phys. Polon. B**36** (2005) 109.

[77] N. Ó Murchadha and J. W. York, *Initial-value problem of General Relativity. I. General formulation and physical interpretation*, Phys. Rev. D**10** (1974) 428; *Initial-value problem of General Relativity. II. Stability of solutions of the initial-value equations*, Phys. Rev. D**10** (1974) 437.

[78] S. Bai and N. Ó Murchadha, *Scaling up the extrinsic curvature in gravitational initial data*, Phys. Rev. D**85** (2012) 044028.

[79] M. Taylor, *Partial differential equations* (Springer, N.Y., 1996).

[80] N. H. Kuiper, *On conformally flat spaces in the large*, Annals of Mathematics, **50** (4), (1949) 916.

[81] G. Perelman, *The entropy formula for the Ricci flow and its geometric applications*, (Preprint, arXiv:math.DG/0211159); *Ricci flow with surgery on three-manifolds*, (Preprint, arXiv:math.DG/0303109); *Finite extinction time for the solutions to the Ricci flow on certain three-manifolds*, (Preprint, arXiv:math.DG/0307245).

[82] G. Rosen, *Unitary Irreducible Representations of $SL(3, R)$*, J. Math. Phys. **7** (1966) 1284.

[83] M. Gell-Mann and Y. Ne'eman, *The Eightfold Way* (W. A. Benjamin, 1964).

[84] R. A. Hulse and J. H. Taylor, *Discovery of a pulsar in a binary system*, Astrophys. J. Lett. **195** (1975) L51–L53; R. A. Hulse, *The discovery of the binary pulsar*, Rev. Mod. Phys. **66** (1994) 699; J. H. Taylor, *Binary pulsars and relativistic gravity*, Rev. Mod. Phys. **66** (1994) 711.

[85] See, for instance, C. M. Will, *The Confrontation between General Relativity and Experiment*, Living Rev. Rel. **17** (2014) 4, and references therein.

[86] See, for instance, D. Kennefick, *Traveling at the Speed of Thought: Einstein and the Quest for Gravitational Waves* (Princeton University Press, 2016); J. L. Cervantes-Cota, S. Galindo-Uribarri and G. F. Smoot, *A Brief History of Gravitational Waves*, Universe **2** (2016) 22.

[87] C. Soo, *Quantum Geometrodynamics with intrinsic time development*, Int. J. Mod. Phys. D**25** (2016) 1645008 (Proc. of Everything about Gravity, 2017).

[88] S. W. Hawking, T. Hertog and N. Turok, *Gravitational waves in open de Sitter space*, Phys. Rev. D**62** (2000) 063502.

[89] B. Bonga, B. Gupt and N. Yokomizo, *Tensor perturbations during inflation in a spatially closed Universe*, JCAP **05** (2017) 021.

[90] Y. H. Chang and C. Soo, *Intrinsic Time Gravity and gravitational waves in closed Robertson-Walker spacetimes*, Master Thesis, National Cheng Kung University, 2021.

[91] A. De Felice and S. Tsujikawa, *Inflationary gravitational waves in the effective field theory of modified gravity*, Phys. Rev. D**91** (2015) 103506.

[92] E. E. Ita III, C. Soo and H. L. Yu, *Intrinsic time gravity, heat kernel regularization, and emergence of Einstein's theory*, Class. Quantum Grav. **38** (2020) 035007.

[93] S. J. Bernau, *The square root of a positive self-adjoint operator*, J. Austr. Math. Soc. **8** (1968) 17; A. Wouk, *A note on square roots of positive operators*, SIAM Rev. **8** (1966) 100; Z. Sebestyen and Z. Tarcsy, *Characterizations of self-adjoint operators*, Studia Sci. Math. Hunger. **50** (2013) 423.

[94] P. M. Morse and H. Feshbach, *Methods of Theoretical Physics*, Part 1 (McGraw-Hill Book Company, 1953).

[95] See, for instance, R. Jackiw, *Fifty years of Yang-Mills theory and my contribution to it*, (Preprint, arXiV:physics/0403109); S. Deser, R. Jackiw

and S. Templeton, *Three-Dimensional Massive Gauge Theories*, Phys. Rev. Lett. **48** (1982) 975.

[96] See, for instance, E. Witten, *Supersymmetry and Morse theory*, J. Diff. Geom. **17** (1982) 661, which featured a Hamiltonian of the form $H = d_g d_g^\dagger + d_g^\dagger d_g$ which is constructed by extending the Laplacian via $d_g = e^{-gh(\vec{\phi})} de^{gh(\vec{\phi})}$ with Morse function $h(\vec{\phi})$.

[97] D. V. Vassilevich, *Heat kernel expansion: user's manual*, Phys. Rept. **388** (2003) 279.

[98] See, for instance, S. Weinberg, *The Quantum Theory of Fields*, Vol. 1 (Cambridge University Press, Cambridge, UK, 1995).

[99] A. D. Sakarov, *Vacuum quantum fluctuations in curved space and the theory of gravitation*, Dokl. Acad. Nauk SSSR **177** (1967) 70.

[100] P. A. M. Dirac, *Generalized Hamiltonian Dynamics*, Can. J. Math. **2** (1950) 129.

[101] T. Maskawa and H. Nakajima, *Singular Lagrangian and the Dirac-Faddeev Method: Existence Theorem of Constraints in 'Standard Form'*, Prog. Theor. Phys. **56** (1976) 1295.

[102] L. D. Faddeev and V. N. Popov, *Feynman diagrams for the Yang-Mills field*, Phys. Lett. B**25** (1967) 29; L. D. Faddeev, *The Feynman integral for singular Lagrangians*, Theor. Math. Phys. **1** (1969) 1; V. N. Popov, *Functional integrals in quantum field theory and statistical physics* (D. Reidel Publishing Company, Dordrecht, Holland, 1983).

[103] J. W. York Jr., *Covariant decompositions of symmetric tensors in the theory of gravitation*, Ann. Henri Poincaré **XXI** 4 (1974) 319.

[104] E. Anderson, J. Barbour, B. Z. Foster, B. Kelleher and N. Ó Murchadha, *The physical gravitational degrees of freedom*, Class. Quantum Grav. **22** (2005) 1795.

[105] A. Hanson, T. Regge and C. Teitelboim, *Constrained Hamiltonian Systems* (Accademia Nazionale dei Lincei, Rome, 1976).

[106] C. Teitelboim, *Quantum mechanics of the gravitational field*, Phys. Rev. D**25** (1982) 3159.

[107] K. Giesel and T. Thiemann, *Scalar material reference systems and loop quantum gravity*, Class. Quantum Grav. **32** (2015) 135015.

[108] V. Husain and T. Pawlowski, *Time and a physical Hamiltonian for quantum gravity*, Phys. Rev. Lett. **108** (2012) 141301.

[109] C. Soo and H. L. Yu, *Geometrodynamics, and the gauging of spatial diffeomorphism and Lorentz symmetries*, Physics Bimonthly, 2014 (幾何動力學, 可微變換及羅倫茲規範對稱。許祖斌、余海禮 物理雙月刊 (36卷5期) 2014年10月 336-340。)

[110] L. N. Chang and C. Soo, *Standard model with gravity couplings*, Phys. Rev. D**55** (1996) 2410.

[111] K. Fujikawa, U. Lindstrom, N. K. Nielsen, M. Rocek and P. van Nieuwenhuizen, *Regularized BRST-coordinate-invariant measure*, Phys. Rev. D**37** (1988) 391.

[112] K. Fujikawa, *Path Integral Measure for Gauge Invariant Fermion Theories*, Phys. Rev. Lett. **42** (1979) 1195; *Path integral for gauge theories with fermions*, Phys. Rev. D**21** (1980) 2848, [Erratum-ibid. D**22** (1980) 1499].

[113] L. N. Chang and H. T. Nieh, *Lorentz anomalies*, Phys. Rev. Lett. **53** (1984) 21; L. Alvarez-Gaumé and E. Witten, *Gravitational anomalies*, Nucl. Phys. B**234** (1983) 269.

[114] H. Georgi and S. L. Glashow, *Gauge theories without anomalies*, Phys. Rev. D**6** (1972) 429; D. J. Gross and R. Jackiw, *Effect of Anomalies on Quasi-Renormalizable Theories*, Phys. Rev. D**6** (1972) 477; C. Bouchiat, J. Iliopoulos and P. Meyer, *An Anomaly-Free Version of Weinberg's Model*, Phys. Lett. B**38** (1972) 519.

[115] L. N. Chang and C. Soo, *Invariant regularization of anomaly-free chiral theories*, Phys. Rev. D**55** (1997) 2410.

[116] E. Witten, *An SU(2) anomaly*, Phys. Lett. B**117** (1982) 324; E. Witten, *Global gravitational anomalies*, Commun. Math. Phys. **100** (1985) 197.

[117] S. P. de Alwis, *Relation between global and U(1) anomalies*, Phys. Rev. D**32** (1985) 2837.

[118] L. N. Chang and C. Soo, *Chiral fermions, Gravity and GUTs*, (Preprint, arXiv:hep-th/9411064), in Quantum classical correspondence: Proceedings of the 4th Drexel Symposium on Quantum Nonintegrability, eds. D. H. Feng and B. L. Hu (Drexel University, Philadelphia, 1994).

[119] A. S. Belyaev and Ilya L. Shapiro, *The action for the (propagating) torsion and the limits on the torsion parameters from present experimental data*, Phys. Lett. B**425** (1998) 246.

[120] C. Soo, *Adler-Bell-Jackiw Anomaly, the Nieh-Yan Form, and Vacuum Polarization*, Phys. Rev. D**59** (1999) 045006; L. N. Chang and C. Soo, *Massive Torsion Modes, Chiral Gravity, and the Adler-Bell-Jackiw Anomaly*, Class. Quantum Grav. **20** (2003) 1379.

[121] S. W. Hawking and C. N. Pope, *Generalized spin structures in quantum gravity*, Phys. Lett. B**73** (1978) 42; A. Back, P. G. O. Freund and M. Forger, Phys. Lett. B**77** (1978) 181; S. J. Avis and C. J. Isham, Comm. Math. Phys. **72** (1980) 103.

[122] See, for instance, G. Senjanovic, *Proton decay and grand unification*, Based on the plenary talks at the SUSY09 and PASCOS09 conferences, Journal reference: AIP Conf. Proc. 1200:131–141, 2010, (Preprint, arXiv:0912.5375).

[123] P. Adamson *et al.*, *Improved Constraints on Sterile Neutrino Mixing from Disappearance Searches in the MINOS, MINOS+, Daya Bay, and Bugey-3 Experiments* (Daya Bay Collaboration, MINOS+ Collaboration), Phys. Rev. Lett. **125** (2020) 071801; M. G. Aartsen *et al.*, *eV-Scale Sterile Neutrino Search Using Eight Years of Atmospheric Muon Neutrino Data from the IceCube Neutrino Observatory*, Phys. Rev. Lett. **125** (2020) 141801; details on the MicroBooNE experiment can be found at https://microboone.fnal.gov/; V. V. Barinov *et al.*, *Results from the Baksan Experiment on Sterile Transitions (BEST)*, Phys. Rev. Lett. **128** (2022) 232501.

[124] W. D. McGlinn, *Problem of combining interaction symmetries and relativistic invariance*, Phys. Rev. Lett. **12** (1964) 467; L. O'Raifeartaigh, *Lorentz invariance and internal symmetry*, Phys. Rev. B**139** (1965) 1052; S. Coleman and J. Mandula, *All Possible Symmetries of the S Matrix*, Phys. Rev. **159** (1967) 1251.

[125] K. S. Stelle, *Renormalization of higher-derivative quantum gravity*, Phys. Rev. D**16** (1977) 953; J. Julve and M. Tonin, *Quantum gravity with higher derivative terms*, Nuovo Cimento B**46** (1978) 137; E. S. Fradkin and A. A. Tseytlin, *Renormalizable asymptotically free quantum theory of gravity*, Nucl. Phys. B**201** (1982) 469.

[126] V. G. Gurzadyan and R. Penrose, *On CCC-predicted concentric low-variance circles in the CMB sky*, Eur. Phys. J. Plus. **128** (2013) 22; D. An, K. A. Meissner, P. Nurowski and R. Penrose, *Apparent evidence for Hawking points in the CMB Sky*, Monthly Notices of the Royal Astronomical Society **495** (2020) 3403.

[127] A. Hajian, *Are There Echoes From The Pre-Big Bang Universe? A Search for Low Variance Circles in the CMB Sky*, The Astrophysical Journal **740** (2010) 52; A. Moss, D. Scott and J. P. Zibin, *No evidence for anomalously low variance circles on the sky*, Journal of Cosmology and Astroparticle Physics 4 (2011) 033; A. DeAbreu *et al.*, *Searching for concentric low variance circles in the cosmic microwave background*, Journal of Cosmology and Astroparticle Physics **12** (2015) 031; D. L. Jow and D. Scott, *Re-evaluating evidence for Hawking points in the CMB*, Journal of Cosmology and Astroparticle Physics 3 (2020) 021.

[128] The context of this work is in 3 spatial dimensions. $(2+1)$-dim. Lorentzian quantum gravity with Chern-Simons terms have been studied in, S. Deser, R. Jackiw and S. Templeton, *Three-Dimensional Massive Gauge Theories*, Phys. Rev. Lett. **48** (1982) 975; E. Witten, *(2+1)-Dimensional Gravity as an Exactly Soluble System*, Nucl. Physics B**311** (1988) 46.

[129] S. Deser, R. Jackiw and S. Templeton, *Topologically massive gauge theories*, Ann. Phys. (N.Y.) **140** (1982) 372; **185** (1988) 406(E); *Three-Dimensional Massive Gauge Theories*, Phys. Rev. Lett. **48** (1982) 975.

[130] M. L. Browne, *A note on the classical groups*, The American Mathematical Monthly, **62** (1955) 424.

[131] See, for instance, C. Soo, *Wavefunction of the universe and Chern-Simons perturbation theory*, Class. Quantum Grav. **19** (2002) 1051.

[132] See, for instance, G. W. Moore, *Introduction to Chern-Simons Theories*, Lecture notes, 2019 TASI School (https://www.youtube.com/watch?v=PSA8QXp7UrM).